T0231276

Assessment of CHEMICAL EXPOSURES

Calculation Methods for Environmental Professionals

Jack Daugherty

CRC Press
Taylor & Francis Group
Boca Raton London New York

CRC Press is an imprint of the
Taylor & Francis Group, an **informa** business

CRC Press
Taylor & Francis Group
6000 Broken Sound Parkway NW, Suite 300
Boca Raton, FL 33487-2742

© 1998 by Taylor & Francis Group, LLC
CRC Press is an imprint of Taylor & Francis Group, an Informa business

No claim to original U.S. Government works

Visit the Taylor & Francis Web site at
http://www.taylorandfrancis.com

and the CRC Press Web site at
http://www.crcpress.com

Dedication

To my children, grandchildren, great-grandchildren,
and future generations:
"God looked at everything he had made,
and he found it very good."

Genesis 1:31

Biography

 Jack E. Daugherty, MSChE, PE, CIH, CHMM, is the Environmental and Safety Engineer for the Vickers, Inc., Aerospace-Marine-Defense Fluid Power Division plant in Jackson, MS.

Jack obtained his training at Auburn University, receiving the B.Ch.E. degree in 1971. He received a M.S. from the University of Mississippi in 1986. He was a monitoring and decontamination team leader for nuclear, biological, and chemical exposures in the Navy, where he also was involved in electromagnetic exposures, noise, and maritime pollution. Since leaving the Navy, Jack has worked as an engineer with increasing responsibilities in the fields of environmental protection and occupation safety and health in the mineral processing, chemical processing, and machinery manufacturing industries. After two years in consulting, Jack began his present assignment in 1987.

Jack is a member of the American Industrial Hygiene Association, a professional member of the American Society of Safety Engineers, president of the Mississippi Chapter of the Academy of Certified Hazardous Materials Managers, and a member of the Air and Waste Management Association. He has been named the Safety Professional of the Year by the Mississippi Chapter of the American Society of Safety Engineers. Besides being a professional engineer, Jack is a Certified Industrial Hygienist and a Certified Hazardous Materials Manager. He is an Honored Member of *Who's Who 1997–1998*.

Jack has authored more than thirty technical articles. He has also been the author, co-author, or contributor to four technical books related to the fields of safety engineering, environmental engineering, and industrial hygiene.

Preface

Exposure assessments for the workplace, the public health, or for ecosystems are a mighty, sometimes overwhelming task. For the first time in history, scientists, business people, and policy-makers are integrating several academic disciplines and job functions into exposure assessment teams. In public policy enforcement agencies, exposure assessments are almost always a team effort. Many businesses, however, still assign the task to one person and too often it just will not fly. Exposure assessments, much less risk assessments, are not a one-man show.

So, why did I write this by myself then? Because when I started out I had a much narrower audience in mind perhaps. The intended audience — industry — grew as the project developed. If I have accomplished what I now want to do, then here is a book that can certainly be used by safety engineers, environmental engineers, and industrial hygienists working alone in industry, perhaps wearing more than one hat, trying to make do. No team is going to assist them. They may not even have a budget. But these men and women are going to roll up their sleeves and do the best they can with what they have. Those professionals in academia and policy-making/policy-enforcing agencies can also benefit from this work. To some specialists, this book may lack the depth needed, but novices and generalists should find enough information and tools herein to begin to practice the art of exposure assessment. The chapters are organized in a manner that would enhance a safety engineering, environmental engineering, or industrial hygiene curriculum, while making it easy for the person in the field to use also.

This is not a new art, though the state-of-the-art is becoming highly developed. No, safety engineers, environmental engineers, and industrial hygienists have been performing exposure calculations for many years. What is new is the expanding scope of exposure assessment and the integration of disciplines. Seeking careers in industry will force some of us to be jack-of-all-trades, while seeking careers in academia or policy-making agencies will force others to specialize. Consulting companies too will be recruiting specialists. Yet, we are all in this together and we are all stakeholders. If fact, a few plank owners are still around. Whether you are novice or veteran, I hope you find my book useful, and good luck in your career.

Jack E. Daugherty
Jackson, Mississippi

Acknowledgments

I would like to acknowledge my engineering and industrial hygiene mentors: Lennis Tollison; Kent Montgomery, PE; E. Corbin McGriff, Ph.D., PE; Jim Dunaway; Dennis Blackledge; and Dan Markiewicz, CIH. I also would like to thank Michael Jayjock, Ph.D., CIH, and Jack Caravanos, Dr.P.H., CIH, for their fine example and leadership in the field of industrial hygiene. Though I met them only once, both have been inspirational to me. Of course I must also thank my wife for her patience, support, and encouragement during this project.

Contents

1 History of Chemical Exposure Assessment

Chemical exposure assessments are estimates of the amount, frequency, duration, and routes of human exposure to chemicals. Modern exposure assessments date back several centuries in the fields of public health, environmental health, toxicology, epidemiology, health physics, and industrial hygiene, though exposure assessments, as we know them today, were not always completed in the format that has evolved.

Public health is that body of medical knowledge and practice that concerns itself with preservation of the health of the community, as opposed to maintaining the health of individuals within the community. The public health official, then, will want to know what diseases or injuries can be expected from public exposure to a chemical. Environmental health is the science of preserving public health and eco-systems, within the context of the environment, which transcends the proximate community of humanity. Epidemiology is the study of disease occurrence and causes, while health physics deals with radiological exposures, and industrial hygiene deals primarily with occupational exposures only. Occupational epidemiology specifically studies diseases incurred in the workplace, whereas environmental epidemiology concentrates on those encountered in our non-occupational, non-residential sur-roundings. Toxicology examines the impact of chemicals on the various systems of the body. Environmental toxicology is the study of the impacts that pollutants have on the structure and function of ecological systems.

Chemical exposure assessment, as a task of interested scientists and engineers, combines elements of all these disciplines, and has become increasingly important since the early 1970s, with the advent of the Environmental Protection Agency (EPA) during the presidency of Richard M. Nixon, and greater public awareness of prob-lems associated with pollution. Within a year of each other, EPA was formed, and, under the Department of Labor, the Occupational Safety and Health Administration (OSHA), and, under the Department of the Interior, the Mine Safety and Health Administration (MSHA) were also established by law to protect the health of the nation's human resources in its citizens in general, workers, and miners. Not only these federal agencies, but those whom they regulate are interested, at one point or another, in chemical exposure calculations. Hence, you could be a professional in any of several occupations, with any of several kinds of employers, and have an interest in chemical exposure calculations.

EPISODES OF EXTREME EXPOSURE

Have you ever wondered how may primitive people sacrificed life and limb to determine what is safe to eat and what is not? Or, given humanity's propensity for painting the body, how many people exposed themselves to harm by means of mercury, lead, uranium, and other colorful, yet hazardous substances found in minerals? It must have taken many hundreds, perhaps thousands, of generations of human experience to accumulate enough knowledge to make correct choices about food and other materials used in everyday life in primitive society. These primitives were making *de facto* chemical exposure assessments arrived at experientially. Of course, physicians and dietitians maintain that, as a group, we still make unwise choices about our foods.

In the fourth century B.C., for instance, Hippocrates, father of medicine, recognized lead toxicity among miners. Pliny the Elder (*Gaius Plinius Secundus*, A.D. 23–79) was a Roman encyclopedist who recorded the use of goat bladders to cover the face of miners to prevent exposure to lead and mercury. Someone had to perform a chemical exposure assessment, in order to determine that the miners required such protection in the first place. The fact that the offending chemicals were isolated indicates that the exposure assessment was successfully completed, nevertheless. A Greek physician named Galen from Pergamum (130–200) wrote about the hazards of mining copper. His exposure assessment stood for many centuries before it was updated. The Romans fully understood the hazards of mining cinnabar, a mercury-bearing ore, and sent slave labor to the mine at Almaden, Spain. Later, they sent volunteer miners who were excused from military duties, if they went to Spain. Someone not only performed an exposure assessment (determined what was killing the miners), but, also, the Roman government made a risk assessment (determined a public policy relative to assuming risk). Whether this public policy decision by the Romans was a foreshadowing of EPA, or OSHA, or both, is inconsequential.

In 1473, Ulrich Ellenbog published a pamphlet about occupational diseases that he had presumably studied. Georg Bauer (1494–1555), who wrote under the Latin pen-name *Agricola*, set forth in his most important work, *De re metallica* (1556), descriptions of illnesses to which miners are subject. As he gave both treatments and prevention methods, someone, probably he, had to have completed a chemical exposure assessment. His book, by the way, was a standard mining text for two hundred years, and a modernized version was printed as late as the 1950s.

By the 16th century, evidence exists that people recognized that no absolutely safe chemical exists. Paracelsus (the adopted name of Philippus Theophrastus Bombastus von Hohenheim), a Swiss physician who lived 1493 to 1541, wrote:

> All substances are poisons; there is none that is not a poison. The right dose differentiates a poison and a remedy.

The quarrelsome, sarcastic Paracelsus learned this firsthand, as he worked for five years in a lead smelting plant making his own personal assessment of lead exposure. Unfortunately, many of his conclusions were wrong, though he understood the basics of dose-response, at least.

French hatters began using mercuric nitrate to make felt in the seventeenth century, and soon presented themselves to physicians with symptoms of nerve disease. A mad hatter even showed up in allegorical literature (see *Alice in Wonderland*).

The Italian Bernardo Ramazzini wrote the first comprehensive treatise on occupational diseases in 1700, *De morbis artificum*, and, also in the eighteenth century, Percival Pott linked scrotal cancer among English chimney sweeps with soot. In the same century, Sir George Baker's exposure assessment of lead in the cider industry solved the appearance of the mysterious disease that came to be called the Devonshire colic. Alice Hamilton, a twentieth century American physician who helped to raise the awareness of the public to occupational illnesses, certainly assisted the progressive development of formal exposure assessments.

Many infamous episodes of acute health effects from pollution have occurred in this century. In fact, many of the situations that led to the current fear of chemical exposure by the general public occurred early in the century and have become part of environmental lore, though few people can give you specifics. Scientists and engineers, in the early decades of this century, who worked on the bench experiments, pilot plants, and full-scale operations of many wonderful chemical products did not, in general, worry about the important details concerning the fate of the residues and wastes. In the early years of the twentieth century, lead-induced deaths were routinely reported, for instance. From 1900 to 1930, in the U.S. alone, thirty-four hundred deaths were attributed to lead poisoning, and the number of children who were affected by lead, but did not die, is unknown, but suspected of being a staggering number.

In the Meuse River Valley, in Belgium, more than sixty people died from an acute exposure to air emissions from steel mills in December 1930. In the 1930s, in the U.S., about sixteen thousand people were affected by tri-*ortho*-cresylphosphate poisoning. In 1934, about four thousand people suffered from lead poisoning in Detroit.

In Donora, Pennsylvania, twenty died during the Halloween of Death in 1948 from exposure to emissions from a group of steel mills and an acid plant. Emissions from the plants, which sit in a valley, were held at the level of Donora's home, which are on the hillside, by a temperature inversion, injuring six thousand. This was the first recorded episode of acute air pollution exposure in this country.

Hydrogen sulfide emissions from a blown-out wellhead in Paza Rica, Mexico killed twenty and made more than three hundred persons ill in 1950. A temperature inversion was implicated in four thousand deaths that occurred during a two week period from smog and emissions of sulfur dioxide and particulate matter in London in December 1952.

Eighteen hundred people were poisoned by parathion in 1952 in Japan. The same year more than twelve thousand Japanese babies were exposed to arsenic. Several thousand actually were received doses of arsenic resulting in about one hundred thirty deaths.

Another temperature inversion, this one in New York City, resulted in two hundred deaths from air pollution in November 1953.

Three years after the first poisoning, another thousand Japanese citizens received methylmercury doses in 1955. Turkey, in 1956, suffered through a hexachlorobenzene exposure episode, with four thousand being poisoned. About eight hundred were poisoned by parathion in India in 1958. Ten thousand Moroccans were exposed to tri-*ortho*-cresylphosphate in 1959. Japan, the same year, had a mercury poisoning episode. Mercury poisoning has come to be called *Minamata* disease, after Minamata Bay, where Japanese pearl divers were exposed by eating contaminated seafood. A thousand Iraqis received ethylmercury doses in 1960. Another seven hundred people died from acute exposure to air pollution in London in December 1962.

More than six hundred Japanese were again poisoned by methylmercury in 1964. Endrin poisoned almost seven hundred Qatari in 1967. Over sixteen hundred Japanese received polychlorinated biphenyl (PCB) doses in 1968. Eleven years after the ethylmercury episode, fifty thousand Iraqis were poisoned by methylmercury in 1971. Milk was contaminated by polybrominated biphenyls (PBB) in Michigan in 1973. Seventy-five hundred Pakistanis were dosed with malathion in 1976. The same year, a lead exposure scare rocked Kellogg, Idaho. Also, in 1976, a reactor ran away and blew up at the ICMESA plant, a Hoffmann–LaRoche subsidiary in Seveso, Italy, near Milan, spreading several pounds of dioxin (2,3,7,8-tetrachlorodibenzo-*para*-dioxin) over the countryside. The town was evacuated after several weeks when the truth of the incident was finally released to the press. The cleanup was completed only recently. The next year brought a 1,2-dibromo-3-chloropropane (DBCP) release in California.

The year 1978 is infamous in the toxic chemical annals: Hopewell, Virginia and Love Canal come to mind. A small chemical supplier in Hopewell dumped kepone into the sewer, which polluted the James River, which flowed into the Chesapeake Bay, a commercial fishing ground. Also, several employees of the plant had symptoms of acute and chronic kepone exposure. Commercial and private fishing were prohibited in the river and its estaurine system for many years, only recently being permitted again. The news about the cover of the old Love Canal dump being violated was released the same year and for weeks the world watched in shock as public officials wrenched their hands, uncertain how to proceed. Finally, public pressure forced Congress to pass the Superfund law in order to get some action and relief for the residents of Love Canal.

Nearly thirteen thousand people were exposed when toxic oil was sprayed on the Spanish tomato crop in 1981. In 1983, we watched astonished as television crews covered the dioxin scares in Times Beach, Missouri and Newark, New Jersey. Then, in 1984, the explosion that released methyl isocyanate in Bhopal, India killed two thousand immediately and for a time afterward, but the total number of deaths resulting from the exposure will perhaps never be known. As late as 1996, the Mississippi Gulf Coast is being rocked by new of the illegal application of the agricultural pesticide methyl parathion in residences. At least one death and several injuries have been alleged as of this writing. The two unlicensed application technicians, operating independently since 1991, face federal and state fines and imprisonment.

Despite that laws and regulations to protect the public health, employee health, and the environment have brought about improvements, the residual deaths are unacceptable. Considering the inestimable value of one human life, one must agree that our current system of government intervention has a basis for existence, and, so far at least, no one has proposed a better way to do things. Estimates of fatalities linked to coal-fired electricity generation, for instance, still average five hundred per year per 1,000 Mwe. Based on the amount of electrical power, in Mwe, produced by coal in the U.S., that amounts to 350,000 fatalities per year. In addition to the human loss, other losses occur from excessive exposures. The good name of the polluting company may be damaged, for instance. Tangible property in the vicinity of the industrial pollution source may be damaged. For this reason, and the aesthetical losses to the community caused by the pollution, the community could also be damaged economically by residents voting with their feet. That is, loss also occurs by potential new residents and industries refusing to move into the polluted area.

Plant life can also be damaged. Deposition of the pollutant onto food crops leads to passing it up the food chain to animals used by humanity as a food source. Upton reminds us of the adage, "you are what you eat." So, the potential effects of chemical exposure are many and varied.

GOVERNMENT INTERVENTION

The Romans (Remember the cinnabar mine at Almaden?) were probably not the first government to intervene in regulating public exposures to dangerous substances, but they certainly have not been the last either. Mostly, until 1970, the U.S. federal government left occupational safety and health to the employers, and environmental emissions to the owners, if no state or local laws applied. Today, in America, we have come to rely on a regulatory approach to control exposures to hazardous chemicals in our lives. Originally, this approach was forced on us because too many industrial managers would not take action quickly enough to protect their workers, the public, and especially the environment from the harmful effects of the myriad chemicals compounds they produced and handled every day. In the 1950s, a series of key technological advances began, which have led to improved safety conditions in most plants today, and cleaned up many of the offensive emissions of the past.

INTERVENTION OF WORKER EXPOSURES

Until 1936, governmental protection of health inside factories had been sparsely regulated by a handful of localities and poorly administered at that. The *National Factory Act* of 1833 had no teeth to it. The Bureau of Mines (BuMines) was established in 1910 to make that one industry safer, but that was a secondary, if not tertiary purpose of the bureau. The Division of Industrial Hygiene and Sanitation was organized by BuMines in 1914. The *Longshoreman's and Harbor Worker's Compensation Act* (33 U.S.C. §§901, 904) was passed in 1927. Then, Congress passed the *Walsh–Healy Act* in 1936, which established standards, for the first time, but only for safety and health of contractors on federal government projects. In 1941, the *Federal Mine Inspection Act* gave BuMines some authority to enforce safety in mines. Maritime safety amendments to the longshoreman's law were passed as Public Law (PL) 85–742 of 1958. The *Longshoreman's Act* of 1959 updated the earlier law and amendments. In 1965, two laws, the *Service Contract Act* and the *National Foundation of the Arts and Humanities Act*, added several categories of persons working indirectly for the federal government to the list of those whose safety was protected by federal law. In 1966, *the Federal Metal and Nonmetallic*

Mine Act, PL 89–577, was an attempt to correct some of the limitations of earlier laws on mine safety. The *Federal Coal Mine Health and Safety Act* of 1969, PL 91 –173. Also, in 1969, the *Construction Safety Act* was passed by Congress, due to the many safety problems and casualties, including fatalities, in the construction industry.

Some progress was made on the state level — in a handful of states — between the passage of the Walsh –Healy Act and 1970. The first real national safety law, though, was the Williams –Steiger Act, named for Senator Harrison A. Williams (D –NJ) and Representative William Steiger (R –WI), which was passed as PL 91 –596 in 1970 and is popularly called the OSHA Act or OSHAct.

The Occupational Safety and Health Administration (OSHA), under the U.S. Department of Labor, was founded by order of Congress under the OSHAct, also known as the Williams –Steiger Act as mentioned above, in 1970 and immediately began developing standards — or regulations — to protect workers in their jobs. President Nixon signed OSHAct into law on December 29, 1970, and it became effective on April 28, 1971. The OSHAct (29 U.S.C. §651) makes it the duty of every employer to furnish employment and a place of employment, which are free from recognized hazards likely to cause serious physical harm. Inasmuch as chemical exposures present such hazards, OSHA is highly interested and involved in chemical exposure assessment.

Not only did OSHAct establish OSHA, but it also established the National Institute of Occupational Safety and Health (NIOSH) as the principle federal agency responsible for researching ways to eliminate workplace hazards. Whereas OSHA was made part of the Department of Labor (DOL), NIOSH was part of, what was then, the Department of Health, Education, and Welfare (DHEW), and is now the Department of Health and Human Services (DHHS). In DHHS, NIOSH is organized under the Centers of Disease Control (CDC) of the Public Health Service (PHS).

The Mine Enforcement and Safety Administration (MESA) was established in 1973 by the Secretary of Interior as part of his department (DOI), which had the responsibility for enforcing the various mine safety laws that were on the books. The BuMines, also part of DOI, was designated the research and standards role with respect to mine safety and health similar to the role of NIOSH with respect to occupational safety and health.

The *Federal Mine Safety and Health Act* of 1977, during President Carter's term, established the Mine Safety and Health Administration (MSHA) in DOL and gave MSHA more enforcement authority than MESA had.

INTERVENTION OF PUBLIC AND ENVIRONMENTAL EXPOSURES

By far, the oldest U.S. environmental law is the *Refuse Act* of 1899 (33 U.S.C. §407), also called the Clean Harbors Act. The quality of water in rivers and lakes was chiefly left to cities and states to maintain or improve. In some cases interstate pacts were formed, but bickering between state governments on how to proceed and the powerful influence of large, wealthy companies, who were well represented on the control commissions, ensured ineffectiveness. In 1965, the Congress passed the *Water Quality Act* to give aid to states and interstate compacts that were trying to establish and enforce water quality standards for surface waters in their jurisdictions.

The same problems plagued the states. They were typically unable to determine whether discharges were in compliance with the standards or not. Many of the standards in the federal law did not apply to particular states, or worse, the proper application of the standard was not obvious at the local level. Most of the states failed to make discharge pollutant load allocations that were enforceable. Finally, the mechanisms of enforcement were so cumbersome that few violations were detected and enforced.

Another attempt at solving the growing problem was the *Clean Water Restoration Act* of 1966. A Ralph Nader commission that had studied the effectiveness of the Water Quality and Clean Water Restoration acts reported to Congress that the nation's water quality was worsening, not improving, and that the implementation of the act was anything, if effective. To worsen the situation, an offshore oil derrick in Santa Barbara Bay blew out and covered the bay with oil. Fishermen were prohibited from plying the trade or sport in the Hudson River, and most of the lakes of South Carolina, due to contamination. Public outrage demanded that government do something. At the time, Congress was trying to get a fairly extensive air pollution control bill passed, and lacked the energy to enact another major environmental law, so it passed, instead, a weakened *Water Quality Improvement Act* of 1970, PL 91 –467.

Of course, severe problems were still encountered, so in 1972, at President Nixon's insistence, a tougher law, the *Federal Water Pollution Control Act* (33 U.S.C. §1251 –1376) was passed, giving EPA authority to set nationwide effluent standards for specific industries based on water pollution control equipment capabilities and cost to industry (called *technology standards*). A permit program was also established. EPA made a decision to concentrate on regulating the oxygen-consuming pollutants, which directly affect wildlife and aesthetics, but do not contribute that much to human toxicity. Consequently, Congress passed the *Clean Water Act* (CWA) (33 U.S.C. §1251 –1387) in 1977 to govern the release of toxic pollutants to water.

The manufacture and use of toxic and hazardous chemicals for pest control is governed by the *Federal Insecticide, Fungicide and Rodenticide Act* (FIFRA)(7 U.S.C. §136 –136y). The law provides EPA with authority to register and regulate (cf. 40 CFR Parts 162 –80) pesticides and herbicides. FIFRA is the second oldest environmental law in the U.S., having been passed in 1947.

Before 1955, each city or town was responsible for its own air quality. Where several cities in close proximity affected each other with air pollution the state would step in. The problem with this arrangement was the most of the states were ineffective in helping the cities maintain a decent air quality and where state boundaries were crossed by pollution plumes, the states were generally totally ineffective. Interstate pacts met with no more success than they did with controlling water pollution. The original U.S. air law, the *Clean Air Act*, was passed in 1955 and was amended in 1963 by the addition of a Title II, the *Motor Vehicle Air Pollution Control Act*. The *Air Quality Act* of 1967 (42 U.S.C. §7401 –7642), which amended the act of 1963, was the first comprehensive federal statute, in the U.S., to authorize regulation of air emissions from stationary and mobile sources by federal agencies. In 1970, Congress determined that changes were needed and a totally restructured federal/state scheme of air pollution was passed and became named the *Clean Air Act* (CAA) amendments. President Nixon asked Congress to approve his plan to establish the EPA, whose chief responsibility, at the time, was to administer the new CAA and the Federal Water Pollution Control Act, neither of which were effectively administered by the commissions that had been formed.

Technical amendments were made to the law in 1973 and 1974. Congress amended the law again in 1977, with provisions for the Prevention of Significant Deterioration of areas that have attained the national standards for air quality, stringent requirements for area that fail to attain these standards, restriction on the use of dispersion by tall stacks, and stiffer enforcement and penalties for noncompliance with the law. Minor amendments were made in 1978, 1980, 1981, 1982, and 1983. In 1990, PL 101 –549, popularly called the Clean Air Amendments (CAAA) made far-reaching changes in the way air pollution control is administered, and will affect industry into the twenty-first century. Many exposure assessments will be required in order to comply with CAAA or to determine the best strategy for compliance.

In 1971, the Environmental Protection Agency (EPA) was also founded, to protect the health of the general public and its environment. Unlike OSHA, no implementing act was passed by Congress to form EPA, only a resolution approving the President's intention to establish a new federal agency to implement and enforce existing laws. Only later did Congress pass new environmental laws. In order to protect public health, the EPA has developed risk-based guidelines.

The manufacture, import/export, processing, use, distribution in commerce, and disposal of chemicals are regulated by the authority of the *Toxic Substances Control Act* (TSCA), PL 94 –469 passed in 1976, codified at 15 U.S.C. §2601 –72. EPA has the authority to require heath and safety and environmental fate testing of chemical products before they are put on the market (cf. 40 CFR 700 –99). TSCA also gives EPA authority to require testing of products already in commercial use, any time information comes to its attention that a potential problem exists.

In another 1976 law, PL 95 –580, the storage and disposal of hazardous wastes is governed by the *Resource Conservation and Recovery Act* (RCRA)(42 U.S.C. §6901 –92k). RCRA exposure details are mainly concerned with worker issues in handling hazardous waste, but the total law is designed to protect public health and the environment. Corrective action requirements at closed RCRA sites are concerned with the protection of onsite workers, the public, and the environment.

The *Asbestos Hazard Emergency Response Act* (AHERA)(15 U.S.C. §2641) governs the removal of asbestos from the nation's schools, but is used for its state-of-the art procedures in many settings.

The *Comprehensive Environmental Response, Compensation and Liability Act* (CERCLA or Superfund) was passed as PL 96–510 in 1980 (42 U.S.C. §9601–75). Superfund deals with the release of hazardous substances to the environment. The also covers abatement of abandoned toxic and hazardous waste sites. CERCLA takes a simple, direct approach. This law authorizes the government to take action to clean up abandoned, uncontrolled hazardous substances (including but not limited to hazardous waste) in the interest of public safety and health, regardless who is to blame. So far, so good. Then, the law authorizes EPA to use law suits to recover the cost of the cleanup, from anyone who contributed hazardous substances to the problem site, or from anyone who contributed to the operation of the site, without regard to blame! You can imagine this law, which is at cross-purposes with English common law (as this nonlawyer understands it), set off lights, whistles, and bells in attorneys' offices from coast to coast, from the Canadian border to the Mexican, and in offshore states and territories, too. Superfund was a great idea, but the un-American law so infuriated the regulated community, which for the most part was not to blame, and presented such a boon to attorneys that ... You guessed it. Most Superfund sites have become so bogged down in lawsuits, to get out of what is perceived as an unfair, guilty no-matter-what law, that nothing tends to happen in the short time frames intended by congress. Some site cleanups have dragged out for years, instead of the 120 days allotted.

The *Hazardous and Solid Waste Amendments* (HSWA) of 1984, PL 98–616, made amendments to RCRA and established EPA authority to register and regulate underground storage tanks (UST). The concern is for public exposure to harmful chemicals via groundwater contamination from leaking USTs.

The *Emergency Planning and Community Right-to-Know Act* (EPCRA) was passed 1986 as Title III of amendments to the Superfund law (42 U.S.C. §11001–50) and governs communications about the storage, use, and disposal of toxic and hazardous chemicals, including the reporting of accidental releases.

The disposal of toxic and hazardous substances into wells is regulated by authority of the *Underground Injection Control Act* (UICA).

The *Safe Drinking Water Act* (SDWA: 42 U.S.C. 300f to j–26) covers the treatment and distribution of potable water and the disposal of marine paints with toxic constituents is controlled by the *Organotin Paint Act*.

EPA AND OSHA COOPERATION

Although the environments OSHA and EPA were chartered to oversee are distinct from each other, a lot of overlap exists and the delineation is not always clear. With respect to chemicals and their effects on people, OSHA regulates the health of workers, usually inside a factory, and EPA regulates the public health, usually outdoors. This last statement is not always true, as OSHA has construction standards, and one which affects workers at hazardous waste cleanup sites, and EPA, under the Toxic Substances Control Act (TSCA) and the Emergency Planning and Community Right-to-Know Act (EPCRA) can come into plants that have no discharges or emissions.

In a memorandum of understanding (MOU) between OSHA and EPA, issued November 23, 1990, the subject of minimizing workplace and environmental hazards is discussed. The MOU was written to establish and improve the working relationship between the two agencies, with the joint goal of improving efforts to protect workers, the public, and the environment at facilities under the jurisdiction of either agency. The MOU establishes a framework for notification, consultation, and coordination between the agencies, and improves information exchange with respect to job-site safety and health, and protection of the public health and environment, thereby reducing the potential for workplace related injury and death, and environmental contamination.

As stated above, and clarified in the MOU, EPA's responsibilities include the protection of public health and the environment, by assuring compliance with federal laws and regulations protecting the environment. OSHA's duties, similarly, are to enforce the OSHAct, and its own safety standards to assure safe and healthful working conditions for every working man and woman in the nation.

Several principal laws, discussed above, pertain to EPA's authority and role.

Tremendous energy and capital have been poured into cleaning up the air, water, soil, and work places of this nation, since the inception of EPA and OSHA. Yet, in some respects, we have a lot more work to do. None of the clean air or water goals, set by Congress in the 70s, have come to fruition. Political conservatives blame political liberals for intrusive standards and regulations, yet laws that were passed by Congress, composed of a blend of political philosophies, are what generates regulations and standards. Some citizens, who are not well-versed in regulations, have the mistaken notion that evil, powerful bureaucrats sit around all day, dreaming up hatefully impossible rules to impose on the good and helpless citizenry of our land, particularly that portion of the citizenry that owns and operates capital producing companies, including their investors. This notion is far from reality. Although regulations are often confusing, duplicative, and overlapping, and in that respect need to be corrected, or repealed, for the most part, regulations are necessary, and perform a needed service for the majority of our citizenry, by protecting them from harmful exposures that they cannot protect themselves from.

In the long run, the fad of bashing the sitting administration and environmentalists will only do great damage to our nation, if it leads to the total dismantling of what has been built, and tolerates self-policing, as predicted by Rene Dubose in the 1970s:

> the greatest danger of pollution may well be that we shall tolerate levels of it so low as to have no acute nuisance value, but sufficiently high, nevertheless, to cause delayed pathological effects and despoil the quality of life.

About six million chemicals compounds are referred to in literature, and about a half million, give or take several tens of thousands, have been registered with EPA as required by the Toxic Substances Control Act. Of these, perhaps a hundred thousand are in active commercial production. Many are food additives, prescription and over-the-counter drugs, cosmetics, and other consumer chemicals that improve the quality of our lives in countless ways we take for granted. Presuming, for the moment, that all these chemicals are considered safe, because FDA, or some other government agency, has approved them, that leaves many chemicals to which we are exposed without respect to regulation. If any of these become industrial waste, they are then heavily regulated by EPA, and those that expose workers in their workplaces are regulated by OSHA.

OTHER EXPOSURE INTERVENTIONS

The *Federal Food, Drug, and Cosmetic Act* (FFDCA) takes the approach that it will not regulate the minutiae of potential exposures within its domain. Rather, this law simply prohibits the intentional addition of any substance to food, drugs, or cosmetics that has been determined to cause cancer in humans or other animals. Therefore, exposure and risk assessments are not needed under this no-risk law. Either a carcinogen has been added, or not. If so, the food, drug, or cosmetic cannot be distributed for commerce. The Food and Drug Administration (FDA) establishes standards for the content of foods, and contamination of food in regulations found at 21 CFR 1–1300.

The Consumer Product Safety Commission (CPSC) also has some exposure and risk assessment responsibilities. The *Consumer Product Safety Act* (CPSA) authorizes CPSC to require specific labeling design, packaging, and composition of products intended for public commerce. CPSC regulations are codified at 16 CFR 1000–1512. The *Federal Hazardous Substances Act* (FHSA) gave CPSC authority to ban or regulate hazardous substances for consumer use. CPSC has labeling authority over consumer products that are defined as being toxic, corrosive, flammable, irritating, or a radioactive hazard.

CHEMICAL EXPOSURE ASSESSMENT AND QUANTITATIVE RISK ASSESSMENT

An extension of chemical exposure assessment, the quantitative risk assessment, is being used by EPA to evaluate and select remediate alternatives at Superfund sites. More and more, environmental consulting firms are using exposure and/or risk assessments to evaluate conditions at their clients' factories, in order to make good engineering and management decisions. Other federal agencies have also used the risk assessment.

From time to time, an incident occurs that reminds us of the potential for disaster that can develop quickly when handling and using chemicals, even beneficial ones. Take for instance, this case of mishandling, which affected people of several states for many years.

> In 1973 a polybrominated biphenyl compound (PBB) used as a fire retardant in the textile industry was inadvertently substituted for a cattle feed supplement. Subsequently, meat but mostly milk was contaminated at a Michigan dairy farm. A large portion of milk consumed in Michigan over the next year or so was contaminated. In 1978, 97% of individuals tested in Michigan were found with PBB in adipose tissue at levels that will remain detectable for the rest of their lives.

Even the courts study exposure assessments and risk analyses that are brought before them.

> *Gulf South Insulation et al. v. Consumer Product Safety Commission* was in response to a proposed ban in 1982 on using urea-formaldehyde foam insulation (UFFI) in schools and residences by the Consumer Product Safety Commission (CPSC). After a six year study CPSC concluded that UFFI presented an unreasonable risk of acute tissue irritation and cancer. The Fifth Circuit Court decided in 1983 that CPSC had not conduct sufficient analysis nor used methodology sufficient to support their

claim of an unreasonable risk. The courts do study exposure and risk analyses brought before them.

The Consumer Product Safety Act (CPSA) requires, before a consumer product safety rule can be affirmed, that:

the Commission's findings ... are supported by substantial evidence on the record taken as a whole.

The substantial evidence test was the issue examined by the D.C. Circuit Court, which wrote in *Industrial Union Department, AFL–CIO v. Hogdson*:

our review basically must determine whether the Secretary carried out his legislative task in a manner reasonable under the state of the record before him.

All things considered, when it comes to a contest between science and public policy, the regulatory agency will always win, if its conflicting or inconclusive evidential information causes it to err on the side of overprotection. Science, decidedly, has the greater burden, as seen in *American Textile Manufacturers Institution v. Donovan*.

In *Citizens to Preserve Overton Park, Inc. v. Volpe*, the court outlined a framework for accommodating science-policy decisions. A regulatory agency assessing health risks has three prime responsibilities. It must (1) adequately evaluate the technical data while (2) following proper administrative procedures to (3) carry out its statutory mandate.

These comments on court cases have been compiled from reviews in the Bureau of National Affairs, Inc. *Environment Reporter*, for the purpose of giving you some history about how exposure (and risk) assessment have become such a factor in the so-called good science vs. public policy dispute. I have tried not to put any legal interpretation on these examples, as I have no legal training. If I have, inadvertently, inserted some amateur lawyering here, beware of my opinions like the plague! Taking legal counsel from me would be like asking a four year old to drive you to the emergency room. Why bother? You won't make it anyway!

The neat way, taught in most schools of industrial hygiene, to conduct an exposure assessment is to take some samples of air, or water, or soil, and analyze for contaminants of concern and compare the results to tabulated data from toxicological studies. In the real world, however, things are rarely, if ever, that neat. Real situations tend to be messy, and lots of data tends to be missing, or otherwise unavailable. The academic discipline that most closely has the skills to model exposures, without benefit of monitoring data, is chemical engineering. Industrial hygiene also has much of the scientific knowledge required, but the mathematical and engineering modeling is another thing. Regardless who does the assessment, the assessor/model developer must:

1. evaluate manufacturing processes to identify potential exposures and release points,

2. estimate the extent of ekposure or release to the environment,
3. evaluate the effectiveness of control alternatives, such as personal protective equipment (PPE), administrative procedures, and engineering controls, for reducing exposure, and
4. recommending effective corrective actions.

The real world, as said earlier, is not neat, and sufficient data on releases rarely exist to establish a full range of releases or the exposures that result. Certainly, not enough data will exist, except in very rare cases, to determine the descriptive statistics of the data base. Missing data and uncertainty abounds. Therefore, the exposures must often be estimated. Thus, this book.

Many disciplines are required to conduct a full-blown public sector exposure assessment, which is typically part of a risk assessment. Some or all of the academic disciplines listed in Table 1.1 are brought into the assessment project. In industry, typically, a much smaller team of generalists prepares the risk assessment, with assistance from specialists consulting with them.

Issues at stake, when people or their environment are exposed to toxic chemicals, are so complex that even a blue-ribbon panel, made up of one or more practitioners of each of the disciplines listed in Table 1.1, cannot, with any certainty, recommend clear policies based on sound science that everyone will be satisfied is the best possible compromise between safety and danger.

In March 1995, EPA Administrator Carol M. Browner issued a memorandum on the EPA Risk Characterization Program, which includes exposure assessment as one of its elements, in which she emphasizes **transparency** in decision-making and **clarity** in communication. Regarding core assumptions and science, when making risk decisions, and the exposure assessment element, she calls for **consistency** and a zone of **reasonableness**.

As you have guessed, from the preceding discussion, exposure assessment is an element out of context, the context being the risk assessment as a whole. Table 1.2 shows the definitions and relationships of the various elements of risk assessment. Risk assessment is the public policy study conducted to make a decision regarding public safety or protection from risk. The risk assessment is based on a series of questions about scientific information, relevant to either public health risks, or environmental risk, or both. Most often, we speak of, and hear about, health risk assessments, the elements of which are discussed in the aforementioned Table. Each question asked, by those conducting the risk assessment, calls for interpretation of available information and data.

The rest of this book will cover the first three elements, though only one bears the title exposure assessment, since the first two are necessarily requisite to the exposure assessment. The scientist or engineer must turn over the process to public policy makers when the exposure assessment is complete, as the risk characterization is in the bailiwick of public servants who are charged with policy making and enforcing. That is not to say that we scientists and engineers cannot make recommendations and render opinions about the matter. We certainly can, but that is not within the scope of this book. To close the loop, risk management is the total of the programs and policies that are implemented to control or manage risk, once

---

TABLE 1.1
Scientific Disciplines in Exposure Assessment

Analytical chemistry
Biochemistry
Biometrics
Chemical engineering
Computer modeling
Ecology
Environmental epidemiology
Environmental toxicology
Evolutionary biology
Industrial hygiene
Kinetics
Landscape ecology
Limnology
Marine biology
Mathematics
Meteorology
Microbiology
Molecular genetics
Oceanography
Occupational epidemiology
Occupational toxicology
Organic chemistry
Pharmacokinetics
Physical chemistry
Physiology
Population biology
Soil science
Thermodynamics
Wildlife biology

---

facts, recommendations, and opinions have been considered and a decision has been made.

CONCLUSION

Humans have found it possible to devise methods for the safe handling of any chemical, no matter how poisonous. Sometimes, these lessons were learned the hard way, though, and people have sacrificed life and limb for this knowledge. You have two basic ways of handling a dangerous chemical safely, you either limit the dose, or control the exposure, to achieve the desired effect — no loss of health. The technology for doing either is outside the scope of this book. Here, we shall concern ourselves with assessing exposure, and predicting dose, such that we can make judgments about the necessity to limit dose or control exposure. Since we will be

TABLE 1.2
Risk Assessment and Its Component Elements

I
Hazard Identification
What is known about the capacity of a chemical for causing adverse health effects?

II
Dose-Response Assessment
What is known about the biological mechanisms and dose-response relationships underlying any effects?

III
Exposure Assessment
What is known about the principle paths, patterns, and magnitudes of human or wildlife exposure and numbers of persons or species likely to be exposed?
In each of the above elements, what are the related uncertainties and science policy choices?

IV
Risk Characterization
A summary of the preceding elements that integrates information and synthesizes an overall conclusion about risk.
Post-Risk Assessment
Risk Communication
The process of exchanging information and opinion with the public, especially the exposed or potentially exposed public.

estimating by mathematical model, rather than measuring actual situations, we must consider, in each case, the reasonable worst case scenario. We may also look at a typical case, if we have sufficient knowledge and data to support it, but we must consider, at the least, the reasonable worst case.

QUESTIONS (ANSWERS BEGIN ON PAGE 401)

1. What is the American approach to controlling exposures to chemical hazards in the workplace?
2. What is the purpose of the Memorandum of Understanding between EPA and OSHA dated November 23, 1990?
3. Discuss the implications of the Rene Dubose prophecy. Do you agree? Why or why not?
4. What is your opinion about the Superfund? Is it a good idea or not? Is it just or not? What would you suggest to make it more palatable to the American public and industry?

REFERENCES

Amdur, Mary O. *Industrial Toxicology.* (Chapter 7) in NIOSH. *The Industrial Environment — its Evaluation & Control.* Washington, D.C.: U.S. Government Printing Office, 1973.

Arbuckle, J. Gordon, *et al. Environmental Law Handbook.* 12th ed., Rockville, MD: Government Institutes, Inc., 1993.

Ayers, Kenneth W. *et al. Environmental Science and Technology Handbook.* Rockville, MD: Government Institutes, Inc. 1994.

Browner, Carol M. *Memorandum re EPA Risk Characterization Program.* March 21, 1995.

Clayton, George D. "Introduction." (Chapter 1) in NIOSH. *The Industrial Environment — Its Evaluation & Control.* Washington, D.C.: U.S. Government Printing Office, 1973.

Crowl, Daniel A. and Joseph F. Louvar. *Chemical Process Safety: Fundamentals with Applications.* Englewood Cliffs, New Jersey: Prentice Hall, 1990.

Daugherty, Jack E. *Industrial Environmental Management: A Practical Handbook.* Rockville, MD: Government Institutes, Inc., 1996.

EPA. "EPA Guidelines on Exposure Assessment." *57 FR 22890.* May 29, 1992.

Hall, John C. "Designing Risk Analysis to Avoid Pitfalls in Cost Recovery Actions: A Legal/Technical Solution." *Hazardous Materials Control Monograph Series: Risk Assessment Volume I.* Silver Spring, MD: Hazardous Materials Control Research Institute, pp. 14–21.

Koren, Herman and Michael Bisesi. *Handbook of Environmental Health and Safety: Principles and Practices, Volume I.* 3d. Ed. Boca Raton, FL: CRC Press/Lewis Publishers, 1996.

Landis Wayne G. and Ming-Ho Yu. *Introduction to Environmental Toxicology: Impacts of Chemicals upon Ecological Systems.* Boca Raton, FL: Lewis Publishers, 1995.

McElroy, Frank E. Ed. *Accident Prevention Manual for Industrial Operations: Administration and Programs,* (Vol. I), 8th Ed. Chicago: National Safety Council, 1981.

Neely, W. Brock. *Introduction to Chemical Exposure and Risk Assessment.* Boca Raton, FL: Lewis Publishers, 1994.

OSHA. "OSHA Preamble and Proposed Rule to Revise Air Contaminants Standards for Construction, Maritime and Agriculture." *57 FR 26002.* June 12, 1992.

OSHA & EPA. "Memorandum of Understanding between The Occupational Safety & Health Administration and The Environmental Protection Agency on Minimizing Workplace and Environmental Hazards." November 23, 1990.

Partridge, Lawrence J. and Arthur D. Schatz. "Application of Quantitative Risk Assessment to Remedial Measures Evaluation at Abandoned Sites." *Hazardous Materials Control Monograph Series: Risk Assessment Volume I.* Silver Spring, MD: Hazardous Materials Control Research Institute, pp. 1–5.

Plog, Barbara A. Ed. *Fundamentals of Industrial Hygiene.* 3rd ed. Chicago: National Safety Council, 1988.

Root, David E., David B. Katzin and David W. Schnare. "Diagnosis and Treatment of Patients Presenting Subclinical Signs and Symptoms of Exposure to Chemicals which Bioaccumulate in Human Tissue." *Hazardous Materials Control Monograph Series: Health Assessment.* Silver Spring, MD: Hazardous Materials Control Research Institute, pp. 31–4.

Schatz, Arthur D. and Michael F. Conway. "OSHA Standards or Risk Analysis: Which Applies?" *Hazardous Materials Control Monograph Series: Health Assessment.* Silver Spring, MD: Hazardous Materials Control Research Institute, pp. 14–7.

Scott, Ronald M. *Introduction to Industrial Hygiene.* Boca Raton, FL: CRC Press, Inc., 1995.

Sheriff, Robert E. "Air Pollution." Chapter 6 in: *Industrial Hygiene Study Guide.* 4th ed. New Jersey Section: American Industrial Hygiene Association, 1989.

Stewart, Patricia A. and Anthony J. Towey. "OSHA." Chapter 21 in: *Industrial Hygiene Study Guide*. 4th ed. New Jersey Section: American Industrial Hygiene Association, 1989.

Upton, Arthur. "Environmental Factors and Human Health." Chapter 1 in: *Report of the Subcommittee on Health Effects, Appendix D: Strategies for Health Effects Research*. Research Strategies Committee, U.S. EPA, Report SAB–EC–88–040D, Revised October 24, 1988.

2 Nomenclature and Terminology

The first thing we have to do is to nail down the language we will use. This is required precisely because so many academic disciplines are involved in exposure assessment and each has its own language. As EPA points out in its *Exposure Assessment Guidelines*, no agreed-upon definitions exist, and terminology in the literature is inconsistent. In preparing its guidelines, EPA had the wisdom to correct its own ways by following the lead of other scientific fields, which have been doing exposure assessments for at least a century.

EXPOSURE TERMS

Scientists generally agree that human *exposure* means contact with the chemical. To some, this means contact with the visible exterior of the recipient, such as the skin, eyes, mouth, or nostrils, while to others this means contact with the exchange boundaries where the chemical is absorbed. Examples of the latter are skin, lung, or gastrointestinal tract. Ambiguity has plagued the measurement of exposure due to these two different approaches to its definition. A third definition of exposure is the occurrence of a chemical in the proximity of a recipient. In terms of this definition, an exposure is an opportunity for contact. The nuclear health physicists use exposure in this manner, and many industrial hygients think of exposure in this way. In its broadest context, exposure subjects a recipient to a potential action or influence. A unwholesome action or influence is implied in the science of exposure assessment. Used in this way, exposure applies to physical and biological hazards as well as to chemical hazards. In ecology, exposure can also mean the potential influence or action of higher levels of biological organization. To illustrate this denotation, EPA, in its guidelines, give the examples of the exposure of a benthic community to dredging, or exposure of an owl population to habitat modification, or exposure of a wildlife population to hunting. The operational definition of exposure, particularly the units of measure, depends on the stressor and receptor (defined below).

Acute exposure is exposure to a high concentration within a short time period, usually less than twenty-four hours. Repeated exposure to a lesser concentration, during the period of a month or less, is called a *subacute exposure*. A *subchronic exposure* occurs over a period of one to three months, at an exposure level less than for subacute exposure. Exposures that are dilute compared to acute exposure, being one hundredth to one thousandth of the level of an acute dose, and which are repeated over a period greater than three months, are called *chronic exposures*.

Exposures are measured in terms of parts per million (ppm), milligrams per cubic meter (mg/m^3), or millions of particles per cubic foot (mppcf). When measured in terms of a volume relative to a volume of air or water, exposure is typically reported in ppm, or the number of volume units of the contaminant to one million volume units of air or water as appropriate. In soil, or other solids, we generally report exposure in terms of a mass ratio, such as milligrams of the contaminant per kilogram of the solid (mg/Kg), which is the number of mass units of the contaminant to one million mass units of the solid matrix. In terms of mass, exposure can also be given in mg/m^3 for gases and vapors in air. For solid particles in air, we often find measurements reported as mppcf (million particles per cubic foot). For any contaminant in aqueous solution or other liquid mixture, concentration measurements of contaminants are given as mg/L, or milligrams per liter, or number of mass units per one million volume units.

The concentration in ppm of gases and vapors in air can readily be converted to mg/m^3 by using the gas laws:

$$mg/m^3 \times 24.5 = ppmxM \qquad (2.1)$$

where M is the molecular weight of the contaminant species. Equation 2.1 is valid when atmospheric pressure is 760 mmHg and the temperature is 25°C (77°F), because one mole of gas occupies 24.5 liters under those conditions. Under other conditions, the equation must be modified accordingly.

Toxic low concentration (TC$_{LO}$) is an exposure, which, according to NIOSH, is the lowest concentration of a substance in air to which humans or animals have been exposed for any period of time, which has produced any toxic effect in humans, or produced tumorigenic or reproductive effects in animals. Another exposure term, called *lethal concentration low (LC$_{LO}$)*, is the lowest concentration of a substance in air, other than the LC$_{50}$, which has been reported to have caused death in humans or animals. The *lethal concentration fifty (LC$_{50}$)*, or the *median lethal concentration*, then, is a calculated concentration of a substance in air, to which an exposure is expected to cause death for fifty percent of a defined animal test population, when exposed for a specified period of time.

DOSE TERMS

Consistent with EPA, and many scientific endeavors, this book shall use human and wildlife exposure to mean contact with the body's external boundaries. Consider the human, or animal, body as a system having an outer boundary that separates outside the body from inside. The skin and body openings make up this outer boundary. Body openings include the mouth, eyes, nostrils, ears, pores of the skin, punctures in the skin, and skin lesions. Contact by a chemical with the outer boundary is a chemical exposure. The amount of chemical that actually crosses the boundary is a *dose* and the amount absorbed into an organ is an *internal dose*. The alimentary canal is outside the body, but passage through it provides plenty of opportunity for absorption into the body. A *dosage* is the dose considered with its frequency and duration.

Exposure and dose are probably the two most confusing terms in the business of exposure assessment. Burmaster and Appling recommend that the definition of dose by an author always be checked when reading an exposure or risk assessment, and this is a good practice, since any of so many disciplines could have prepared the assessment report and each defines dose and exposure in its own way. Many reporters assume no dilution of concentration between source and receptor; therefore, to them, exposure equals dose. Burmaster and Appling define *exposure dose* as the mass of chemical entering a person's body via ingestion, inhalation, or dermal contact, with no allowance for excretion or exhalation before the chemical is absorbed or metabolized.

Absorbed dose is the mass of chemical that is absorbed or metabolized by the body. Absorbed dose is always less than the exposure dose because some of the chemical is exhaled or excreted before being absorbed or metabolized. Absorption of chemicals in the body is hard to measure quantitatively. *Biological effective dose* (BED) is the mass or concentration of chemical that reaches the target organ or tissue, causing physiological or genetic damage. BED is so much more difficult to measure than absorbed dose that it is rarely used.

NIOSH defines *toxic dose low (TD$_{LO}$)* as the lowest dose of a substance, introduced by any route other than inhalation, over a given time period, which produces any toxic effect in humans or tumorigenic or reproductive effects in animals. *Lethal dose low (LD$_{LO}$)* is the lowest dose, other than LD$_{50}$, of a substance, introduced by any route other than inhalation, which causes death in humans or animals. The *lethal dose fifty (LD$_{50}$)* is a calculated dose of a substance, introduced by any route other than inhalation, which is expected to cause the deaths of fifty percent of a defined animal test population.

EXPOSURE ASSESSMENT

An *exposure assessment* is the quantitative or qualitative evaluation of the potential for contact of a chemical (or physical or biological agent) with the outer boundary of a recipient, or with an ecosystem. The assessment consists of establishing an *exposure profile* and a *scenario*. The exposure profile summarizes the magnitude and spatial and temporal patterns of exposure for the scenarios described in the conceptual model. The exposure scenario consists of a set of assumptions concerning how an exposure may take place, including assumptions about the exposure setting, stressor characteristics, and activities that may lead to exposure. A proper exposure assessment not only describes the intensity, frequency, and duration of contact, as well as routes of exposure, but also evaluates the rate at which the chemical crosses the outer boundary into the body, known as chemical intake or uptake rate. The assessment also evaluates the route by which the chemical crosses the boundary or routes of exposure, which are dermal, oral, or respiratory. Ideally, the assessment defines a range of potential releases from a source, the mean or median of estimates made, descriptive statistics of the exposures or releases, and the uncertainty of the estimates. Unfortunately, such a full analysis is rarely made due to insufficient data on releases and exposures. If little or no data are available, a *reasonable worst case* estimate, called the *outer bound estimate*, is the only option. Conservatism is traded for hard data.

Any physical, chemical, or biological entity or agent that induces an adverse response is a *stressor*. The term *stress regime* has been used in three distinct ways.

Stress regime has been used to mean an exposure to multiple chemicals, or to both chemical and nonchemical stressors. In this usage, a stress regime is a multiple exposure, complex exposure, or exposure to mixtures. The term has also been used as a synonym for exposure in an attempt to avoid overemphasis on chemical exposures. Finally, the term has been used to describe the series of interactions of exposures and effects resulting in secondary exposures, secondary effects, and, finally, ultimate effects. This series of interactions has been referred to also as a risk cascade by Lipton [cited in: EPA Proposed Guidelines for Ecological Risk Assessment]. Andrewartha and Birch call it a causal chain, pathway, or network [*loc. cit.*].

An entity, or action, that releases a chemical, physical, or biological stressor to the environment is a *source*. As applied to chemical stressors, *source term* means the type, magnitude, and patterns of chemical(s) released. The *stressor–response profile* summarizes data on the effects of a stressor and how the data relates to the assessment endpoint, the environmental value that is to be protected. An assessment endpoint includes both an ecological entity and specific attributes of that entity. If the spotted owl is valued ecological entity, then the reproduction and population maintenance of the spotted owl is the assessment endpoint. The characterization of ecological effects evaluates the ability of a stressor to cause adverse effects under a particular set of circumstances. The characterization of exposure evaluates the interaction of the stressor with one or more ecological entities.

Engineering estimates of releases and exposures are made by analogy or model. Analogy is used when similar chemicals are used in similar circumstances. For instance, if we have a case study involving trichloroethylene, a common industrial cleaning solvent, we can extrapolate results to perchloroethylene, another solvent, used under similar conditions. Modeling techniques are based on the laws of physics involving parameters such as vapor pressure or solubility in water.

Exposure *routes* or *pathways* are the means by which a chemical reaches the boundaries of the body. An evaporating chemical may diffuse through the air of a still room, or be carried by air currents, to inhaled by the receptor. A chemical that dissolves easily in water may be carried many hundreds of yards, sometimes miles, to a receptor. Chemicals that end up in the soil either remain there indefinitely, leach into groundwater, or get carried off by soil water into a body of surface water. Those that remain in the soil are potentially ingested with food or directly by small children or indirectly from dust particles. The mechanisms of transfer in each of these cases are the topic of subsequent chapters. A living organism, such as a rat, that transfers a chemical contaminant from a source to a receptor is called a *vector*.

A *receptor* is the person or animal or plant or community exposed to the chemical contaminant or stressor. The three-way link to have an exposure to a stressor is source–pathway–receptor. Think of them as a triangle — not unlike the fire triangle — and you see that if any are removed or interdicted the triangle fails to exist. Does that suggest some strategies for dealing with exposures?

ECOLOGICAL TERMS

A *population* is the aggregate of individuals of a species within a specified location in space and time. A *community* is an assemblage of populations of different species within a specified location in space and time. The *biotic* community and *abiotic* environment within a specified location in space and time is an ecosystem. *Trophic* levels are functional classifications of *taxa* within a community that is based on feeding relationships. As we shall see in Chapter 4, aquatic and terrestrial green plants comprise the first trophic level and herbivores comprise the second. Carnivores have several trophic levels. A *conceptual model* consists of a series of working hypotheses about how the stressor might affect ecological entities. The conceptual model also describes the ecosystem potentially at risk, the relationship between measures of effect and assessment endpoints, and exposure scenarios. An *ecological entity* may be a species, a group of species, an ecosystem function or characteristic, or a specific habitat. Or, an ecological entity can be one component of an assessment endpoint. Any event, or series of events, that disrupts an ecosystem, community, or population structure and changes resources, substrate availability, or the physical environment is a *disturbance*.

TOXICOLOGICAL TERMS

Toxicity is an adverse effect on an organism. *Acute toxicity* is the effects resulting from a brief exposure while *chronic toxicity* is the adverse effects that follow a long term exposure. *Cumulative toxicity* are adverse effects due to repeated doses occurring as the result of a prolonged action. *Subchronic toxicity* is the adverse effects from repeated daily exposure for approximately ten percent of life time (about seven years).

Adverse effects from chronic exposures are observed when absorption of the contaminant in body tissues is greater than biotransformation and excretion combined. If the excretion rate is less than the rate of absorption (mass out < mass in) the contaminant accumulates in the body. Steady state occurs when absorption = excretion (mass in = mass out). *Biotransformation* is a specific phase of metabolism during which the contaminant (called a *xenobiotic* in the body) is transformed to another chemical by a biotic system, typically an enzyme. Besides biotransformatiom, metabolism processes include absorption, distribution, and excretion. A *primary effect* is one where the stressor acts on the ecological component of interest, or the human target organ, itself, not through effects on other components of the ecosystem. A *direct effect* is synonymous with primary effect. A *secondary effect* of a stressor acts on supporting components of a body or an ecosystem, which in turn affect the physiological or ecological component of interest. A synonymous term is *indirect effect*. *Recovery* is the rate and extent of return of an individual or a population or community to a condition that existed before the introduction of a stressor. Due to the dynamic nature of physiological and ecological systems, the attributes of a recovered person or system must be carefully defined. Though the exposure assessor is not typically involved in the recovery phase, he or she may be called upon to provide exposure data, modeling expertise, or other information.

Toxicologists use the terms NOAEL, NOEL, LOEL, and LOAEL as follows. The maximum dose level (mg/Kg day) in an experiment that does not produce a statistically or biologically significant adverse effect is the *No Observed Adverse Effect Level* (NOAEL). The maximum dose that does not produce any statistically or biologically significant effect at all is the *No Observed Effect Level* (NOEL). The *Lowest Observed Adverse Effect Level* (LOAEL) is the lowest dose (mg/Kg day) used in a toxicity test that produced an observed adverse effect that is statistically or biologically significant. If the effect observed is not considered adverse, the latter level is called the LOEL, *Lowest Observed Effect Level.*

Either of these dose levels are calculated as follows:

$$D = \frac{C \times I}{W \times F} \qquad (2.2)$$

where: D = dose, NOEL, NOAEL, LOEL, LOAEL, as appropriate
        C = concentration in contaminated media, mg/Kg or mg/m^3, as appropriate for air, water, or food
        I = intake, Kg/day or m^3/day, as appropriate
        W = body weight, Kg
        F = interspecies scaling factor.

The interspecies scaling factor for dose rate extrapolation from rats and other laboratory animals to humans is taken as unity (1) if human data is available (no scaling is required). However, when required, the factor is composed of three subfactors

$$F = F_1 \times F_2 \times F_3 \qquad (2.3)$$

The first of these, F_1, accounts for potential *interspecies* variation in response sensitivity to the same exposure. A value from one to ten is selected for animal data depending on whether biokinetics and mechanism of toxicity match between the animal species and humans. If these match, $F_1 = 1$, and $F_1 = 10$ if no match. The second subfactor, F_2, accounts for the potential *intraspecies* variation in human sensitivity to the same exposure. Some people get a headache from chlorinated organic solvents, before others can even smell them, as an example. Again, the values range from one to ten and the extremes are the only numbers ever used. Typically, $F_2 = 1$, unless some data exists showing that sensitivity varies among humans. Finally, F_3 depends on the quality of data match to the situation studied. If a lifetime NOEL or NOAEL is desired, but less than lifetime data (NOEL, NOAEL, LOEL, LOAEL) is available, $F_3 = 10$. If LOEL or LOAEL must be derived from short-term data, an additional factor of ten is used, and $F_3 = 100$.

One must keep reality in mind when using these numbers (NOEL, NOAEL, LOEL, and LOAEL). It is easy to think of the NOEL or NOAEL as being some threshold below which no one will be injured. That is untrue. By way of example, consider this. If you take a basket of 10,000 marbles of the same color and replace 1,000 of them (10%) with black marbles, you would rightly expect, on the average, to select a black marble from the basket one out of every ten selections. Can we say that p = 0.1 is our NOEL? On the average, yes. But, we know that we could possibly draw one thousand colored (nonblack) marbles (p = NOEL = 0) or all black marbles (p = NOEL = 1). The two extremes are less likely that p = NOEL = 0.1, but just as

TABLE 2.1
Sample Size Required to Detect Differences
(95% Confidence Level)

Animals per Test Group	Detects this % Difference
1,237	2
309	5
137	8
77	12
49	16
34	20

After Neeley.

possible. So, we could conclude that all the marbles are colored and none of them are black, which we know to be false. We could also conclude that all the marbles are black, and none are colored, which is equally false. Yet, our laboratory test, consisting of drawing marbles from the basket, could lead us to either of those false conclusions or to the equally false conclusion that p = NOEL is anything other than 0.1. Now, a standard laboratory test uses forty animals. What if we draw only forty marbles from the basket? We have a 28% chance of selecting no black marbles at all, and, thus, drawing an erroneous conclusion about NOEL. Therefore, for any animal test considered, chances are that the NOAEL is something other than reported.

Neeley points out the sample sizes required to determine whether a statistically significant difference exists between any control and treated group of animals. As the difference desired becomes smaller, the greater the number of test subjects is required. Table 2.1 shows these requirements.

As you can see, huge percentage differences between a control group and its treated counterpart are necessary when testing on the order of thirty to forty animals. The cost of testing twelve hundred animals is prohibitive in most cases. The point is this: The next time you hear someone carelessly bandy about an NOAEL, as if it were a hard and fast threshold, correct them. It is not so.

Two common dose terms in exposure and risk assessments are *maximum daily dose (MDD)* and *lifetime average daily dose (LADD)*. The MDD is simply the maximum expected dose for any given day. The LADD is a term frequently used in carcinogenicity risk studies. It is the MDD averaged over the fraction of a life span that includes exposure to the chemical of concern (COC), considering factors such as age, sex, state of health, tobacco use, and occupational exposures, among others. *Acceptable daily intake (ADI)* is a unit of risk based on NOELs and a safety factor based on variation of response in order to protect most people.

EPA has established a protocol for developing a *reference dose* (RfD) for any COC. The highest dose of a chemical that will be safe is usually greater than the RfD, which is a dose that is unlikely to cause effects. Test animal studies and human epidemiology studies, if available, are reviewed for study quality, and to establish LOAEL and NOAEL. Uncertainty factors are applied to the LOAEL or NOAEL, depending on the reliability of data, source of data, and variability of response. When the LOAEL or NOAEL is derived from human exposure data, they are used without modification. Otherwise the exposure level may be divided by a number from 1 to 10 in order to account for variations among persons. Also, a factor varying from 1 to 10 is divided into the exposure level, if animal data is used to protect humans. The animal exposure level may also be divided by an uncertainty factor, if the test lasted less than the lifetime of the animals, yet the results will be used to establish a lifetime limit for humans. While having its limitations, this approach does provide the public with some degree of protection, albeit, at great expense.

Dose response is the mathematical relationship between dose and magnitude of biologic effect. However, the way that most sampling is conducted, the result is an *exposure response* curve instead, although called a dose response curve. When the body's natural defenses cannot be accounted for, and if no controls are available, always equate dose with exposure as a worst case situation.

Carcinogenicity is the ability of a chemical to initiate or promote malignant tumors, especially carcinomas (cancer). However, cancer is not a single disease with several variations, as measles, for example. Cancer, in reality, is a group of diseases and must, therefore, be treated differently than other adverse health effects. For instance, when the liver is damaged by disease, a certain amount of damage must have happened before any practical difference in liver health is even a concern. That is because the body can live with or replace many damaged cells. Cancer, on the other hand, is thought to be caused by a single mutation in DNA, which is theoretically possible with the presence of a single molecule of a carcinogen. A single molecule, therefore, can theoretically cause cellular damage leading to cancer. While scientists and medical doctors believe it is not always true that a single molecule of a carcinogen can lead to cancer, the current practice in exposure and risk assessment is to treat it that way. *Oncogenicity*, as a chemical characteristic, is the formation of tumors, period, whether benign (noncancerous) or malignant.

To test for carcinogenicity, animals (such as rats, mice, or dogs) are administered either the highest dose or half the highest dose of the COC that will give no adverse effect other than cancer, when given over the test life. For rats and mice, a test life is two years. At the end of the test, the animals, both test and control, are *necropsied* (killed and autopsied) with emphasis on examination for tumors. When test groups have more tumors than their control groups do, test data is used to predict cancer development rates in human beings. However, two differences must be accounted for when comparing test animals and humans: 1) the much longer lifetime of humans, and 2) the much lower anticipated exposure. Consideration of these factors led EPA to develop the *cancer slope factor*, or CSF, to estimate cancer risk in humans. The CSF is a multiplier that assumes a linear relationship between dose and carcinogenic response. Typically, CSF is expressed in units of $(mg/Kg\ day)^{-1}$; although, any consistent set of units could be used. The factor is used thus:

$$R = LADD \times CSF \qquad (2.4)$$

Two databases are available from EPA for the serious assessor of carcinogen exposure or cancer risk. The Integrated Risk Information System (IRIS) is a database that is updated monthly and available on-line through several wide area services. Fixed (not updated once obtained) versions of IRIS can be purchased on disk from several sources. If you choose this route, keep in mind that the information is updated often and purchase a new disk set at least annually, if not more often. The Health Effects Assessment Summary Tables (HEAST) is a database that is updated quarterly or semiannually and available in print from the National Technical Information Service in Springfield, Virginia.

Noncancerous toxicological effects on the body or its organs include a variety of manifestations. Two, in particular, deserve special mention, because they are becoming more and more the focus of interest in health and lifestyle tabloids. *Mutagens* are chemicals that cause somatic or genetic mutations at conception. *Teratogens* cause birth defects in newborns by their action on fetal development.

When reversible changes are caused in the eyes or skin after exposure, we speak of *irritation*. However, we have a situation of *dermal* or *eye corrosion*, if the tissue damage produced by the exposure is irreversible.

GENERAL RISK TERMS

With the foregoing discussion in mind, let us examine some terms that cause much confusion: safety, hazard, danger, and risk. *Safety* is a condition where harm, danger, injury, or damage are not only perceived as being remote possibilities, but, indeed, are remote potentials. Obviously, then, true safety and apparent safety are two different things entirely. Often we live in apparent safety, when, in fact, we have merely been lucky. In the past, people considered this lucky condition as divine favor. The gods are pleased and have blessed us with no catastrophes. Even today, the lack of harm or danger is often considered as evidence that nothing will change, that the way things are is somehow blessed. In fact, the law of probability works in safety as well as it works on the gaming tables. Sooner or later, we have to pay for our streaks of good luck. Sooner or later, the bill comes due and our lucky streak ends with a bang! Harm, injury, and damage should be obvious. Those are things we spend our lives trying to avoid in many ways.

A *hazard*, then, is the potential for some activity, condition, or circumstance to produce harmful effects, such as injury or damage. The chance, or probability, of injury, damage, or other harm, and the severity of the loss, is *risk*. The term risk implies an uncertainty about a possible event that may have serious consequences. Manuele would have us ask four questions about the risk of an event. Can it happen? If yes, what is exposed to harm or damage? What will be the consequences? How often can it happen? Relating the terms hazard and risk, the source of risk is one or more hazards. So, Manuele would have us ask these two questions. What is the source of risk? What lays out for us the probability of the occurrence of events that could have adverse effects?

As a society, we have to decide what risks are acceptable to us, but complicating this decision is the personal acceptance of risk by millions of citizens. Some people will jump out of an airplane and call it fun, yet others will not even get into an airplane to travel from point A to point B. Some people smoke two packs of cigarettes every day, yet will protest a planned hazardous waste site a few miles from their homes. Some people want preindustrial age air quality, yet they consistently drive fifteen to twenty miles per hour over the legal speed limit. The difference is that people willingly accept great risks if they perceive a derived personal benefit (the thrill of skydiving; the pleasure of nicotine; getting somewhere a few minutes sooner), yet reject lesser risk as unacceptable if some external agency seems to force it on them when they do not perceive a benefit for themselves (airlines — they can go by bus; industry — they do not see the connections between the disposal site and their material wealth; the LAW — why should their freedom be restricted when it is the other drivers who are unsafe).

Therefore, as scientists and technologists, we cannot cling to our exposure assessments as if they are somehow holy. The fact is that once we turn our exposure assessment over to the decision- and policy-makers, anything can happen and will. Our study may be used in ways we never intended or imagined. The best we can do is to stand and be heard, but do not be surprised if we are not heard. Policy-makers will use exposure studies to get reelected. Decision makers will use them to get budgets approved. Manufacturing lobbyists will use them if they support the needs of industry, but ignore them or refute them if they do not. Environmentalists will use them to oppose the lobbyists, or else refute them.

Yet we must not be deterred. An exposure assessment ought to be completed to the best of our abilities, regardless of outcome, and we should be willing to defend our studies with integrity. Our goal can only be to protect the employee in the work place, or the citizen in the community, or the environment itself. If our assessment says take no action, fine. If it says the situation is too close to call, that is what we tell management and/or the public policy-makers. If it says, someone needs protection, that is our stand.

PHYSICAL STATE OF CONTAMINANT

Contaminants, or chemicals of concern, or stressors, may manifest themselves in any of several states of matter, which too often confuses discussion. If literature spoke only of the three physical states of matter — solid, liquid, gas, the issue would be simplified in one respect (we all think we understand the three states), but would be more complicated in another (some manifestations do not fit well into those three categories).

For instance, scientists and engineers can agree that a *gas* is a formless fluid, which at standard temperature and pressure (STP) will completely fill any container it may be put into. A *liquid* is a fluid that takes the shape of its container at STP, and may overflow it or fail to completely fill it, depending on the comparative volumes of the liquid and the container. A *solid* at STP is a crystalline structure with molecules, or atoms, packed so tightly in a formal organized fashion that it retains its shape, regardless of the shape and volume of its container.

Now, those definitions, though simplistic, are easily agreed to. The following terms are sometimes misused, however: vapor, aerosol, dust, fog, fume, mist, and smoke. A *vapor* is the gas phase of a liquid or solid. Therefore, when we speak of vapor, the environment cannot be at STP conditions. At STP, a vapor will condense back to its original liquid or solid form. The next six terms are all forms of *aerosol*, which is a dispersion of microscopic particles in a gas, usually air. The particles may be solids, in which case they are either dust, fume, or smoke, or they may be liquid, that is, either mist or fog.

We have known *dust*, airborne solid particles, since we were children and watched dust motes in the sunlight. Dust ranges from 0.1 to 50 microns (μ). Actually, dust particles can be much larger, but for our purposes in exposure assessments, the larger particles are not very interesting because they drop out of suspension in air almost immediately after leaving the source. The smaller end of the range (0.1–10 μ), called *respirable* dust, is much more relevant to exposure assessment. The naked eye can see particles that are 50 μ or greater in diameter. A microscope is required in order to examine particles less than 50 μ in diameter. The smaller the diameter, the longer the particle remains in suspension. *Fume* is a term that is widely misused. Many use fume interchangeably with vapor or gas. Few nontechnical persons realize that fumes are more related to dust than to vapors. A true fume is an aerosol of solid particles condensed from a gaseous state formed when metals melt. Fine particles, typically less than 1.0 μ in diameter, make up fume. *Smoke* is an aerosol formed of soot and carbon particles less than 0.1 μ in diameter. These microscopic, but solid, soot and carbon particles are generated from incomplete combustion of carbonaceous materials, such as coal, wood, or oil. Not all of smoke is solid particles, some droplets of liquid also may be found in smoke, mostly from condensing vapors.

Liquid dominated aerosols exist in nature, as well as being anthropogenic. A *fog* is a visible liquid aerosol caused by condensing vapors. A *mist* is composed of suspended liquid droplets formed either by vapor condensing to liquid, or by the mechanical dispersion of a liquid into the air. Examples of the latter case include atomizing, foaming, or splashing. The term mist implies that the liquid droplets have somehow been finely divided and suspended in air.

RISK ASSESSMENT

As stated several times, expect several more mentions, the work product of the exposure assessment is the raw material of the risk assessment. Several terms used by EPA, OSHA, and others overlap to varying degrees with the broad concept of a risk assessment: hazard assessment; comparative risk assessment; relative risk assessment; cumulative ecological risk assessment; and environmental impact statement. Hazard assessment as used by EPA means either the evaluation of the intrinsic effects of a stressor or the defining of a margin of safety or safety quotient by comparing the effects of a toxicologically significant concentration with an exposure estimate. The OSHA use of the term differs only in the setting of interest and the receptor is always an employee or group of employees.

A comparative risk assessment is a process that uses expert judgment to evaluate the relative magnitude of effects and set priorities among a wide range of environmental problems. Sometimes this process is similar to the problem formulation portion of an exposure assessment, the outcome of which may help select topics for further evaluation and help the risk manager focus limited resources on areas having the greatest risk reduction potential. Otherwise, a comparative risk assessment is essentially a preliminary risk assessment. Either way, the comparative risk assessment is considered the purview of the EPA and state counterparts, though industrial risk assessors could use it as tool similar to the preliminary design phase in the construction of a grassroots plant or in the selection of new process equipment. The relative risk assessment process is similar to comparative risk assessment, involving the estimation of the risks associated with different stressors or management actions. Relative risk connotes, to some people, the use of quantitative risk techniques, while comparative risk approaches more often rely on expert judgment. Others, as the EPA points out in its *Ecological Risk Guidelines*, do not make this distinction.

The cumulative ecological risk assessment considers the aggregate ecological risk to a target entity caused by the accumulation of risk from multiple stressors. An environmental impact statement (EIS) is an assessment required under the National Environmental Policy Act (NEPA) to fully evaluate environmental effects associated with proposed major Federal actions. Like ecological risk assessments, the EIS typically requires a scoping process analogous to problem formulation, an analysis by multidisciplinary teams, and a presentation of uncertainties. By virtue of special expertise, EPA may cooperate with other agencies by preparing EISs or otherwise participating in the NEPA process.

CONCLUSION

An entire book could be dedicated to the discussion of these and other terms used in exposure assessments. Let this preliminary discussion of these terms suffice to begin our discussion of calculating chemical exposures on the same game board, at least.

QUESTIONS (ANSWERS BEGIN ON PAGE 401)

1. Explain the difference between exposure and dose. Why is exposure used in lieu of dose?
2. Differentiate acute and chronic exposure from acute and chronic health effects.
3. The molecular weight of ethyl acrylate is 100.1; what is the exposure in ppm, if sample taken in the workplace at STP conditions yields a concentration of 4.16 mg/m^3?
4. The molecular weight of phenyl glycidyl ether is 150.1; what concentration would be measured in mg/m^3, if the exposure were 5 ppm?
5. Why is the biological effective dose not used?
6. If a 100 Kg person eats 1 Kg of contaminated food per day and receives a dose of 0.1 mg/Kg–day, what concentration was he exposed to if the interspecies scaling factor is 1.0?
7. Discuss the meaning of NOAEL in a study population of forty. What implication does this have on dosing of the study animals? How would you respond to someone who ridicules a study because the study animals were dosed with a much larger amount of the study substance than would be reasonable to consume otherwise?
8. Who decides what risk is acceptable? Why is the acceptability of risk not consistent from person to person? From hazard to hazard for one person?

REFERENCES

Ayers, Kenneth W., *et al*. *Environmental Science and Technology Handbook*. Rockville, MD: Government Institutes, Inc., 1994.

Brauer, Roger L. *Safety and Health for Engineers*. New York: Van Nostrand Reinhold, 1990.

Burmaster, David E. and Jeanne W. Appling. "Introduction to Human Health Risk Assessment, with an Emphasis on Contaminated Properties." *Environment Reporter*. April 7, 1995. pp. 2431–40.

Caravanos, Jack. *Quantitative Industrial Hygiene: A Formula Workbook*. Cincinnati: American Conference of Governmental Industrial Hygienists, 1991.

Cockerham, Lorris G. and Barbara S. Shane. *Basic Environmental Toxicology*. Boca Raton, FL: CRC Press, 1994.

EPA Proposed Guidelines for Ecological Risk Assessments.

Hallenbeck, William H. *Quantitative Risk Assessment for Environmental and Occupational Health*. 2d. Ed. Chelsea, MI: Lewis Publishers, 1993.

Manuele, Fred A. "Learn to Distinguish between Hazards and Risks." *Safety & Health*. November 1994. pp. 70–4.

Neey, W. Brock. *Introduction to Chemical Exposure and Risk Assessment*. Boca Raton, FL: Lewis Publishers, 1994.

OSHA. *OSHA Preamble and Proposed Rule to Revise Air Contaminants Standards for Construction, Maritime and Agriculture*. 57 FR 26002. June 12, 1992.

Piantanida, Lillian G. and Thomas J. Walker. "Industrial Hygiene Aspects of Occupational Chemical Exposure." In *Proctor and Hughes' Chemical Hazards of the Workplace*. 3d. ed. New York: Van Nostrand Reinhold, 1991.

Proctor, Nick H. "Toxicological Concepts — Setting Exposure Limits." In *Proctor and Hughes' Chemical Hazards of the Workplace*. 3d. ed. New York: Van Nostrand Reinhold, 1991.

Talbott, Evelyn O. and Gunther F. Craun. *Introduction to Environmental Epidemiology*. Boca Raton, FL: CRC Press/Lewis Publishers, 1995.

3 Basic Chemical Hazards to Human Health and Safety — I

Chemicals, whether natural or anthropogenic, present hazards to humans. Properly stored, handled, and used, chemicals present minimal hazards, but rarely, if ever, can the risk be said to be zero. The probability of an undesirable exposure occurring is something greater than zero. In exposure assessment, we cannot speak in absolute terms such as **always** or **never**. Fallible environmental engineers and industrial hygienists hope, at best, to reduce the risk of an undesirable exposure to something approaching zero. The whole premise of this book is that, for whatever reason, the hazards of chemical exposure are not minimized, and, therefore, we need to find out exactly what hazards exist, so someone may take the exposure information we give them and make a policy decision about making the workplace, or outside environment, safer. If the decision belongs to a higher authority, the exposure assessment is taken one step further by analyzing the probability of any specific person, or community, being harmed, and the entire project becomes a risk assessment. For our purposes, we will stop with the exposure assessment proper. We will discuss risk assessment along the way, with respect to exposures. Before learning to predict exposures, then, let's examine the specific hazards of chemical exposure to humans.

The identification of hazards is the starting point for all exposure assessments, and involves the determination of whether a chemical or contaminant of concern (COC) is causally linked to particular health effects or not. Generically speaking, four hazards of chemicals exist: fire, irritation and burning of tissue, violent reactions, and toxic effects. Typically, the immediacy of exposure to one of these general hazard classes is shown in Table 3.1.

While not precisely true for all cases, Table 3.1 represents how we need to deal with chemicals in general. Some poisons (toxic hazard) present acute hazards, for instance. Would you be worried about the long term health effects of cyanide, if you spilled a beaker of undiluted acid onto a sack of sodium cyanide? No, of course not, you would get out the room as quickly as you could, or you would quickly become deceased.

Can you think of an ignitable that is a chronic hazard? If you drink a thimble full of gasoline, daily, for several years, expect some internal damage. However, we do not normally drink any amount of gasoline. If we did, it would be accidentally, and an event that happens every day for thirty years is not accidental. However, a gas-station attendant might breathe gasoline vapors, for that length of time, and, given the trace amount of benzene in gasoline, we expect him or her to develop liver cancer, eventually. So, as you see, Table 3.1 should be taken generically, because it

---

TABLE 3.1
Immediacy of Exposure

Acute			Chronic
Ignitable	Caustic	Reactive	Toxic

TABLE 3.2
General Chemical Properties and Information

Constituents by % wt. or % vol.

Impurities
Molecular weight
Total quantity of material
Color
Odor
Hygroscopicity
Appearance
Physical state
Solubility
Vapor pressure
Viscosity

After IRI.

---

may not fit every specific example you might think up. By the way, Table 3.1 represents the logic EPA used in writing hazardous waste regulations under the authority of the Resource Conservation and Recovery Act. While having to force fit some of the discussion below, I have used this logic for presenting this chapter, and the next one, on human health hazards in order to demonstrate how exposure assessments help people live more healthful lives, but also how they fit into the grand scheme of complying with regulations. While air, water, toxic substance, emergency release, and uncontrolled release regulations do not use the same logic, all hazardous or toxic substances can generally be fit into one or more of these four categories.

Some chemicals present multiple hazards. Phenol, for instance, is ignitable, highly corrosive, and toxic too. Therefore, we expect an overexposure to phenol to potentially have both acute and chronic effects on our body. Some general properties and information you should know about a chemical are listed in Table 3.2.

Obviously, knowing the constituents of a mixture is important, because we rarely find data on mixture, but on pure chemicals. Preferably, you want to know each constituent's contribution on both a weight and volume percent basis, but, if you know either, you can calculate the other. Weight and volume are related by the molecular weight.

Example 3-1. You are dealing with a mixture that contains, on a volume basis, the following:

Toluene	50%
Xylene	25%
Methyl Ethyl Ketone	25%
Total Mixture	100%

What is the weight fraction of each?

Toluene	$= CH_3C_6H_5 = 92$ lb./lb.-mole
Xylene	$= C_6H_4(CH_3)_2 = 106$ lb./lb.-mole
Methyl Ethyl Ketone	$= CH_3COC_2H_5 = 72$ lb./lb.-mole

On a 100 mole basis:

50 mole toluene	= 4,600 lb.
25 mole Xylene	= 2,650 lb.
25 mole MEK	= 1,800 lb.
100 mole mixture	= 9,050 lb.

Therefore, the weight fraction of each component is:

Toluene	= 50.8%
Xylene	= 29.3%
MEK	= 19.9%
Total	= 100.0%

If you had elected to use the volume fraction data, as is, to represent weight fraction, and you were interested only in the toluene constituent, your results would be nearly accurate, accidentally. If you were interested in all the components, or in Xylene or MEK individually, you would have been off by five percent. The Material Safety Data Sheet (MSDS) for the mixture should give you the volume or weight percentage of the components, and most manufacturers are good about specifying which. Call the number on the MSDS, if either weight or volume is not specified. Most report in volume percentage, but you cannot count on this, so do not assume.

Impurities are important if they impact the behavior of the mixture, or if they add their own set of health problems. Molecular weight is important for the reason you have just seen demonstrated; it related weight and volume. The total quantity of material tells us how much is available to spill into the ground, or to evaporate and release into the air. Color is an identifying factor and, while it has no direct health effect, can be important information. Odor is another identifying characteristic that can be important or not. *Hygroscopicity* is the readiness of a substance to absorb moisture from the atmosphere. This has implications on the potency of a chemical, especially a water reactive one that undergoes violent reaction when mixed with water. Hygroscopicity can also be an important factor with a chemical that reacts with water nonviolently to produce an inert mass. While appearance of the mixture has no effect on health, it is another identifying characteristic and, therefore, is important information.

The physical state of the mixture and its components is very important to know. Think of water. It behaves completely different as liquid, solid (ice), or vapor (steam), and each state has its own peculiar health hazards. While the physical state can often be guessed intuitively, this is not always the case.

Solubility of each component is important from a partitioning standpoint. Where will the component go when released into the environment? Solubility is a chief factor in the answer to this question. Water soluble components may be carried great distances by flowing streams, surface runoff, or groundwater. Insoluble components generally stay close to the release point, unless something else is available for transportation.

Vapor pressure is another critical transport property. Vapor pressure varies greatly with temperature, so the temperature should always be specified with the pressure at standard atmospheric conditions. Remember, if the MSDS gives only one vapor pressure point, assuming it is not at the boiling point of the material, then you automatically have two points. That is because the vapor pressure of any compound at its boiling point is 760 mmHg or 1 atmosphere.

> **Example 3-2.** If the vapor pressure of trichloroethylene is given as 69.0 mmHg at 25°C, and its boiling point is given as 87°C, what is its vapor pressure at 105°F?
>
> Since vapor pressure is proportional to temperature, the most direct approach would be to interpolate the desired pressure.
>
> 105°F = 40.6°C
>
> From 25 to 87°C, the vapor pressure changes from 69.0 to 760 mmHg, or 11.15 mmHg/°C. Therefore, from 25 to 40.6°, the change is 173.86 mmHg.
>
> So, at 40.6°C, the vapor pressure is 243 mmHg.

Viscosity is an important transport property where a large amount of material is released or in determining flow through conduits.

Now, with these preliminaries out of the way, let's examine some of the health effects of chemicals on humans. This is, necessarily, a precursory and nonexpert look at physiology and biochemistry, but adequate for the purposes of an exposure assessment.

CHEMICALS WHICH IGNITE EASILY

When evaluating the flammability of any material, the properties listed in Table 3.3 must be determined. Flammability (OSHA), or ignitability (EPA), presents an imminent hazard to workers on site and also the local population, in the event of an explosion, or a spreading fire. Essentially these chemicals release vast amounts of energy, either slowly in the form of fire, or quickly in the form of an explosion, which has the potential for mass — that is widespread — destruction.

Generically, *flammable* substances burn readily, whereas *combustible* materials require more energy input to sustain combustion than flammable materials do. Hence,

TABLE 3.3
Properties Related to Flammability

Flash point
Fire point
Explosive limits
Specific gravity, liquid phase
Vapor density
Vapor pressure
Heat of vaporization
Boiling point at pressure of concern
Ignition temperature
Autoignition temperature
Spontaneous heating
Dielectric constant
Melting point
Flow point
Percent volatile
Products of decomposition
Heat of fusion

After IRI.

OSHA and DOT distinguish between flammable and combustible materials. EPA refers to *ignitable* waste materials as hazardous waste, and avoids the terms flammable and combustible. That is because *ignitability* is a characteristic of a liquid, describing the ability of its vapors to be ignited in the presence of an ignition source. Ignitability is also a characteristic of solids that catch fire due to friction, or ignite when water comes into contact with the substance. Some easily ignited compressed gases and oxidizers are also classified by EPA as ignitable. Ignitability, for liquid wastes, is defined as having a flash point less than 140°F. Two subclasses of combustible materials (Class II; Class IIIA), as defined by NFPA, OSHA, and DOT, break at 140°F, so ignitable wastes include all DOT/OSHA flammable liquids (Class I) plus some combustible liquids (Class II).

Knowledge of flash point is essential to understanding ignitability. Essentially, molecules of a liquid cross back and forth between the bulk liquid and the air over it. After a short time, a vapor cloud exists over the liquid, and unless drawn away by sufficient ventilation, can become substantially concentrated. At some point, this vapor cloud may become concentrated enough to form a combustible mixture with air, if sufficient heat is present. The temperature at which this combustible cloud of vapor ignites is the *flash point*. Several devices are available to measure flash point, so sometimes it is important to know the device because of variability in measurements, however, precision is not usually a factor in exposure assessments. Precision is more typically required in determining whether a waste material is a hazardous waste by characteristic of ignitability.

Whereas the flash point is the lowest air temperature at which a vapor will ignite when an ignition source is introduced, the *fire point* is the lowest bulk liquid temperature at which the liquid evolves enough vapor to support continuous combustion. The two characteristics are numerically close together, but not the same.

Explosive limits define the range of vapor concentrations in air that constitute an explosive mixture. The *upper explosive limit*, or UEL, is the highest concentration of the vapor that will explode. Any concentration of vapor above the UEL will not explode and the atmosphere is said to be too *rich*. The *lower explosive limit*, or LEL, is the lowest concentration of the vapor in air that will explode. Any concentration of the vapor less than LEL will not explode and is said to be to *lean*.

The specific gravity of the liquid phase is important to know, in order to convert from mass to volume or from volume to mass. Specific gravity is the unitless ratio of the density of a liquid to the density of water. While temperature affects the specific gravity, under ordinary environmental conditions the density, or specific gravity of most liquids do not change much. Hence, the imprecision introduced by ignoring temperature difference from standard conditions does not affect the precision very much. Often, a question asked is, if you have a given volume, such as a drum, of material, how much does it weigh?

> **Example 3-3.** One gallon of pure ethyl bromide, with a specific gravity of 1.46, spills from a drum. In order to enter the reportable quantity table, you need to determine the amount in pounds that spilled.
>
> Gallon, being a volumetric measure, is constant regardless of the material. Therefore, a gallon of ethyl bromide and a gallon of water have the same volume, but weigh different amounts in proportion to the specific gravity. A gallon of water weighs 8.34 pounds.
>
> 1 gal × 8.34 lb. water/gal. × 1.46 lb. ethyl bromide/lb. water = 12.2 pounds.

Vapor density is another characteristic we need to know about flammable or ignitable materials. The vapor density can be reported as a true density, weight of vapor per volume of air, or as a specific gravity related to the density of air. Temperature is critical with vapor density, because the volume of vapor may change drastically with a small increase or decrease in temperature; thus the density is vastly changed, too. Charles' law can be used to determine volume and, hence, vapor density changes with temperature. The law of Charles states:

> If pressure remains constant, the volume of a given mass of gas is directly proportional to the absolute temperature.

$$V/T = K; \therefore$$

$$V_1/T_1 = V_2/T_2 \tag{3.1}$$

where: k = a constant
         V = volume
         T = absolute temperature.

The density is determined using the new volume.

$$n/V = P/RT \qquad (3.2)$$

where: P = pressure
       n = moles
      R = universal gas constant.

Ideal Gas Constant R

1,545 ft-lb.$_f$/lb.-mole °R	8.313 kJ/K kg-mole
0.7302 atm ft³/lb.-mole °R	8.313 kPa m³/K kg-mole
10.73 psia ft³/lb.-mole °R	82.06 atm cm³/K g-mole
21.9 inHg ft³/lb.-mole °R	0.08206 l atm/K g-mole
1.987 Btu/lb.-mole °R	1.987 cal/K g-mole

Or more simply, the density ratio is:

$$D_1/D_2 = T_2/T_1 \qquad (3.3)$$

where

$$D = \frac{n}{V} \times M \qquad (3.4)$$

where M = molecular weight. Standard temperature and pressure (STP) vary according to usage. Scientists use one number and engineers another. Be certain you understand which STP is used:

Standard Temperature	Standard Pressure
Scientific, metric	
0°C	1 atm
0°C	760 mmHg
0°C	1,013.25 mb
Scientific and Engineering, English	
32°F	29.92 inHg
32°F	14.7 psia
Natural Gas Industry	
60°F	14.7 psia
Heating Ventilation Air-conditioning (HVAC)	
77°F	14.7 psia

We discussed Dalton's law and vapor pressures above. Obviously, the greater the vapor pressure of a liquid the more of it will evaporate in a given time period. This is important because liquids and solids do not burn, vapors do. What appears to be the log burning in your fireplace is actually the vapors generated by the superheated woody substance. This is an oversimplification, a complex series and amount of chemical reactions are taking place, which we summarize by labeling the whole process combustion. Charring begins with dehydration caused by radiant heat from a flame or hot surface. As steam is released, it also damages the partially charred material. Some hydrogen is generated in the process and burning begins at the surface of the material, which not shows visible signs of charring, causing structural changes within the material, in this case, the skin. The hydrogen that is being released, fatty material that vaporizes, and other combustibles are releasing more heat as they burn, and fracturing appears on the char, allowing deeper levels to participate in the process. Again, this is a gross oversimplification, but covers what we need to know in a nutshell.

Heat of vaporization becomes important for calculating evaporation and BLEVE (boiling liquid, expanding vapor explosion) potential of flammable liquids. Typically, the pressure of concern is atmospheric pressure, so the boiling point can be obtained from standard handbooks, if not given on the MSDS for the chemical product or the profile sheet of the waste material.

The *ignition temperature* is the lowest surface temperature of a material that will allow ignition to occur. *Pilot ignition* occurs when the combustible material surface temperature has reached the ignition temperature, and some external source, a pilot, causes ignition. Pilots include embers, flames, hot air (convection heat), radiant heat, or a mechanical or electrical spark. If ignition occurs without an external pilot, then it is called autoignition and the *autoignition temperature* is somewhat higher than the ignition temperature. *Spontaneous heating* is often the result of a heat buildup from oxidation of organic material, such as bacterial decomposition.

A dielectric is a poor conductor of electricity that sustains the force of an electric field passing through it, a property not exhibited by conductors. Typically, dielectric properties are caused by the polarization of the substance. When the dielectric is placed in an electric field, the electrons and protons of its constituent atoms reorient themselves, and in some cases molecules become similarly polarized. As a result, the dielectric is under stress, and stores energy that becomes available when the electric field is removed. Dielectric effectiveness is measured by the relative ability to store energy compared to a vacuum. This effectiveness is expressed in terms of the dielectric constant, with the value for a vacuum taken as unity. With respect to flammability and ignitability, the dielectric constant is a measure of the substances ability to retain a static charge, which could act as a pilot when released to ground. The ability of a dielectric to withstand electric fields without losing insulating properties is known as its dielectric strength. If two materials are good conductors, the static buildup between them, resulting from physical separation, is small because the electrons easily pass between the surfaces. However, if one or both materials are dielectrics, electrons become trapped on one of the surfaces and the magnitude of the static charge is greater. The streaming current, the flow of electrons from one surface to another, is directly proportional to the dielectric constant. Table 3.4 gives dielectric constants for several materials.

The melting point, of course, tells us when to expect melting, and the flow point when flowing will occur and may not be relevant to the exposure assessment in every case. The heat of fusion, sometimes referred to as the equivalent heat, or the latent heat of fusion, is the amount of heat that must be absorbed by a solid to change one gram of the solid to its liquid state with no change of temperature. The percent

TABLE 3.4
Dielectric Constants

Material	Dielectric Constant
Air	1.0
Benzene	2.3
Cellulose	5.7
Ethanol	25.7
Heptane	2.0
Hexane	1.9
Isopropanol	25.0
Methanol	33.7
Paraffin	2.1
Pyrex	4.8
Rubber	3.0
Slate	6.8
Teflon	2.0
Toluene	2.4
Water	80.4
Wood	3.0
Xylene	2.4

volatiles in a liquid mixture will be germane in most assessments. Once again, make sure you understand whether the percentage is expressed in volume or mass. As demonstrated above under component percent, it makes a difference how the percentage is expressed.The products of decomposition are typically listed on an MSDS for the benefit of firefighters and first responders when a fire is involved, but where these products are persistent in the environment, chronic effects can be expected. Also, in some cases products of decomposition may cause delayed effects, such as phosgene.Many flammable materials are also classified as explosives and those which are devices, such as ammunition, explosive projectiles, grenades, bombs, mines, torpedoes, rockets, missiles, blasting caps, dynamite, and fire works, are not considered part of the context of this book. However, solid, liquid, and gaseous explosive substances are very much a part of our concern as exposure assessors. Table 3.5 defines the explosive subgroups per DOT/OSHA. The explosive devices might also be considered for exposure assessment, but the knowledge and skills required to evaluate them are left to military and other specialists.

OSHA and the Department of Transportation (DOT) define flammable liquids as having a flash point less than 100°F according to the National Fire Protection Association method. The test for flash point is specific and its protocol is found in DOT regulations at 40 CFR 173.115, except that any liquid having a vapor pressure greater than 40 psi at 100°F is considered to be flammable. Mixtures having one or more components with a flash point of 100°F or higher that make up at least 99% of the total volume of the mixture are not flammable liquids. Another exception is

TABLE 3.5
Explosive Subgroups

Class A: A detonating or otherwise maximum hazard.
Type 1:   Can detonate by contact with sparks or flame.
Type 2:   Solid explosives that contain liquid explosive ingredient, which can detonate by a blasting cap.
Type 3:   Solid explosives that contain no liquid explosive ingredient, but can detonate by a blasting cap.
Type 4:   Solid explosives can detonate with sparks or flame.
Type 5:   Desensitized liquid explosives that may detonate separately or when absorbed in cotton, by
           blasting cap.
Type 6:   Liquid explosives that explode by impact in a drop of less than 10 inches.
Type 7:   Blasting caps.
Type 8:   Any explosive not specifically included above.
Class B: A flammable explosive.
Class C: An article containing small quantities of A or B.

any alcohol-water solution containing 24% or less alcohol by volume, if the remainder of the solution does not meet the definition of hazardous material (DOT). Wines, beers, and ales meet the definition terms of the latter exception. Note that most whiskeys are not an exception from the definition of flammable liquid.

A combustible liquid has a flash point from 100° to 200°F as measured per DOT protocol. Mixtures having one or more components, with a flash point of 200°F or higher, that make up 99% or more of the total volume of the mixture are not combustible liquids.

The U.S. Department of Transportation (DOT) has developed evacuation tables for different compounds for fire and explosive hazards and downwind air pollution hazards. The table gives suggested distances for isolating unprotected persons from spill areas involving a listed hazardous material, if the material is not on fire. Materials selected for inclusion in the Initial Isolation and Protective Action Distances Table have the potential to produce poisonous effects by means of vapor diffusion or advection.

Thermal burns injure the skin and deeper tissues. Skin exposed to temperatures as low as 120°F is burned in about five minutes. The severity of depends on the depth of the burn, extent or surface area covered by the burn, and the age of the victim. Burns are classified by depth as first, second, or third degree. First-degree burns cause redness and pain, affecting only the outer layers of the epidermis. Second-degree burns are marked by blisters. The epidermis is destroyed at the site of the burn, which extends into the dermis. Blisters of formed by damaged cells yielding fluid. The burn is extremely painful after the incident, due to the nerves of the dermis being irritated by the products of cellular destruction. In third-degree burns, both the epidermis and dermis are destroyed, and underlying tissue may also be damaged. Typically, no epidermis remains and underlying layers of skin are charred. The destruction of nerves in the dermis and subcutaneous tissues renders the remaining skin in the burned area nonfeeling. Inhalation of smoke significantly increases mortality during a fire. For fire survivors, however, thermal destruction of

---

TABLE 3.6
Symptoms of Irritation

Stinging, itching, burning eyestearing (lacrimation)
Burning nasal passages, throat
Nasal inflammation (rhinitis)
Cough
Sputum production
Chest pain
Wheezing
Breathing difficulty (dyspnea)

---

the skin exposes the victim to infection, which is the most common cause of death for extensively burned persons. Body fluids and minerals are lost through the wound and have to be replaced to prevent dehydration. The lungs, heart, liver, and kidneys are also affected by infection and fluid loss.

CHEMICALS WHICH IRRITATE AND BURN TISSUE

Damage to the skin and eye tissues, and mucous membranes, which are reversible after the exposure to a chemical, is defined as *irritation*. *Corrosion* is the irreversible tissue damage to skin or eye. Irritant effects do not go unnoticed for long; the affected individuals soon become aware of the problem, and seek to cease the exposure. However, when the irritation is slight, sometimes stopping the exposure receives such a low priority that the exposed individual learns to tolerate it. Irritants and corrosives (typically the difference depends more on pH than on chemical species) not only can damage the external tissues of the body, but the internal tissue as well. Inhalation of an irritant leads to sore throats, bronchitis, and may, in extreme cases such as high concentration or extremely low pH, cause pulmonary edema, which is fluid in the lungs (pneumonia). Irritants may also cause skin rashes. Corrosives, in the extreme case, can cause destruction of tissue (chemical burns). Table 3.6 lists some symptoms of irritation.

IRRITATION OF THE RESPIRATORY SYSTEM

Respiratory irritants are either primary or secondary irritants. Primary irritants act directly at the point of contact. These type irritants are discussed immediately below. As a class, primary respiratory irritants have no systemic effects either because the products formed by the tissues of the respiratory tract are nontoxic or because the irritant damage is far in excess of any systemic toxic action that may have been produced. Secondary respiratory irritants do have systemic effects and are discussed in the section on toxicity.

Respiration begins and ends with the nose or the nasal cavity, where irritants and corrosives may affect the nasal passage membranes and sinuses. The nasal concha are turbinate bones that create turbulence in the inspired air to aid the mucous membranes of the concha to remove particulate matter and water soluble vapors from the air. The nasal hair at the entrance of the nose removes larger particles and the mucous membranes in the concha remove finer particles.

Eight sinuses, four on each side of the nose, help equalize air pressure in the nasal cavity, and, because they are linked with the nasal passage, are easily irritated. The pharynx and larynx are also easily irritated by either inhaled or ingested irritants. On either side of the pharynx, behind the mouth, tonsil tissue can be susceptible to irritants. The pharynx serves as a passageway for air, but also for food and other ingested materials. At the lower end of the pharynx is the glottis, a narrow slit covering the opening into the larynx. Covering the glottis is the epiglottis a small flap that prevents swallowed materials from entering the larynx. This is where the pharynx separates into the larynx and esophagus. Next, on the respiratory side, are the trachea and bronchi, then the bronchioles and alveoli are attacked by irritants in the worst exposures.

Damage to the epithelial cells in the upper respiratory tract (URT) produces inflammation causing the permeability of blood cells to increase and an accumulation of defense cells from the immune system that can occur quite rapidly. In the worst case, the irritant causes necrosis of URT cells (cell death). Blood vessel permeability is proportional to the accumulation of fluids in a specific area of the body. Sulfur dioxide, formaldehyde, and other water soluble irritants create dyspnea, or difficult breathing, due to swelling and edema of the URT that narrow the URT airways. Secretion of mucus may complicate the matter.

Sputum production increases with irritation, and the nasal passages become inflamed, as well as the trachea and upper bronchial passages. Development of bronchitis means that the cilial clearance mechanism is compromised. The accumulation of mucus increases the risk of secondary bacterial infection. Wheezing may develop where the exposed person has a history of hyperactive airway disease.

Intense exposure of water soluble irritants that make it past the URT, can produce bronchioconstriction by causing the smooth muscles, which surround the bronchioles, to contract. Sometimes the irritant does not act directly, but increases the sensitivity of the smooth muscles to other substances, which cause the contraction. Asthma and other preexisting respiratory diseases make a person especially susceptible to irritants and the swelling and bronchioconstriction caused by them. Such materials may burn the parenchyma and fluid may suddenly collect in the interstitial spaces and alveoli, which is pulmonary edema.

Solubility is the primary factor that determines whether the irritant will act in the upper or lower respiratory systems. Alkaline materials, ammonia, hydrogen chloride, hydrogen fluoride, and other highly water-soluble substances irritate the upper respiratory tract, but get washed down the esophagus by mucus and swallowing. Substances that are partially soluble may affect both the upper and lower respiratory systems. These substances include diethyl sulfate, dimethyl sulfate, halogens, ozone, and phosphorus chlorides. Arsenic trichloride, nitrogen dioxide, and phosgene, which are insoluble in water, are primary irritants of the lungs. When the irritant reaches the alveoli, and, typically, the less water soluble species, such as nitrogen dioxide and ozone, do reach the alveoli, an accumulation of fluids creates an interference with the gas exchange between the lungs and the blood. The fluid in the alveoli acts as a barrier to diffusion. Also, oxidizers such as ozone may damage the membranes of the alveoli, or inhibit enzyme activity within the cells by lipid peroxidation. If the exposure continues for a long period of time, some chronic effects may be noticed in the lungs.Respiratory irritants induce changes in the biomechanics of respiration. Pulmonary flow-resistance increases, for example, with irritation. Compliance, a measure of the elasticity of the lungs, decreases with irritant insult. Irritants such as acetic acid, acrolein, formaldehyde, formic acid, iodine, sulfur dioxide, and sulfuric acid increase flow-resistance, decrease compliance only slightly, and decrease the frequency of breathing at higher concentrations. The nitrogen oxides and ozone have little effect on flow-resistance, but decrease compliance, and increase the respiratory rate. The potency of irritant aerosols is indirectly proportional to their average particle size. Small particle size aerosols are more potent than large particle size aerosols. Long-term exposure to respiratory irritants results in increased mucus secretion and a bronchitis-like pathology without the infection.

IRRITATION OF THE EYE

The sclerotic coat, or outside of the eyeball, is tough, fibrous tissue, but many nerve endings make irritant attacks quite painful. The layer just inside the sclerotic coat is the choroid, where many blood vessels nourish the eye tissues. The conjunctiva, the smooth layer protecting the cornea, is also susceptible to irritants.

The mechanism of eye irritation is believed to be the stimulation of nerve endings in the cornea. Active sites on the irritant molecule bind to sulfhydryl groups of eye tissue protein, inhibiting cellular respiration and cholinesterase.

Damage to the eyes can go beyond irritation of tissue. The cornea, lens, retina, ganglion cells, or optic nerve can suffer transient damage, temporarily disabling damage, or permanently blinding damage. The degree of damage depends on exposure severity, susceptibility of the victim, and the specific stressor.

Delayed symptoms are presented for painful corneal injuries by the vapors of certain chemicals. Not only are the symptoms delayed, but they persist painfully for hours. The vapor of methyl silicate is among this group. Typically, no pain is experienced during the exposure, but several hours later burning begins, vision blurs, conjunctival hyperemia occurs, as well as tears, photophobia, and uncontrollable squinting. Such delayed reaction chemicals inhibit enzymes, denature certain proteins, alter DNA, and interfere with the mitotic process of the cells of the eye. After exposure, affected cells die, although the damage is typically only a few layers of cells and regeneration repairs the damage. However, sometimes the damage involves the corneal stroma and endothelium (internal boundary layer), and, when this happens, scarring, vascularization, opacity, and loss of vision are the consequences. Equally unfortunate, these substances give no warning by causing immediate eye pain. By the time the eye becomes painful, the damage has been done.

The cornea, choroid, and conjunctiva are the eye's front line of defense against chemical attack. However, the cornea must be transparent in order to admit light in all frequency bands, so any scar formation after an injury renders the eye incapable of seeing at least some frequency range. The transparency of the cornea is maintained by thin inner and outer boundary layers. The death of cells in these boundary layers results in loss of transparency. Sometimes the outer boundary layer (corneal epithelium) can regenerate, if the depth of necrosis is slight.

Another group of chemicals causes a painless edema of the corneal epithelium, accompanied by delayed onset of visual haloes. That is, the victim sees a halo around his or her field of vision, especially around light sources. N-ethylmorpholine, for instance, causes swelling of the epithelia cells, diffracts the light entering the cornea, and places a colored halo around light sources.

The lens is a transparent, avascular tissue that is surrounded by a thin, collagenous capsule. Most of the lens is made of long, thin fibers forming closely packed, onion-like layers. Transparency of the lens depends on the highly ordered organization of cells; the size, shape, and uniformity of the fiber bundle; molecular structure of the fibers; and the regularity of the packing. COCs that interfere with lens metabolism, interrupt transport across cell boundaries, or interfere with the integrity of the lens capsule, can cause the lens to lose transparency and ultimately cause loss of visual acuity. When the lens transparency changes, the victim is said to have cataracts.

IRRITATION OF THE SKIN

Three layers of tissue in the skin offer natural protection from chemical and biological attack and provide our sense of touch or feel, as well. Each of these main layers is sub-layered. The top layer of skin is the *epidermis*, the deepest levels of which are constantly producing new layers of skin. The top of the epidermis is the horny layer, which are nonliving cells. The living part of the epidermis is found deep, where blood from the dermis interfaces. The separation of the epidermis from the dermis, due to excessive rubbing and friction between the two layers, produces blisters, with the epidermis providing the blister bubble and fluid from damaged dermal cells to fill it. The true skin is the dermis and provides the thickness of the body, contains blood vessels, nerves, nerve receptors, hair follicles, sweat glands, and oil glands. The subcutaneous layer, which is not part of the skin, contains fat lobules, dense connective tissue, blood vessels, and nerves. The dermis is anchored to the subcutaneous layer by collagenic fibers.

Skin irritants are often primary irritants, that is, they cause irritation at the point of contact. Concentrated aqueous solutions of acids and alkalis can be severely corrosive and irritating to tissue. Ongoing exposure to dilute acidic or alkaline compounds leads to dermatitis, if more immediate burns are not experienced. Acids cause irritation and burning almost immediately upon contact, but an alkaline solution may produce delayed burning as the attacking solution slowly dissolves the fatty layer of the skin. Alkaline solutions often do not burn upon contact, but give a slippery, or soapy, feeling to the skin caused by extracted fat. Alkalis damage the keratin layer by softening and removing cells. Simultaneously, alkalis dehydrate the keratin layer, leading to drying and cracking of the skin. The eruption of dermatitis is typically on the hands and arms when the COC is liquid or solid, and on the face and neck, if the COC is a gas or vapor. Dermatitis on other parts of the body indicates the COC has penetrated clothing. Sometimes irritating vapors and gases will cause an eruption of dermatitis at points of friction, such as under a belt, or where the sock elastic binds the leg. However, in cases where a sensitivity or allergy is developed, the dermatitis may be generalized, rather than local. The COC may simply stain the skin.

Contact dermatitis is a general condition that may have any of several causes. However, dermatitis may also be caused by the body's immune system response. In this case, the dermatitis is due to the presence of antigen peptides in major histo-compatibility complex (MHC) class II context in the afferent phase with $CD4^+$ cells (another type of antigen).

Some irritants and skin sensitizers cause contact-type acute eczematous derma-titis, the symptoms for which are reddening (erythema), edema, sores of various kinds (papules, vesicles, bullae, and crusts), and peeling or flaking (desquamation). Symptoms for chronic eczematous dermatitis, of the contact type, include erythema, lichenification, and cracking (fissuring). Punctate ulcers (holes) are uniquely a symp-tom of chromic acid exposure. Purpura is common with hepatotoxins. Folliculitis and boils are typical symptoms of oil and grease exposure. Acne is the most common form of dermatitis caused by chlorinated compounds. Coal pitch and its derivatives cause light sensitivity (photosensitization) upon contact with the skin. Keratosis and cancer are the common results of chronic exposure to coal tar, pitch and its deriv-atives, and soot. Beryllium causes skin granulomas to develop. A condition called hyperpigmentation, or leukoderma, is caused by the benzyl ether of hydroquinone. Hair damage and loss (alopecia) can also be forms of dermal damage.

IRRITATION IN GENERAL

Irritation and corrosion of tissue are a threshold phenomenon, in that, for any given irritant, some concentration exists for which few persons, if any, develop symptoms of exposure. This is the sensory irritation threshold for that chemical. An increase in exposure means that more and more people will begin to feel the effects of the irritant, which will become increasingly severe, and, at some exposure con-centration, all exposed persons will feel the effects, although intensity of responses will vary from person to person. By the time mucus membrane irritation produces objective signs of irritation, such as coughing, wheezing, conjunctivitis, or tearing, medical treatment may be required, and the exposed individuals are at risk from increased absorption of the substance and decreased resistance to infections. Repeated exposures may cause the exposed persons to become inured to the warning properties of the irritant, and increase the risk of exposing themselves to greater concentrations, at which exposures other health effects may come into the picture.

CHEMICALS WHICH REACT DANGEROUSLY

For those chemicals that have dangerous reactions, the properties listed in Table 3.7 need to determined.

Many dangerous reactions require the presence of one or more catalysts to proceed at normal temperatures and pressure. Therefore, a good management prac-tice is to keep these materials separated until the reaction is needed and conditions

---

TABLE 3.7
Reactive Properties

Catalysts
Degree of dilution with nonreactive media
Degree of mixing
Heat of reaction
Physical form of reactants
Pressure of container
Qualitative analysis of gases/vapors released
Rate of mixing
Rate of reaction
Required energy of activation

---

are under control. Another way to control violent reactions is to dilute the reactants with nonreactive media. This is not an alternative in every case, and water, the general dilutent, reacts vigorously with some chemicals. The degree of mixing can be an important factor in the violence of a reaction. A certain degree of mixing of reactants may be required to start the reaction. Again, the prevention is to avoid mixing the reactants until the proper time, if ever. The heat of reaction is important for two reasons. First, the exposure assessor must know how to calculate potential thermal exposure from a violent reaction. On the other hand, the heat of reaction may be high enough that it is nearly impossible to accidentally provide adequate heat to start the reaction.

The physical state of the reactants (solid, liquid, or gas) typically matter when it comes to predicting the sudden release of energy from a violent reaction. Pressure is often another important factor. Sometimes, one or more of the reactants must be stored at a high pressure to prevent the initiation of the dangerous reaction, but other times the reactant must be kept at atmospheric pressure until the reaction is desired at some higher pressure. Mostly, the former case applies. It is hard to have an accidental pressurization, whereas an accidental depressurization is too easy. Often, the products of violent reactions are noxious gases and vapors.Therefore, where uncertainty exists, a qualitative analysis of the volatile reactions products is recommended, if this information is not provided on MSDSs or elsewhere. The rate of mixing sometimes a controlling factor in the kinetics of a reaction. A literature search of standard handbooks and research papers often turn up rate of reaction, or kinetics, data. Nevertheless, many reactions have never been studied. Finally, the required energy of activation provides a hint as to how readily the reaction will initiate itself. Table 3.8 lists properties contributing to dangerous reactions.

Reactive substances are ill-defined in regulations, but generally have a tendency to undergo violent chemical change. RCRA defines reactives as chemicals that are unstable or undergo rapid or violent reactions. In a sense, explosives are included in this category, and some regulations, including RCRA regulations, list explosives as examples of reactive chemicals. Properly speaking, explosives are among the flammable/combustible group, but with respect to chemical reaction, however, who can argue that the effects of explosives are not violently reactive. Pyrophoric, or

---

TABLE 3.8
Properties Relative to Dangerous Reactions

Differential thermal analysis
Impact test
Thermal stability
Detonation with blasting cap
Drop weight test
Thermal decomposition test
Lead block test
Influence test
Self-acceleration decomposition temperature
Card gap test
Critical diameter
Limiting oxygen value
Hazardous decomposition products
Incompatibility
Self-reactivity
Instability
Shock/friction sensitivity
Decomposition temperature
Specific heat
Gas evolution
Adiabatic temperature rise
Heat of reaction

---

fire-starting, materials are generally included as reactive substances. Water reactive materials are certainly included, as are cyanide- and sulfide-bearing materials that can release poisonous gases upon contact with acids.

Certain chemical structures are associated with violent reactivity. Strong oxidizing or reducing agents are considered reactive substances. Oxidation is the increase in oxidation number, or the loss of electrons, and reduction is the decrease of oxidation number, or the gaining of electrons. Oxidation and reduction are coupled reactions. That is, for every oxidation, an opposite reduction exists. The oxidizing substance causes the oxidation of another and gets reduced in the process. That means the oxidizer causes another substance to lose electrons, and gains electrons itself. The reducing substance causes reduction of another and gets oxidized in the process. That means the reducer causes another substance to gain electrons and loses electrons itself.

Some substances, such as the halides fluorine (UN1045), chlorine (UN1017), and bromine (UN1744), are self-oxidizing, acting as both an oxidizer and a reducer. Therefore, interhalogen compounds, such as BrCl (UN2901), BrF_5 (UN1745), BrF_3 (UN1746), ClF_5 (UN2548), ClF_3 (UN1749), ICl (UN1792) or IF_5 (UN2495), whether combined with anything else or not, are very reactive because they can undergo self-oxidation-reduction. These compounds react with water to produce the halide acids (HBr, HCl, HF, HI) and elementary oxygen (O). These powerful corrosives burn tissue through acidification, oxidation, and dehydration.

Single compounds containing both oxidizing and reducing units are called *redox* compounds and other examples of these compounds include nitrogen halides, halogen oxides, halogen azides, and hypohalites. Nitrogen halides are NCl_3, NI_3, or NF_3.

The reaction of NCl_3 and water is a strong redox type reaction as shown in Equation. 3.5.

$$NCl_3 + 3H_2O \rightarrow NH_3 + 3HOCl \qquad (3.5)$$

BrO_2 and ClO_2 are examples of halogen oxides. These compounds are highly unstable and toxic. The azide ion is N_3^- and ClN_3 is an example of a halogen azide. The hypohalites are very powerful bleaching agents, such as hypochlorite, ClO^- (UN1791). These compounds are strong irritants and toxic. Nitrous oxide, N_2O (UN1070 as a compressed gas; UN2201 as a cryogenic liquid), is another strong oxidizer.

Allene is an unstable organic gas having two double carbon bonds, $C=C=C$. Another unstable double carbon bond is the diene structure such as 1,3-butadiene, $C=C-C=C$, 1,2-butadiene, $C-C-C=C$, and 1,4-butadiene, $C-C=C-C$. Unsaturated organic molecules such as these can undergo violent polymerization. Azo-compounds and triazenes are also unstable and reactive. The azo group, $-N=N-$, is a common component, as a chromophore, in organic dyes. The triazene structure, not to be confused with triazine pesticide, is an organic containing $-N=N-N$.

CHEMICALS WHICH HAVE TOXIC EFFECTS ON HUMANS

In this age, people are vitally interested in the effect natural and *anthropogenic* (that is, human-made) substances have on the body, its health, and well-being. Civilization has advanced in many wonderful ways, but at what cost? The question begs an answer. The task of the exposure analysts is to determine a reasonably accurate answer about the cost.

Scientists and technologists involved with chemical exposure must keep three questions in mind at all times:

- What is the extent of human exposure to this chemical?
- What health effects are anticipated from such exposure?
- What magnitude of health problem does the chemical cause?

Table 3.9 lists properties necessary for evaluating toxicity. Some of these properties were discussed in the chapter on nomenclature, the remainder are discussed below. Toxicological ratings have been developed per Table 3.10.

If the substance causes *systemic damage*, no single organ can be identified as the target of low doses. Sax and Lewis assign a hazard rating to chemicals according to the scheme in Table 3.11.

Several organ systems may be affected simultaneously. A variety of adverse signs and symptoms may be indicated, when exposed to a systemic toxin. Nonspecific indicators of systemic poisons are dizziness, respiratory irritation, blood in the

TABLE 3.9
Properties Relative to Toxicity

Threshold limit value (TLV)
Time-weighted average (TWA)[1]
Lethal concentration (LC50)
Lethal dose (LD50)
Lowest lethal concentration (LCL)
Lowest lethal dose (LDL)
Lowest toxic concentration (TCLo)
Lowest toxic dose (TDLo)
Chronic effects
Acute effects
Cumulative effects
Carcinogenicity
Mutagenicity
Teratogenicity
Neoplastic effects

[1] TWA is not a separate property, rather
the time-basis for exposure limits proposed by various organizations such as
the TLV-TWA by ACGIH.

After IRI.

TABLE 3.10
Toxicological Rating Scheme of Sax and Lewis

U Unknown	Unknown toxicology
0 No toxicity	No toxic effects
1 Low toxicity	
Acute local	Slight irritation
Acute systemic	Slight toxic effects
Chronic local	Slight, reversible damage
Chronic systemic	Slight, usually reversible toxic effects
2 Moderate toxicity	
Acute local	Moderate irritation, tissue corrosion
Acute systemic	Moderate toxic effects
Chronic local	Moderate corrosive damage
Chronic systemic	Moderate, perhaps reversible effects.
3 High, severe toxicity	
Acute local	Life-threatening, disfiguring burn
Acute systemic	Life-threatening, toxic effects
Chronic local	Permanent or irreversible damage
Chronic systemic	Death or serious physical impairment

TABLE 3.11
Sax & Lewis Hazard Ratings

HR: Hazard Rating	LD50
1	4,000–40,000 mg/kg
2	400–4,000 mg/kg
3	<400 mg/kg

urine, chest tightness, weight loss, decreased rate of weight gain in children, lethargy, loss of appetite, nervousness, or gastrointestinal disturbance. Other systemic poisons have well-defined effects. For instance central nervous system depression is an effect observed in *beta*-chloroprene exposure. Another example is methemoglobinemia in anisidine exposure.

The duration of exposure plays a key role. A brief, but relatively large, dose may be less harmful than a much lower total dose administered over a longer time period. For instance, when I was young, I remember the veterinarian giving a large dose of arsenic to a horse to cure it of some ailment or another, yet the ladies in *Arsenic and Old Lace* administered small doses of arsenic to the tea cup of their hapless victim, over a long period of time, while the poor man's doctor merely scratched his head at the frustrating, but crippling, stomach condition of his patient, and, alas, the poor man eventually died. In occupational settings, we tend to focus on exposures that occur eight hours per day, for a forty year working life, while in the environmental setting we focus on twenty-four hours per day exposures, for a seventy year life time. Sooner or later, we may have to increase this life time to eighty years due to the ever increasing average life span, which already approaches eighty years for women.

Acute assessments are rarely a concern in the environment, emergency releases notwithstanding, but acute exposure may present a first aid or emergency room case in the workplace. The type and severity of health effects, then, depends on (1) the duration, or how rapidly the dose is received, and (2) frequency, or how often the dose is received. To rephrase what we said earlier, in these terms, an acute exposure is typically a single incident, or a relatively short duration, ranging from minutes to days. Chronic exposures involve frequent doses at relatively low levels over a period of time, ranging from months to years. If a low dose is administered so slowly that the rate of elimination or the rate of detoxification keeps up with intake, no toxic response may be observed at all. The same total dose administered once, or more rapidly, may produce a health effect.

A given amount of a chemical elicits a specific type and intensity of response. One person does not respond one way to a dose, X, and another person respond in a different way to the same dose. A slight variation may be noted in the degree of response, but the response will be the same. Dose-response is fundamental to toxicology, and the basis for comparing the harmfulness of one COC to another. The dose-response relationship is a consistent mathematical and biologically plausible correlation between the number of individuals responding and a known dose, over

TABLE 3.12
Units of Dose

mg/Kg (mass COC/mass of body)	Ingestion
mg/cm2 (mass COC/surface area of skin)	Dermal absorption
ppm (volume parts per million volumes)	Inhalation
mg/m3 (mass/volume of air)	Inhalation

TABLE 3.13
Toxicity Rating

Toxicity Rating	Oral Acute LD50 for rats	Typical Substance
Extremely toxic	Up to 1 mg/Kg	Dioxin, botulinum
Highly toxic	1–50 mg/Kg	Strychnine
Moderately toxic	50–500 mg/Kg	DDT
Slightly toxic	0.5–5 g/Kg	Morphine
Practically nontoxic	>5 g/Kg	Whisky

some exposure period. Table 3.12 lists the units of dose typically encountered in literature. The period of time over which the dose is administered is significant, and must be assumed, if not known or otherwise specified.The health risk due to exposure to noncarcinogens can be evaluated by comparing average daily dose (ADD) of a chemical with its corresponding acceptable daily intake (ADI) which is published by EPA:

$$ADD = C \times E \times I \qquad (3.6)$$

where E is the exposure duration in hours. Noncarcinogens can cause such effects as mutagenicity, teratogenicity, reproductive system toxicity, and systemic toxicity. The latter group of health effects includes such target organs as the skin, eyes, central nervous system, peripheral nervous system, sensory nervous system, motor nervous system, respiratory system, blood and lymphatic systems, digestive system, the kidneys, the liver, the endocrine system, as discussed above.

Table 3.13 gives the toxicity rating used by NIOSH with some common substances to compare. Some LD_{50} values for common substances are summarized in Table 3.14. Refer to the literature for other numbers or recent studies involving the Table 3.14 substances.

Narcotic effects are indicated by alcohols, aliphatic hydrocarbons, and chlorinated hydrocarbons. Narcosis is a general depression of the central nervous system. As the narcotic depression increases, awareness of consciousness is affected, diminishing more with increasing dose. Initially, narcosis produces drowsiness, decreased ability to concentrate, and mood changes. Progressive narcotic symptoms include

TABLE 3.14
LD_{50} for Rats

Substance	LD_{50} mg/Kg
Table sugar (sucrose)	29,700
Whisky (ethanol)	14,000
Table salt (sodium chloride)	3,000
Vitamin A	2,000
Vanilla extract	1,580
Aspirin	1,000
Chloroform	800
Copper sulfate	300
Caffeine	192
Sodium phenobarbital	162
DDT	113
Sodium nitrite	85
Nicotine	53
Aflatoxin B1 (found on peanuts)	7
Sodium cyanide	6
Strychnine	3

slurred speech, dizziness, and loss of coordination. If the dose is extreme, coma and death can occur. The depression of the central nervous system due to narcosis is reversible in all but the most extreme cases, if the exposure ceases. The extreme case is called alcohol poisoning in the case of whisky. Narcotic exposure is pertinent in the workplace where exposure may adversely affect concentration and coordination, leading to increased risk of injuries and accidents caused by slower reactions to sensory input, incoordination, reckoning mistakes, and errors in judgment.

Liver (*hepatotoxicity*) and kidney (*nephrotoxicity*) toxicity are two of the more common toxicities encountered in the industrial setting. Hepatotoxicity is generally regarded as a graded response to a substance. That means that the severity of the lesion produced by the hepatotoxic substance or its metabolites on the liver is directly proportional to the intensity and duration of exposure. Many effects on the liver are reversible, as in mild alcohol toxicity, but some exposures produce lasting damage. Hepatotoxicity is complicated because liver damage is typically not manifested as any one thing you can lay your finger on. The manner in which liver damage presents itself depends on the dose, duration, and COC involved. Acute exposure may accumulate lipids in the liver cells — fatty liver. Cell death may also be produced by acute exposure, but hepatobiliary dysfunction is more common. Chronic overexposure leads to cirrhotic changes and the development of neoplasms. Fatty accumulation or necrosis may be either localized or general in the liver under attack. Lesions produced from COC attack can cause noticeable changes of the entire liver (noticeable, that is, to the pathologist or biologist examining the liver specimen). Early in the attack, the biochemical liver functions change and enzyme production is altered. These changes may be noticed in the blood, however, other reasons for enzyme changes exist and the examining physician must be familiar with the patient's history in order to determine the probable cause of the enzyme changes.

Enzymatic changes may be accompanied by changes in the morphology of specific organelles in hepatocytes. Relatively low doses of halogenated aliphatic hydrocarbons cause an increase in the activity of microsomal mixed-function oxidase enzymes, which is ordinarily accompanied by proliferation of the endoplasmic reticulum. Many COCs cause abnormal accumulation of microscopic vacuoles of fat, especially triglycerides, in liver cells. The only grossly detectable manifestation of this effect is increased liver size, which a physician can determine by palpitation of the organ. At some high dose, however, hepatotoxins kill cells, which results in general tissue necrosis, clinically manifested as gangrene, which may be initially localized, but, at sustained high doses, may involve the entire liver. Moderate to severe necrosis is indicated by increased concentrations of the marker enzymes, such as glutamate-pyruvate transaminase or glutamate-oxaloacetate transaminase, in the blood serum. Thus, blood work, reviewed by a physician familiar with the patient's history, is a useful diagnostic tool.

The development of liver and kidney damage by COCs is typically progressive. Beginning with subcellular changes, progressing to cellular damage, and the damage finally involves the whole organ. As stated before, as dose increases cellular necrosis spreads and eventually causes organ dysfunction. Since the liver can regenerate itself, minor cellular damage, even transient diseased areas of the liver, can regenerate, if the exposure ceases. However, if the exposure does not cease, the dose may exceed the liver's capacity for regeneration.

The mechanism for kidney damage by chemicals is not understood; however, it is known that certain chemicals (1,3-dichloropropene, dicyclopentadiene, ethyl silicate, hexachlorobutadiene, methyl isobutyl ketone, and soluble uranium compounds, for example) are selectively toxic in the renal tubules. The kidney, of course, is an important organ in the elimination of waste from the body, but another, equally important, function of the kidney is the regulation of total body *homeostasis*. The kidney controls homeostasis by regulating extracellular volume, controlling electrolyte concentration, balancing pH, and secreting hormones that control systemic metabolism.

Nephrotoxins, then, do one of five things that damage the body. They either interfere with hydration of tissue, interrupt the proper elimination of waste materials, disturb the electrolyte balance, interfere with metabolism, or alter the pH of the body. The least severe lesions in the kidney are both graded and reversible. The earliest changes during a chemical insult involve the alteration of enzyme activity in the tubular cells. Minor morphological cellular changes may be observed under an electron microscope. Higher or more sustained doses of the nephrotoxin begin to cause cellular necrosis that is observable under light microscopy. Fortunately, the kidneys have a large reserve capacity. Unfortunately, that same reserve capacity masks the damage, sometimes until the effects are irreversible. Therefore, significant tubular cell necrosis may occur before any loss of kidney function is noticeable. Because the kidney can handle the damage at first, the worst damage can blind-side you, precisely due to the capacity of the kidney to function despite lesser damage.

Indicators of impaired kidney function — proteinuria, glucosuria, and increased BUN — give no warning of early kidney damage. Increased kidney weight, swelling of the tubular epithelium, fatty degeneration of the tubular epithelium, and the presence of tubular casts in the urine indicate significant damage has occurred. Kidney damage is progressive at the early stages, as well as reversible. Damage becomes more extensive as exposure continues, until it reaches the point beyond which the damage is irreversible.

METABOLISM

Chemicals that enter the body undergo the same *biotransformations* that food does. Food, after all, is another species of chemicals, though more complex in both molecular and architectural structure than the typical pollutant or toxic chemical. This biotransformation process, or *metabolism*, is nothing more than a series of chemical reactions that transform energy for living, besides performing other functions. The reactions are initiated, controlled, and terminated by specific *enzymes*, or catalysts.

Two metabolic processes are anabolism and catabolism. *Anabolism*, or constructive metabolism, is the process of synthesis required for the growth of new cells and the maintenance of all tissues. *Catabolism*, or destructive metabolism, is a continuous process concerned with the production of the energy. Catabolism also maintains body temperature and breaks down chemicals into simpler substances, called metabolites, that are then eliminated as waste through the kidneys, intestines, lungs, or skin.

Metabolic reactions follow linked pathways to achieve their end. Anabolic pathways begin with simple, thinly distributed chemical intermediates that absorb energy from enzyme-catalyzed reactions. These processes then build large, complex molecules: carbohydrates, proteins, and fats. Catabolic pathways use different enzyme sequences and take the opposite direction. Catabolic pathways break down complex molecules into simpler compounds, called metabolites, for use as building blocks or waste elimination.

The chemical reactions in tissues undergo degradation in catabolism and resynthesis in anabolism. *Exergonic* reactions occur during catabolism, and release energy from within the reaction system. *Endergonic* reactions occur during anabolism, and require external energy. After the reactant and product substances involved in an endergonic reaction absorb energy, they sometimes start an exergonic reaction going. Oxidation sets off endergonic reactions within cells. When an exergonic reaction drives an endergonic reaction, the two are said to be coupled.

Metabolism, as a total process, takes place through many such energy-yielding reactions, linking up and forming an intricate, interrelated network within the cell. Chemical energy is exchanged by way of adenosine triphosphate (ATP), which contains three high-energy phosphate bonds. ATP transfers energy by losing one or two of its phosphate groups, becoming either adenosine diphosphate (ADP) or adenosine monophosphate (AMP) depending on how many phosphate groups were lost. ADP and AMP are reconverted to ATP by way of absorbing chemical energy.

Enzymes. Rate-limiting enzymes regulate metabolic reactions. The molecule of an enzyme has an active site, which is an unmasked reactive group (-NH$_2$, -OH, -SH, or -COOH), that fits a particular compound, called the substrate, with which the enzyme combines to form an end-product. Enzymes and substrates join together so precisely that other potential reactions are inhibited from occurring indiscriminately. Minute amounts of an enzyme cause profound changes in cellular metabolism.

Hormones. Metabolism is also regulated by the nervous system, the pancreas, and the pituitary and adrenal glands of the endocrine system. Hormones are secreted into the bloodstream, which they use as a highway to reach target tissues. Hormones alter the permeability of cell membranes, therefore altering the amounts of other substances that get into and out of cells. In this way, hormones control metabolism and other cellular processes.

BIOCHEMISTRY OF METABOLISM

The biochemistry of breaking down and excreting any chemical in the body depends on the chemical itself, and whether its structure can serve as a substrate for enzyme attachment. The two most important reactions in the body, with respect to breaking down chemicals, are hydrolysis and oxidation.

Hydrolysis. The general reaction in which ions react with water to form the products H^+ and OH^- is called *hydrolysis*. This is a chemical reaction in which a molecule of water, HOH, reacts with a molecule of a compound R•R'. The water molecule breaks into H^+ and OH^- and the molecule R•R' breaks into R^+ and R'^-. These four ions then join to give the final products ROH and HR'. The generic reaction is also called a double decomposition or, sometimes, an exchange reaction.

Ester. An *ester* is an organic compound formed from an alcohol by dehydrating an acid in a process called *esterification*. The breaking down of the ester to reform an alcohol and an acid is a hydrolysis reaction. Contact with water breaks esters down into their component acids and alcohols. The reaction rate of esterification is greatly enhanced by the presence of acids. The ester "ethyl acetate" becomes acetic acid and ethyl alcohol. The reaction between an ester and a metallic base is *saponification*. Since the hydroxyl ions attack electrophilic carbons, hydrolysis of esters is favored under alkaline conditions. This ester linkage is called a *carboxylester* and is found in common drugs, such as aspirin, but also in pesticides, such as malathion and pyrethrum. Such chemicals are easily hydrolyzed in the body, whereas chemicals that do not have a carboxylester branch will accumulate in fat, if they are lipophilic.

Alcohol. Alcohols are organic compounds that include the hydroxyl group, OH, attached to a carbon atom. Alcohols are monohydric, dihydric, or trihydric, depending on whether they have one, two, or three attached OH groups. Methanol and ethanol are monohydric alcohols. Alcohols are further classified as primary, secondary, or tertiary according to whether one, two, or three other carbon atoms are bound to the COH. Although analogous to inorganic bases, alcohols are neither acid nor alkaline. They are normal by-products of digestion, and processes within cells, and are found in the tissues and fluids.

Fat. Fats are hydrolyzed, or decomposed, into component glycerol and fatty acids, which are synthesized to neutral fats, cholesterol compounds, and phospholipids. Fat is either synthesized into body structure or stored in the tissues until needed, when it is catabolized to carbon substances that break down into carbon dioxide and water.

Amides (RCNH$_2$), hydrazides (H$_2$N•NH$_2$), and nitriles (-C::N-) are also hydrolyzed in this fashion.

Hydrolysis Enzymes. *Hydrolases* are enzymes that catalyze hydrolysis, including *amidases* and *peptidases*, which help digest protein, and *lipases*, which split fatty acid esters and glycerides. Amidase reactions are:

$$RC\overset{O}{\overset{\|}{-}}NHR' \longrightarrow RC\overset{O}{\overset{\|}{-}}OH + R'NH_2$$

A-esterases, or *arylesterases*, hydrolyze phosphotriesters and are not inhibited. *B-esterases*, or *serine hydrolases*, are inhibited by paraoxon, an organophosphate pesticide. *Cholinesterases* are B-esterases that catalyze the hydrolysis of choline esters. Other B-esterases include the *carboxylesterases* and amidases and *monoacylglycerol lipases*. The liver has several carboxylester hydrolases, which detoxify important internal and foreign carboxylesters, and which also function as lipases in digesting fats. Carboxylester hydrolysis proceeds:

$$RC\overset{O}{\overset{\|}{-}}OR' + H_2O \longrightarrow RC\overset{O}{\overset{\|}{-}}OH' + HOR$$

Besides the liver, B-esterases occur in the kidneys, brain, muscle, serum, and saliva. *C-esterases* hydrolyze acetyl esters.

Organophosphorus pesticides are composed of phosphorus esterified to some organic compound such as glucose or sorbitol. Organophosphorus pesticides such as Azodrin, DDVP, Diazinon, Malathion, Methyl Parathion, Parathion, Phosdrin, and TEPP, bind up, or inhibit, cholinesterases that are important to the nervous system at the synapses resulting in paralysis. [Note: Some of these are registered trademarks.] Though they are highly toxic, they are low in persistence. The pyrethroids, Bioresmethrin, Cypermethrin, Pyrethium, and Resmethrin, also affect the nervous system, leading to paralysis. A similar group of cholinesterase inhibiting pesticides are the carbamates, which are neutral esters of carbamic acid: Baygon, Carbaryl (Sevin trademark), Maneb, Matacil, Temik, Zectran, and Zineb. Some of these compounds may be detectable in a fat biopsy or as an irregular liver function indicted in a SGPT blood test. Yet, neither of these indicator tests is as simple nor direct as delta cholinesterase testing.

Krebs Cycle. *Serum Glutamic-Pyruvic Transaminase* (SGPT) is one of two enzymes that catalyzes a reversible amino group transfer reaction in the Krebs cycle, which fuels the energy production in living tissue. The Krebs, or citric acid or tricarboxylic acid, cycle is a series of chemical reactions that breakdown food molecules to form carbon dioxide, water, and energy. The process is controlled by seven enzymes. At the beginning of the cycle, food is broken down into acetyl groups

(CH$_3$CO), which combine with four-carbon oxaloacetate molecules to yield citric acid. Citric acid molecules eventually lose two carbon atoms in the form of carbon dioxide. Four electrons are also released and transported along a nearby series of carrier molecules in the cell, called the electron transport chain, where they produce energy in the form of an energy-rich ATP before combining with oxygen to form water. Another energy-containing molecule called guanosinetriphosphate (GTP) is also produced in the Krebs cycle. A fuel molecule called creatine phosphate provides extra energy to brain and muscle cells. At the end of the cycle, oxaloacetate is regenerated, which can then recombine with acetyl groups to start the cycle over again, producing more energy.

SGPT. The enzyme SGPT primarily appears in hepatocellular cytoplasm, with lesser amounts found in the kidneys, heart, and skeletal muscles. Therefore, SGPT is a fairly accurate indicator of acute hepatocellular damage and of chemical toxicity as one of the potential causes of liver damage. SGPT levels in adult males range from 10 to 32 units/liter, while the range is from 9 to 24 units/liter in women. Infants normally range twice as much as adults.

The cholinesterase test measures the amounts of two enzymes that hydrolyze acetylcholinesterase. *Pseudocholinesterase*, which is also called PCHE or serum cholinesterase, is produced primarily in the liver, but small amounts are found in the pancreas, intestine, heart, and white matter of the brain. Even though physicians and scientists do not fully understand the function of pseudocholinesterase, when a chemical inhibits cholinesterase it also affects pseudocholinesterase, which can be observed. *Acetylcholinesterase*, the second choline enzyme, is present in nerve tissues, red cells of the spleen, and the gray matter of the brain and transmits impulses across nerve endings to muscle fiber. Hence, paralysis occurs when the enzyme is inhibited. Organophosphates inactivate acetylcholinesterase directly. When organophosphate poisoning is suspected, either cholinesterase enzyme may be measured, but pseudocholinesterase is usually chosen. Pseudocholinesterase levels range from 8 to 18 units/ml as determined by a kinetic colorimetric technique.

Dehalogenation. Hydrolytic dehalogenation metabolizes the haloethanes and chlorinated pesticides in this fashion: $RCH_2Cl \rightarrow RCH_2OH$.

Oxidation. Proteins, carbohydrates, and fats differ in chemical composition and proceed on independent biochemical pathways, yet, in metabolism, they ultimately form the same carbon compounds by means of oxidation-reduction reactions that yield carbon dioxide and water for excretion from the body. Each step in this process involves highly complex, simultaneous biochemical reactions.

Protein and Amino Acids. Complex *proteins* break down into amino acids needed for cellular anabolism. Amino acids, in turn, form hormones and digestive enzymes. Excessive amino acids are catabolized themselves, in two steps. In *deamination*, the nitrogen-bearing part of the molecule is removed and united with carbon and oxygen to form urea, ammonia, and uric acid. The remaining amino acids form other compounds, which are further catabolized, often by pathways common to the catabolism of carbohydrates and fat. The products of these reactions are carbon dioxide and water. Amines, as opposed to amino acids, are deaminated to aldehydes. Flavin adenine dinucleotide (FAD) is a cofactor in monoamine oxidase deamination of catecholamines and serotonin to stop their actions and reduce their levels.

TABLE 3.15
Isomers of Lactic Dehydrogenase

LDH in heart tissue
  LDH₁   HHHH
  LDH₂   HHHM
LDH in lung tissue
  LDH₃   HHMM
LDH in liver and skeletal muscle
  LDH₄   HMMM
  LDH₅   MMMM

H represents enzyme units that most consistently appear
in heart tissue. M represents units that most consistently
appear in skeletal muscle tissue. The specific isomer
indicates where the damage has occurred.

Microsomal FAD monooxygenase is the catalyst in oxidative desulfuration. Monoamine oxidase (MAO) is the catalyst for the oxidative deamination of a number of amines.

Carbohydrates. Carbohydrates are sugars, chiefly glucose, that are maintained in the blood at an approximately constant level. Glucose is readily catabolized to provide energy for the body. The glucose molecule breaks down into carbon compounds that are oxidized to carbon dioxide and water, then excreted. If not used immediately, glucose is converted to glycogen and stored in the liver and muscles. Excess glucose is converted to fat and deposited in adipose tissue.

Oxidation-Reduction Couples. Biochemical oxidation-reduction reactions require the presence of *oxidoreductase* enzymes, which transfer electrons. *Isomerases* are enzymes that form isomers of molecules by group transfer creating *cis-* and *trans-* forms and the reverse, which do all sorts of tasks in the body systems. For instance, lactic dehydrogenase (LDH) has five isomers with differing molecular details, such as electrophoretic migration, yet have the same basic identity. Table 3.15 lists the isomers of LDH, which, along with the cytochrome enzymes, are among the enzymes that are more susceptible to excess levels of xenobiotics.

Transport Enzymes. *Transferase* enzymes move chemical groups from a substrate molecule such as an acetyl, amino, or methyl group, to a recipient molecule. *Gamma Glutamyl Transferase* (GGT), for instance, transfers amino acids across cellular membranes. *Lyase* enzymes remove chemical groups, or segments of molecules, without hydrolysis, and form or split C = C bonds, or add other species to double carbon bonds. *Aldolases* and *carboxylases* are typical lyases. *Ligase* enzymes and the high-energy ATP link other molecules together by forming carbon-to-carbon covalent bonds (C-C) or carbon-sulfur covalent bonds (C-S). Ligases also form C-O-, and C-N bonds.

Energy Production. ATP is the instantaneous source of energy for living cells, built up in special compartments called mitochondria. The energy-exchanging function of ATP and the catalytic function of enzymes are intimately connected, so ATP is referred to as a coenzyme. The adenosine part of the molecule is made up of adenine, which is a nitrogen-containing compound that is one of the principal components of the gene, and ribose, which is a five-carbon sugar. Three phosphate units, each made of one phosphorus and four oxygen atoms, $-PO_4$, are attached to the ribose. The two bonds between the three phosphate groups are high-energy bonds that are relatively weak and readily yield energy when split by enzymes. When the end phosphate group is split, seven kilocalories become available for work and the ATP molecule becomes adenosine diphosphate (ADP). The energy-consuming reactions in cells that are powered by the conversion of ATP to ADP include the transmission of nerve signals, the movement of muscles, the synthesis of protein, and cell division. ADP quickly regains the third phosphate unit through the action of cytochrome, a protein that builds it up by using food energy. In muscle and brain cells, excess ATP joins with creatine to provide a reserve supply of energy. The release of two phosphate groups from ATP by the enzyme adenyl cyclase forms adenosine monophosphate (AMP), a nucleotide component of nucleic acids, the material of DNA. AMP is also important in many of the body's reactions. One form of AMP called cyclic AMP, or cAMP, is created by the action of the enzyme adenyl cyclase and is instrumental in many hormones, such as epinephrine and ACTH.

Cytochromic Enzymes. Monooxygenases are *cytochrome P_{450}* enzymes, which include the heme group in which molecular oxygen is bound to iron. These enzymes get their name from the fact that carbon monoxide, one of the more potent inhibitors of these enzymes, has a characteristic peak of light absorbence at 450 nm. Whole families of cytochrome enzymes perform various functions involving oxidation-reduction reactions. They actually form an oxidase system of metallo proteins with three major components: the cytochrome P_{450}, NADPH-P_{450} reductase, and dia-acylglyceryl-3-phosphorcholine. As a system, these enzymes depend on vitamins B_2, B_3, B_6, copper, iron, and several amino acids, including choline, and are the chief catalysts in the detoxification of COCs through oxidation, reduction, and conjugation.

Primarily, the monooxygenases catalyze carbon oxidation. Aromatic methyls, RCH_3, for instance, are oxidized very rapidly to RCH_2OH. A methylene group located between two aromatic rings, or a methyl radical attached to an aliphatic ring, undergoes a similar reaction. A terminal methyl on a longer aliphatic chain, RCH_2CH_3, oxidizes to RCH_2CH_2OH. Similar penultimate methyls, RCH_2CH_3, oxidize to $RCH(OH)CH_3$. Nitrogen oxidation by microsomal P_{450} follows RNHR $\rightarrow$ RNOHR, RN = O, or R = NOH. Microsomal FAD monooxygenase reactions oxidize R_3N to $R_3N = O$. Other reactions catalyzed by P_{450} enzymes are hydroxylation, dealkylation, and epoxidation, which are oxidation reactions.

Liver oxidases have a remarkable capacity for breaking down toxic substances by adding a hydroxyl radical to the basic structure. Aliphatic, side-chain, aromatic, and heterocyclic compounds are hydroxylated to form alcohol derivatives, which allows the metabolite to be further converted to a water soluble compound that can be excreted in the urine. Monoamine oxidases oxidize $RCH_2NH_2 \rightarrow RCHO + NH_3$. Xanthine oxidase, a soluble molybdenum-based flavoprotein oxidizes xanthine and purine analogs to uric acids, and oxidizes metabolites of trimethylxanthines and some aromatics. Aliphatic and aromatic dehydroxylation, where the compounds are reduced, are the reverse of hydroxylation. Alkene oxidation breaks the double carbon bond.

Dealkylation of N-, O-, and S- radicals are also oxidation reactions. N-dealkylation proceeds $RNHCH_3 \rightarrow [RNHCH_2OH] \rightarrow RNH_2 + HCHO$. The compound in brackets [ ] is a short-lived intermediate. O-dealkylation of ethers is the reaction $ROCH_3 \rightarrow [ROCH_2OH] \rightarrow ROH + HCHO$. O-dealkylation of phosphate esters is $(RO)_2POOR' \rightarrow (ROOH)POOR'$. The O-dealkylation of dioxymethylenes proceeds $(RO)_2CH_2 \rightarrow (ROH)_2 + HCHO$. S-dealkylation proceeds $RSCH_3 \rightarrow RSH + HCHO$. A variation of N-dealkylation is oxidative deamination, which cleaves a larger group from the nitrogen: $RCH_2NH_2 \rightarrow RCHO + NH_3$.

Epoxidation consists of adding an oxygen atom between two carbon atoms in an unsaturated system. For instance, the insertion of an oxygen into an olefin, $-C=C-$ $+ O \rightarrow -C-O-C-$, is an epoxidation. Olefin epoxidation breaks down such COCs as Aldrin, trichloroethylene, and vinyl chloride. Epoxide hydratase catalyzes:

$$RC\overset{O}{\overset{/\backslash}{=}}CR \longrightarrow R-\underset{\underset{H}{|}}{\overset{\overset{OH}{|}}{C}}-\underset{\underset{OH}{|}}{\overset{\overset{H}{|}}{C}}-R$$

Epoxidation of aromatics results in the formation of arene oxides.

Epoxides are extremely nucleophilic and chemical reactive. Many are carcinogens. Epoxide hydrolases detoxify epoxides. Amines, $R-NH_2$, are oxidized to aldehydes and acids and conjugated to hydrophilic derivatives. Industrial nitro-derivatives, $R-NO_2$, are reduced by hydroxylation. Aromatic hydrocarbons, halogenated aromatic hydrocarbons, and polycyclic hydrocarbons are detoxified by reaction with acetyl mercapturic acid, $-SCH_2CHCOOH$. Inorganic and organic cyanides are neutralized by producing thiocyanate, $RCNS$. Glycine, $-NHCH_2COOH$, metabolizes aromatic acids, aromatic-aliphatic acids, furane carboxylic acids, thiophene carboxylic acids, and polycyclic carboxylic acids (the Bile acids). Primary, secondary, and tertiary aliphatic and aromatic hydroxyl compounds are metabolized by glucuronate. Hydrazine derivatives are neutralized by glucose hydrazone.

Reductive dehalogenation of halothane is $F_3CHBrCl \rightarrow F_3CCH_3 + Cl^- + Br^-$. DDT is reduced to DDD in this manner. The oxidation-reduction of pentachloroethane is $CCl_3-CHCl_2 \rightarrow \ldots Cl_2C=CHCl$. Nitroreduction occurs in a chain reaction: $RNO_2 \rightarrow RNO \rightarrow RNHOH \rightarrow RNH_2$. Azo reduction proceeds $RN=NR' \rightarrow RNH_2 + R'NH_2$. The transhydrogenases catalyze $RSSR' \rightarrow RSH + R'SH$. Finally, the reduction catalyzed by peroxidases is $ROOH \rightarrow ROH$.

Chlorinated hydrocarbon pesticides cannot be detected by cholinesterase monitoring. These persistent COCs include Aldrin, Chlordane, DDT, Heptachlor, Lindane, and Toxaphene. Although they too affect the nervous system and may result in paralysis, it is by another mechanism than cholinesterase inhibition.

Conjugation. Conjugation reactions are synthesis processes, usually following one or more of the hydrolysis or oxidation-reduction reactions discussed above. Xenobiotics are conjugated by nutrient derived enzymes, called cofactors, and amino acids. These conjugations are called acetylation, gluconation, methylation, peptide conjugation, or sulfonation. Conjugation of cyclic compounds with sulfuric acid forms water soluble ethereal sulfates, $-OSO_3H$, for example. Conjugation with the sugar acids forms a soluble derivative of the form $RO(CHOH)_n COOH$. Conjugation with amino acids, such as glycine or cysteine, also forms soluble products, which are cleared through the kidney and eliminated via the urine.

Typically, conjugated COCs are more water-soluble than unconjugated COCs and, with very few exceptions, biologically inactivated. Glucuronic acid, sulfuric acid, acetic acid, glycine, and glutathione promote conjugation. Hepatic enzymes of the smooth endoplasmic reticulum are stimulated by highly lipophilic COCs. As the density of the smooth endoplasmic reticulum increases, the detoxification of the COC is enhanced by enzyme induction, though sometimes the metabolites are toxic too. Enzymes that catalyze conjugate hydrolysis are deacetylases, glucosidases, glucuronidases, and sulfatases.

The glycoside conjugation of alcohol, ROH, yields RO-glucuronic acid. Carboxylic acids, RCOOH, become RCOO-glucuronic acid. Conjugated amines, RNH_2, yield RNH-glucuronic acid. Sulfate conjugation uses phenol sulfotransferase to break down phenol but not steroids: $ArOH \rightarrow ArOSO_3H$. Alcohol sulfotransferase, enhanced by the activity of other sulfotransferases, catalyzes alcohol and polyol, ROH, synthesis to $ROSO_3H$.

Methylation. Methylation, as a conjugation process, detoxifies inorganic compounds, such as those of arsine and tellurium, ring nitrogen compounds, and complex aromatics (phenols) by an attaching a methyl radical, $-CH_3$, to the compound. However, methylation renders lead and mercury more toxic. S-, N-, and O-methyl transferase enzymes and the amino acids methionine and ethionine also catalyze methylation.

Catechol O-transferase acts in the soluble fraction of tissues where catecholamines exert effects, especially in the liver and kidneys. The reaction is $Ar(OH)_2 \rightarrow Ar(OH)OCH_3$. Other O-methyl transferase species are hydroxyindole O-methyl transferase, which acts in the soluble portion of the pineal gland and retina mostly during the hours of darkness, microsomal catechol O-methyl transferase, and microsomal phenol O-methyl transferase. The latter catalyzes $ArOH \rightarrow ArOCH_3$.

N-methyl transferase is nonspecific, acting in most tissue, especially the lungs, and is soluble, catalyzing $RNHR' \rightarrow RN(CH_3)R'$. Another soluble enzyme, phenylethanolamine N-methyl transferase (PNMT), is also nonspecific though it is found at high levels in the adrenal gland, and catalyzes $RCH_2(OH)NH_2 \rightarrow RCH_2(OH)NHCH_3$. Histamine NMT is a highly specific enzyme, on the other hand, but acts in many tissues especially in the brain.

Thiol S-methyl transferase yields this reaction: $RSH \rightarrow RSCH_3$. The last reaction is deadly when the R is hydrogen, $H_2S \rightarrow CH_3SH$ (toxic) $\rightarrow CH_3SCH_3$.

The microbial biomethylation of heavy metals and metalloids, which are solid elements resembling metals, increase the lipophilicity of the metals and metalloids, with the implication that they are stored in fat until a later time when they can be retrieved and continue their effect on their target organ. Elements in this category include arsenic, lead, mercury, thallium, tin, selenium, and sulfur.

TABLE 3.16
Urine Monitoring for Chemical Toxicity

Chemical	Toxicity indicated, units
Antimony	>1.0 mg/l
Arsenic	>850 µg/l
Benzene (as phenol)	>75 mg/l
Boron	>0.3 mg/100 mL
Cadmium	≥10 mg/dl
Cyanide	>0.5 mg/100 mL
Fluoride	>10 mg/l
Lead	>120 µg/l
Mercury	>150 µg/l
Nickel	>350 µg/l
Phenol (benzene metabolite)	>75 mg/l
Selenium	>400 µg/l
Thallium	>50 µg/l

After Hamilton.

Acetylation. In *acetylation*, a CH$_3$CO- group is added to aromatic amines, sulfur amines, or amino acids to detoxify them. Carcinogenic aromatic amines and aliphatic amines do not detoxify with acetylation, however. The enzyme N-acetyl transferase acetylates and detoxifies many COCs, including potentially carcinogenic aryl amines. The rate of acetylation varies with race. As the body's pollutant load increases, more and more acetylation enzymes and acetyl molecules are tied up in detoxification rather than food metabolism, resulting in inefficient use of energy and nutrition.

Elimination. Soluble metabolites of COCs are eliminated from the body in saliva, sweat, milk, tears, bile, mucus, feces, and urine. Most elimination is with urine, followed by feces, and thirdly by bile. In exposure assessments, this has two implications. Though we do not measure and account for elimination, our assessments are, therefore, typically worst cases. Second, elimination provides an opportunity, in many cases, to verify receipt of dose.

Mercury compounds affect kidney function but kidney monitoring by urine testing is usually effective. Urinary mercury from 0.15 to 0.5 mg/l is significant. Table 3.16 shows some urine monitoring tests for chemical toxicity.

Water soluble materials pass through the body quickly, so it is important to sample soon after the exposure, or the urine test may give a false indication of low exposure. Spot collection of urine may yield different results than a twenty-four hour sample. Also, some urine tests, such as the one for arsenic, may be falsely positive due to recent ingestion of foods that normally contain measurable levels of the material. OSHA has, in some cases, established numbers lower than those indicated in the table. Refer to the appropriate standard for the specific chemical of concern.

TOXIC IMPACT ON THE EYE

The retina is the compact neural structure in the eye that is responsible for converting an ocular light image to a neural impulse that the brain can interpret. On the inside of the eye, the retina is normally protected from exposure to chemicals. Internally absorbed COCs can affect retina, causing lesions and other changes such as retinal edema or hemorrhage. A few COCs cause acute narrowing of the retinal arteries, leading to damage to the optic nerve with ultimate loss of vision. Below the retinal surface layer lies the ganglion cell layer, composed of neuron cell bodies extending to the mid-brain via the optic nerve. Some COCs act directly on the neural cell bodies, while others attack the optic nerve, such as methyl alcohol. The degree of loss of visual acuity depends on the severity of the exposure.

TOXICITY IN THE RESPIRATORY SYSTEM

Primary respiratory irritants damage the cell linings in the airways of the respiratory system. This effect was discussed under sensory irritation effects. Chemical damage increases the permeability of the cell membranes and lead to edema, hemorrhage, and localized necrosis. Chronic inhalation of irritating chemicals destroys alveolar septa, resulting in emphysema. Whether the cellular damage is localized or diffuse depends on how the toxic substance is distributed in the lung. Solubility is the primary factor in determining distribution. Secondary irritants are those that have systemic effects. These are discussed here.

Simple Asphyxiation. Asphyxiants do not damage the lungs, but deny the body of oxygen needed to sustain life. The simple asphyxiants are otherwise inert physiologically. All they do is replace oxygen, by having a partial pressure sufficient to reduce the partial pressure of oxygen. For instance, helium is a simple asphyxiant that is encountered in everyday life as the gas of choice in party balloons. Helium is a good example of how familiarity with a chemical removes any fear of it. How many people do you know, who, in clowning around, inhale helium from a balloon in order to talk with the high-pitched voice of a lovable cartoon character? No one is afraid of helium, though it is potentially dangerous, even fatal. Fortunately, helium is typically inhaled on purpose in settings where plenty of oxygen is available. Another simple asphyxiant is nitrogen. How many people have you seen play with liquid nitrogen, which is a cryogenic hazard as well, by pouring it over some object that freezes and breaks into many pieces? Yet, liquid nitrogen expands over six hundred times its volume as it evaporates. Fortunately, ventilation is usually adequate to prevent suffocation, but the potential is there. Other common asphyxiants include carbon dioxide and nitrous oxide, as well as the flammable gases: hydrogen, methane, ethane, and other aliphatic hydrocarbon vapors.

The ability of hemoglobin to carry oxygen is directly proportional to the partial pressure of oxygen in the air we breathe. Our bodies have adapted to work efficiently when the volume percent of oxygen in the atmosphere is about twenty-one percent. Dalton's law of partial pressures states that in any mixture of gases, the total pressure is the sum of the partial pressures of the component gases. The partial pressure of any component gas in the mixture is proportional to the percentage of its molecules in the total mixture, or, in other words, the volume fraction of the gas component. In the case of oxygen in air at sea level:

$$P_{O_2} = P_{air} \times \%O_2 = 760 \text{ mmHg} \times 0.21 = 160 \text{ mmHg} \tag{3.7}$$

Ordinarily, the percent of red blood cells transporting oxygen to organs is 97%. As this transport number decreases, the body breathes faster and the heart rate increases. The lack of oxygen in living tissue is called *hypoxia*. When the lack is caused by respiration, it is called *anoxic hypoxia*. *Anemic hypoxia* means the lack of oxygen is due to the inability of blood to carry oxygen. If the blood can carry oxygen efficiently, but blood flow is restricted, thus preventing oxygen from reaching tissue, the deficiency is called *hypokinetic hypoxia*. If the tissue is unable to use the oxygen as delivered, the condition is called *histotoxic hypoxia*.

Chemical Asphyxiants. Substances that render the body incapable of transporting or using oxygen are chemical asphyxiants. The two best-known chemical asphyxiants are probably carbon monoxide and hydrogen cyanide gas. The percent of hemoglobin carrying carbon monoxide is directly proportional to the ratio of the partial pressure of carbon monoxide to the partial pressure of oxygen. Wadden and Scheff refer to a 1965 journal article by Coburn, Forester, and Kane to give us this relationship:

$$\frac{\alpha[COHb]_t - \beta \dot{V}_{CO} - P_i(CO)}{\alpha[COHb]_0 - \beta \dot{V}_{CO} - P_i(CO)} = \exp(-at/\beta V_B) \tag{3.8}$$

where: t   = duration of ex[osure, min,
         [COHb] = concentration of COHb in blood, mL/ml,
         V_{CO} = endogenous CO production rate, mL/min,
         $P_i(CO)$ = inspired CO partial pressure, mmHg,
         V_B = blood volume, mL.

α and β are determined as follows:

$$\alpha = \bar{p}c(O_2)/M[O_2Hb]' \tag{3.9}$$

where: $pc(O_2)$ = mean pulmonary capillary oxygen pressure, mmHg,
         M   = Haldane constant, dimensionless,
         $[O_2Hb]'$ = concentration of oxyhemoglobin, mL/ml.

$$[O_2Hb]' = 1.38(m_{Hb}) - (COHb)_t \tag{3.10}$$

where: m_{Hb} = hemoglobin concentration in blood, gHb/ml blood.

TABLE 3.17
Human Physiology

	Typical	Range	95% Range
Alveolar Ventilation, l/min	6.4		1.9–10.9
Blood, female, mL/Kg	62		
Blood, male, mL/Kg	69		
Creatinine Excretion, male,			
mg/Kg day	1.8		1.1–2.5
female, mg/Kg day	1.2		1.1–1.3
Diffusion Capacity, CO, lung,			
ml/min mmHg	22.4	10.5–45	
Endogenous CO Production,			
ml/min	0.007		
Endogenous COHb%, normal		0.4–0.7	
hemolytic anemia patient		4–6	
smoker		5–20	
Hemoglobin in Blood, g/100 mL	15.85		13.4–17.3
Pulmonary Capillary Oxygen			
Pressure, mean, mmHg	100	95–105	
Urine Excretion, male, l/day	1.02	0.51–2.0	
female, l/day	0.99	0.50–1.88	
Urine specific gravity		1.010–1.025	

The Haldane constant represents the ratio of the affinity of hemoglobin for CO to that for O_2. Wadden and Scheff report that M is 220 for a 7.4 blood pH. Scott reports that it is 245 at normal blood pH.

$$\beta = 1/D_L + P_L/\dot{V}_A \qquad (3.11)$$

where: D_L = diffusion capacity of the lungs for CO, mL/min mmHg,
P_L = partial pressure of gas in the lungs, mmHg, and
V_A = alveolar ventilation rate, mL/min.

$$P_L = P_b - P_{wv} \qquad (3.12)$$

where: P_b = barometric pressure, mmHg,
P_{wv} = partial pressure of water vapor at body temperature.

P_{wv} is 47 mmHg at 37°C. The barometric pressure, P_b, is typically 760 mmHg at sea level, but can be determined directly from a barometer or by consulting local weather information. The concentration of carboxyhemoglobin expressed as a percent of saturation is:

$$(COHb\%) = [COHb]_t \cdot 100/(1.38\,mlO_2/gHb) \cdot m_{Hb} \qquad (3.13)$$

Edema. Pulmonary edema is the release of fluid in the open spaces of the airways (lumen) or the alveoli. Several species of organic chemicals cause pulmonary edema, but particulate matter, such as asbestos, can also lead to edema. If the edema takes several hours to develop, the victim may not realize at the time of the exposure that a life-threatening exposure has taken place. Also, pulmonary edema is far more serious when the victim is suffering from the effects of a cold or flu virus.

Tissue Necrosis. Necrosis, or cell death, reduces the functional surface area of the lungs. Benign granulomas are a type of lesion that is common in victims of respiratory toxins. These localized masses in the lungs form when the immune system isolates a foreign material by sequestering it.

Emphysema is the gradual destruction of the alveolar septa, causing a loss in elasticity. Most of the adult human population has slight emphysema anyway, yet no functional impairment is noticed. As emphysema progresses, whether due to attack by chemicals or for other reasons, serious and life-threatening reduction of lung capacity becomes apparent. Once emphysema is apparent its effects are irreversible.

Pulmonary fibrosis has long been recognized as an occupational phenomenon. Although some people refer to pulmonary fibrosis and pneumoconiosis interchangeably, pneumoconiosis is a general term meaning that a foreign substance is present in the lungs as diagnosed by radiographic, or X-ray, methods. Functional damage is not implied in the term pneumoconiosis, and a variety of conditions may be indicated. Fibrosis is a specific and debilitating group of diseases. For instance, interstitial fibrosis is a type of pneumoconiosis that is characterized by deposition of fibrous tissue in the interstitial spaces between the alveolar membrane and the pulmonary capillary membrane. The diffusing capacity of the lungs is vastly reduced by interstitial fibrosis, thus the body is deprived of oxygen. Fibrosis is mostly irreversible, except at the earliest stages when very minor damage has occurred, and once the disease is well established it progresses on its own without further exposure. Silicosis is a specific type of interstitial fibrosis that is caused by respirable particles of silica. Coal dust causes another form of pneumoconiosis with fibrosis that is particularly debilitating. Graphite, mica, carbon black, and grain dust also initiate forms of fibrosis.

CARDIAC TOXICITY

Any toxic material eventually affects the heart. After all, at death, the ultimate health effect, the heart stops. Generally, effects on the heart are secondary problems caused by inattention to a primary problem, or else the primary toxicity problem simply overwhelms attempts to mitigate the effects of exposure. Some chemicals, however, cause one of three direct effects on the heart and the cardiac system: cardiac sensitization, vasodilation, and atherosclerosis. These effects are material impairments of health and functional capacity, because they are potentially disabling and have a life-threatening potential too.

Cardiac Sensitization. Cardiac sensitization makes the heart sensitive to biochemical compounds called *sympathomimetic amines*. Such hormonal compounds, one being adrenaline or epinephrine, make the heart beat faster. When the body anticipates an increase in physical exertion, as when terror causes panic, adrenaline is secreted into the bloodstream. In order to increase the heartbeat rate, the dose of adrenaline must equal or exceed its no-effect level. Cardiac sensitizers lower the no-effect level of adrenaline so that the heart is stimulated by ever lower quantities of adrenaline. They do this by affecting the pacemaking and conduction center of the heart, which determines the rhythm and rate of the heartbeat. Interference with this region leads to arrhythmia with various consequences. Young persons generally have no serious consequences, though fatal arrhythmias are not unknown in healthy young people. Also, either cerebral or myocardial ischemia may result from arrhythmia. Shock may occur. Or, in older people or those who suffer from compromised cardiac systems already, congestive heart failure may occur. Organic solvents are well known as cardiac sensitizers. Fatal cardiac arrhythmias following organic solvent exposure are due to cardiac sensitization to circulating catecholamines.

Leukotriene C_4 (LTC$_4$), leukotriene D_4 (LTD$_4$), prostaglandin I_2 (PGI$_2$), and prostaglandin E_2 (PGE$_2$), metabolites of arachidonic acid (AA), are potent vasoconstrictors in the circulatory system. AA is a fatty acid that is a principle substrate for several enzymatic pathways. When AA is diverted to the lipoxygenase pathway, leukotrienes are released. LTC$_4$ and LTD$_4$ are not only powerful vasoconstrictors, but bronchoconstrictors as well, adding to the severity of cardiac sensitization. It has been suggested by Gardner that the prostaglandins interfere with hormone secretions, receptor sites, glucose transport, and other metabolic processes having indirect, if not direct, effects on the heart and the circulatory system.

Vasodilation. The vasodilators cause blood vessels to expand, lowering the blood pressure (*hypotension*), and, thus, decreasing the supply of blood to the various organs. Acute hypotension causes shock. Chronic hypotension causes lethargy, general weakness, fatigue, dizziness, and faintness. Calcium channel blockers are well-known vasodilators and are used as medicines to counter *hypertension* (high blood pressure). Gardner discusses the interrelationship of cAMP, prostaglandins, and calcium. When the heart is subject to hypoxia and ischemia, PGI$_2$ and adenosine are released as vasodilating agents.

Atherosclerosis. Atherosclerosis is a serious disease involving the progressive degeneration of the arteries. Plaques, composed of lipids, complex carbohydrates, blood products, and calcium, form on the interior walls the arteries, usually the major vessels, which affect several organ systems or areas of the body. The plaques, called atheromas, narrow the arteries until they cause any of several health effects: renal hypertension, stroke, and myocardial ischemia. Increased levels of polyunsaturated fatty acids in the diet reduce atherosclerosis, and hypertension too.

CHOLINESTERASE INHIBITION

Some toxic substances are cholinesterase inhibitors as discussed above. Organophosphate and carbamate pesticides, for instance, inhibit acetylcholinesterase at cholinergic synapses in the central and peripheral nervous systems. The effect of this inhibition is an accumulation of acetylcholine at the effector sites, which arouses signs and symptoms of excessive cholinergic activity such as bronchioconstriction, increased bronchial secretions, salivation, and lacrimation. Other symptoms are nausea, vomiting, cramps, constricted pupils, muscular weakness, and cardiac irregularities. Increasingly severe cholinesterase inhibition causes coma, irreversible central nervous system damage, and death.

Carbamates bond ionically with the cholinesterase enzyme, deactivating it. Organophosphates form a covalent bond with the enzyme, and also deactivate it. Usually, the inhibition is reversible. Carbamate-cholinesterase complexes dissociate, regenerating the active enzyme. Cholinesterase inactivated by an organophosphate species is replaced by the *de novo* synthesis of active enzyme. Unless the inhibition of the enzyme caused brain damage or death, the effects are reversible and the patient recovers without *sequelae*. A large portion of the endogenous cholinesterase activity must be inhibited before signs and symptoms appear, though the inhibited fraction of the total enzyme varies per individual. Intensity and duration of exposure also matter. The lack of warning signs at low exposures means the health effects can blind side you.

Blood Toxicity

Other COCs reduce the ability of the blood to carry oxygen. Carbon monoxide bonds with hemoglobin in red blood cells with greater affinity than oxygen does, and the carbon monoxide also alters the dissociation characteristics of the oxygen-hemoglobin complex. This reduces the oxygen carrying capability of the blood. Aromatic amines and nitro compounds also react with hemoglobin to reduce it to methemoglobin, which will not bond with oxygen. Therefore, these substances reduce the ability of the blood to carry oxygen and result in tissue anoxia. The overt symptoms of these COCs are neurobehavioral disturbances, dizziness, cardiac irregularities, cyanosis, unconsciousness, and death. People who have existing cardiovascular disease are at greater risk when exposed to carbon monoxide. Even healthy persons, who are engaged in heavy physical labor may be at greater risk when five percent or more of their hemoglobin is tied up with carbon monoxide. Whether carboxyhemoglobin or methemoglobin is formed, the condition is reversible in the absence of carbon monoxide. Carboxyhemoglobin dissociates in time to carbon monoxide and hemoglobin. Methemoglobin is gradually oxidized to hemoglobin by endogenous mechanisms. However, the synthesis of fresh hemoglobin is the chief factor in recovery, for which time may be lacking if the exposure was severe.

Alcohol Sensitivity

Some COCs make the body sensitive to alcohol. A prescription trademark medicine called Antabuse does exactly that and these particular COCs may be thought of as Antabuse analogs. If alcohol is consumed after being exposed to calcium cyanamide, cyanamide, or disulfiram, the face becomes flushed. The victim may also present nausea, vomiting, hypotension, and a faster heart rate. In very severe exposures, the victim may have convulsions, cardiac arrhythmia, or a heart attack, or in the worst case, the victim may die. Even though the symptoms are disabling, the majority of times, the reaction is reversible.

ALLERGY RESPONSES

When a person suffers an allergic reaction to the second or subsequent exposure to a chemical or to a structurally similar chemical, a sensitization reaction has taken place. In this case the immune system has reacted to the COC. A cross-sensitization occurs when subsequent exposure to the allergen causes reactions not only to itself, but to some other chemical as well, although the other chemical will usually be structurally similar. Common target organs for sensitizers are the skin and eyes resulting in dermatitis and conjunctivitis accordingly. With some chemicals the respiratory system is sensitized, leading to asthma or bronchitis. Allergic reactions are mediated by two immunoglobulins, IgD or IgE. Immunoglobulin D (IgD) mediated reactions lead to delayed contact dermatitis. Reactions mediated by immunoglobulin E (IgE) cause severe, potentially fatal effects, such as acute asthma, urticaria, and anaphylactic shock. Sensitization reactions are unpredictable. Sensitivity typically persists for the remainder of the lifetime of the victim, though a few cases have been reported where the effects disappeared over time. Avoidance of the COC is the most prudent treatment and prevention strategy. Isolation may be the only way to restore good health to the victim, however in many cases, residual respiratory symptoms persist after exposure is discontinued. Workers sensitized to toluene 2,4-diisocyanate (TDI) persisted with respiratory symptoms such as impaired pulmonary function and chronic bronchitis for as long as three and one-half years after the exposure ceased.

Several occupational dermatoses can present allergic-type reactions. A small percentage of dermatoses cases are allergic reaction, of the cell-mediated-type hypersensitivity. The remainder of cases are caused by primary irritants or toxic effects, presenting such symptoms as rash, itching, and scaling.

QUESTIONS (ANSWERS BEGIN ON PAGE 401)

1. What test should be used to identify malathion intoxication?
2. What is an effective test to monitor for mercury poisoning?
3. Comment on the use of a delta cholinesterase test for DDT exposure.
4. What would cause an arsenic in urine test to be false?
5. What metabolite of benzene is measurable in urine?
6. Is a urine lead level of 90µg/l acceptable?
7. Why would spot levels of a COC in urine be different from twenty-four hour measurements?
8. Concerned about potential exposure to a cardiac sensitizer, you send a worker to see a physician. The blood levels of lactase dehydrogenase are mostly LDH_5, some LDH_4, but no LDH_1 through LDH_3, what might the physician conclude?
9. A mixture is being evaluated and the components are listed as weight percent. What is the volume fraction of each?
10. If the vapor pressure of Dalapon, a pesticide, is 0.12 mmHg at 25°C, and its boiling point is 190°C, what is its vapor pressure at 88°F?

11. The volume of carbon disulfide is 2,600 mL at STP (scientific) and weighs 5.148 g. What is its density at STP?

12. If the density of helium is 0.1782 g/l at STP, what is its density at 31.1°C?

REFERENCES

American Conference of Government Industrial Hygienists. *Guide to Occupational Exposure Values – 1991*. Cincinnati: ACGIH, 1991.

Anderson, H.V. *Chemical Calculations*. New York: McGraw–Hill Book Company, 1955.

Bartok, William, and Adel F. Sarofim. Eds. *Fossil Fuel Combustion: A Sourcebook*. New York: John Wiley & Sons, Inc., 1991.

Brauer, Roger L. *Safety and Health for Engineers*. New York: Van Nostrand Reinhold, 1990.

Burmaster, David E. and Jeanne W. Appling. "Introduction to Human Risk Assessment, with an Emphasis on Contaminated Properties." In *Environment Reporter*. April 7, 1995. pp. 2431–40.

Burrell, Robert, Dennis K. Flaherty, and Leonard J. Sauers. *Toxicology of the Immune System: A Human Approach*. New York: Van Nostrand Reinhold, 1992.

Campbell, Reginald L. and Roland E. Langford. *Fundamentals of Hazardous Materials Incidents*. Chelsea, MI: Lewis Publishers, 1991.

Cockerham, Lorris G. and Barbara S. Shane, eds. *Basic Environmental Toxicology*. Boca Raton, FL: CRC Press/Lewis Publishers, 1994.

Cornett, Frederick D. and Pauline Gratz. *Modern Human Physiology*. New York: Holt, Rinehart and Winston, Publishers, 1982.

Daneshyar, H. *One-Dimensional Compressible Flow*. New York: Pergamon Press, 1976.

De Serres, Frederick J. and Arthur J. Bloom. *Ecotoxicity and Human Health: A Biological Approach to Environmental Remediation*. Boca Raton, FL: CRC Press/Lewis Publishers, 1996.

Dufour, James T. *Hazardous Waste Management*. Sacramento: California Chamber of Commerce, 1992.

Federal Register, Vol. 57, No. 114, Friday June 12, 1992. "OSHA Preamble and Proposed Rule to Revise Air Contaminant Standards for Construction, Maritime, and Agriculture." pp. 26002–26601.

Gardner, Robert W. *Chemical Intolerance: Physiological Causes and Effects and Treatment Modalities*. Boca Raton, FL: CRC Press, Inc., 1994.

Hamilton, Helen Klusek, *et al. Nursing Reference Library: Diagnostics*. Springhouse, PA: Springhouse Corp., 1984.

Hansen, Doan J. ed. *The Work Environment, Volume III: Indoor Health Hazards*. Boca Raton, FL: Lewis Publishers, 1994.

Hartman, Catherine E., George J. Schewe, Leslie J. Ungers and Michael J. Petruska. "Waste Oil Risk Assessment Study." *Hazardous Materials Control Monograph Series: Health Assessment*. Silver Spring, MD: Hazardous Materials Control Research Institute, pp. 35–8.

Hathaway, Gloria J., Nick H. Proctor, James P. Hughes, and Michael L. Fischman. Eds. *Proctor and Hughes' Chemical Hazards of the Workplace*. 3d. ed. New York: Van Nostrand Reinhold, 1991.

Himmelblau, David M. *Basic Principles and Calculations in Chemical Engineering*. 5th ed. Englewood Cliffs, NJ: Prentice Hall, 1989.

Kay, Robert L., Jr. and Chester L. Tate, Jr. "Public Health Significance of Hazardous Waste Sites." *Hazardous Materials Control Monograph Series: Health Assessment.* Silver Spring, MD: Hazardous Materials Control Research Institute, pp. 65–71.

Koren, Herman and Michael Bisesi. *Handbook of Environmental Health and Safety, Volume I: Principles and Practices.* 3d. ed. Boca Raton, FL: CRC Press/Lewis Publishers, 1996.

Landis, Wayne G. and Ming-Ho Yu. *Introduction to Environmental Toxicology: Impacts of Chemicals upon Ecological Systems.* Boca Raton, FL: Lewis Publishers, 1995.

Levesque, Joe. "Fire Protection." In *The Work Environment: Volume I, Occupational Health Hazards.* Doan J. Hansen, ed. Chelsea, MI: Lewis Publishers, 1991.

Luster, Michael. "Immunotoxicology." Section 5 of Chapter 3, "Newer Basic/Long-term Research with Application to Environmental Health Problems." In *Report of the Subcommittee on Health Effects, Appendix D: Strategies for Health Effects Research.* Research Strategies Committee, U.S. EPA, Report SAB-EC-88-040D, Revised October 24, 1988.

Mahaffey, Kathryn. "Research Advances in the Toxicology of Lead." Chapter 3 in: *Report of the Subcommittee on Health Effects, Appendix D: Strategies for Health Effects Research.* Research Strategies Committee, U.S. EPA, Report SAB-EC-88-040D, Revised October 24, 1988.

Manahan, Stanley E. *Environmental Chemistry.* 6th. Ed. Boca Raton, FL: CRC Press/Lewis Publishers, 1994.

Mattison, Donald, and Alan Wilcox. "Human Chorionic Gonadotropin (HCG)." Section 6 of Chapter 3, "Newer Basic/Long-term Research with Application to Environmental Health Problems." In *Report of the Subcommittee on Health Effects, Appendix D: Strategies for Health Effects Research.* Research Strategies Committee, U.S. EPA, Report SAB-EC-88-040D, Revised October 24, 1988.

Neely W. Brock. *Introduction to Chemical Exposure and Risk Assessment.* Boca Raton, FL: CRC Press/Lewis Publishers, 1994.

Office of Solid Waste and Emergency Response. U.S. EPA. Draft Memorandum. "Implementation of Exposure Assessment Guidance for RCRA." Washington, D.C. September 24, 1993.

Overview: A Total Management System for Loss Prevention and Control. Hartford, CT: Industrial Risk Insurers. 1986.

Partridge, Lawrence J. and Arthur D. Schatz. "Application of Quantitative Risk Assessment to Remedial Measures Evaluation at Abandoned Sites." *Hazardous Materials Control Monograph Series: Risk Assessment Volume I.* Silver Spring, MD: Hazardous Materials Control Research Institute, pp. 1–5.

Philp, Richard B. *Environmental Hazards and Human Health.* Boca Raton, FL: CRC Press/Lewis Publishers, 1995.

Plog, Barbara A. ed. *Fundamentals of Industrial Hygiene.* 3d. ed. Chicago: National Safety Council, 1988.

"Polymerization." In *Chemical and Process Technology Encyclopedia.* Considine, ed. New York: McGraw–Hill Book Company, 1974.

Preuss, Peter W., Alan M. Ehrlich and Kevin G. Garrahan. "The U.S. Environmental Protection Agency's Guidelines for Risk Assessment." *Hazardous Materials Control Monograph Series: Risk Assessment Volume I.* Silver Spring, MD: Hazardous Materials Control Research Institute, pp. 6–13.

Rea, William J. *Chemical Sensitivity: Volume I, Principles and Mechanisms.* Boca Raton, FL: CRC Press/Lewis Publishers, 1992.

Reiter, Lawrence. "Neurotoxicology." Section 3 of Chapter 3, "Newer Basic/Long-term Research with Application to Environmental Health Problems." In *Report of the Subcommittee on Health Effects, Appendix D: Strategies for Health Effects Research.* Research Strategies Committee, U.S. EPA, Report SAB-EC-88-040D, Revised October 24, 1988.

Rochow, Eugene G. *The Metalloids.* Boston: DC. Heath and Company, 1966.

Rose, Arthur, and Elizabeth Rose. Eds. *The Condensed Chemical Dictionary.* 7th ed. New York: Van Nostrand Reinhold, 1966.

Sax, N. Irving and Richard J. Lewis, Sr. *Hazardous Chemical Desk Reference.* New York: Van Nostrand Reinhold, 1987.

Schatz, Arthur D. and Michael F. Conway. "OSHA Standards or Risk Analysis: Which Applies?" *Hazardous Materials Control Monograph Series: Health Assessment.* Silver Spring, MD: Hazardous Materials Control Research Institute, pp. 14–7.

Schewe, George J., Joseph Carvitti and Joseph Velten. "Human Exposure Estimates Using U.S. EPA Guideline Models: An Integrated Approach." *Hazardous Materials Control Monograph Series: Health Assessment.* Silver Spring, MD: Hazardous Materials Control Research Institute, pp. 9–13.

Scott, Ronald M. *Introduction to Industrial Hygiene.* Boca Raton, FL: Lewis Publishers, 1995.

Smith, S.L. "Nervous about Neurotoxins." *Occupational Hazards.* December 1990, pp. 37–40.

Solomon, T. W. *Organic Chemistry.* 5th ed. New York: John Wiley & Sons, 1992.

Staab, Robert J. "General Concepts of Toxicology." Chapter 1 in: *Industrial Hygiene Study Guide.* 4th ed. Phil Roets, ed. New Jersey Section American Industrial Hygiene Association, 1989.

Stine, Karen E. and Thomas M. Brown. *Principles of Toxicology.* Boca Raton, FL: CRC Press/Lewis Publishers, 1996.

Talbott, Evelyn O. and Gunther F. Craun. *Environmental Epidemiology.* Boca Raton, FL: CRC Press/Lewis Publishers, 1995.

Talty, John T. *Industrial Hygiene Engineering: Recognition, Measurement, Evaluation and Control,* 2d. Ed. Park Ridge, NJ: Noyes Data Corporation, 1988.

U.S. Department of Transportation, Research and Special Programs Administration. *Emergency Response Guidebook.* DOT P 5800.5. The latest is the 1996 version.

Upton, Arthur. "Environmental Factors and Human Health." Chapter 1 in: *Report of the Subcommittee on Health Effects, Appendix D: Strategies for Health Effects Research.* Research Strategies Committee, U.S. EPA, Report SAB-EC-88-040D, Revised October 24, 1988.

Wadden, Richard A. and Peter A. Scheff.

Wagner, Travis. *In Our Backyard: A Guide to Understanding Pollution and Its Effects.* New York: Van Nostrand Reinhold, 1994.

Weast, Robert C. ed. *CRC Handbook of Chemistry and Physics.* 56th ed. Cleveland: CRC Press, 1975.

Wiesner, Paul J. and Stephen Margolis. "The Epidemiologic Approach to Hazardous Waste Problems." *Hazardous Materials Control Monograph Series: Health Assessment.* Silver Spring, MD: Hazardous Materials Control Research Institute, pp. 18–20.

4 Basic Chemical Hazards to Human Health and Safety — II

CANCER

Dread of cancer is a primary emotional stressor in the late Twentieth Century. Some people apparently live in near-constant fear of the disease. Take for instance, four plaintiffs won a case against a major industrial company in California, who happened to dispose of a Class I toxic chemical in a Class II landfill. The foursome was awarded $800,000 for fear of cancer, $269,500 for psychological damages, $142,975 for medical monitoring costs, $108,100 for disruption of their lives, and $2.6 million for punitive damages. The state court of appeals upheld a lower court, saying that the plaintiffs did not have to establish a physical injury or even a reasonable certainty that cancer would occur in order to recover for fear of cancer as long as the fear itself is genuine, serious and reasonable. However, the higher court did set aside the award for medical monitoring, saying that it was unwilling to open a new cause of action against the defendant. The plaintiffs not only did not have to establish a physical injury, they did not even have to show that cancer or some other future injury was more likely than not.

Apparently, in disagreement with the appellate court, the state supreme court ruled that a plaintiff with a fear of developing cancer from exposure to hazardous waste, but without a current physical injury, must prove it is more likely than not that the feared cancer will develop in the future, due to the toxic exposure. Experts must provide evidence that the plaintiff has a serious fear that the toxic exposure was of such a magnitude and proportion as to likely result in the feared cancer. However, this proof is not required in the situation where plaintiff's distress is caused by the defendant's despicable act. The court required also that damages for medical monitoring requires an expert's assertion that medical surveillance is a reasonably certain consequence of plaintiff's toxic exposure.

Increasing evidence in studies of women patients and laboratory animals suggests that women who have high concentrations of chlorinated pesticides and other chlorinated chemicals in their blood and fat are up to ten times more likely to develop breast cancer than women with lower concentrations. This is the induced conclusion of examining available scientific studies for a possible link between chlorine and cancer. Cohorts of women who have higher exposure to chlorine chemicals than normal have elevated rates of breast cancer. Genetic inheritance, reproductive, and hormonal factors account for twenty-five percent of breast cancer cases.

An oft-maligned chemical with respect to cancer is ozone. The National Toxicology Program recently completed some long term studies in which they concluded that ozone exposure indicates no evidence of carcinogenic activity in rats, equivocal evidence in male mice, and some evidence of carcinogenic activity in female mice. Increased incidences of pulmonary tumors were observed in the mice.

Cancer is abnormal tissue growth that is insidious and life-threatening. Perhaps we would all develop cancer if we lived long enough. We certainly develop tumors throughout our lives. The cells in our bodies at the present moment did not exist, as our bodies, when we were born. Statistically, some of our countless cells, as they multiply and replace themselves during our lifetimes, are going to go awry. If they are altered in such a way that the new, mutant cells start multiplying rapidly, and replacing healthy cells in the process, we have cancer. If new cells grow that do not functionally replace old cells that have died off, but which do not grow out of control, we simply have a *neoplasm* or tumor. Cancer is a neoplasm, or series of neoplasms, that grow out of control and attack other cells. At present life expectancies, about one in seven of us will have certain body cells alter in such a way as to cause unrestricted growth of unhealthy cells sometime in our life span. The unrestricted growth progresses to the point of interfering with the normal functions of organs, and the whole process can be quite painful. Without medical intervention, death ensues. With medical intervention, the patient may die anyway, or require recurring medical treatment, but some patients do recover. The point is that cancer is typically lethal and the approaching death is usually unpleasant with respect to pain and the quality of life towards the end. In the case of colon cancer, breast cancer, and a few others, life can be prolonged with chemotherapy (treatment by a selective poison to kill the cancer cells), radiation treatment, surgery, or any combination of these options. Lung cancer is one type where little hope of survival is offered. Hence, cancer is an emotional topic under the best of circumstances.

Carcinogens. A *carcinogen* is a cancer-causing agent, which may also be called a *tumorigen, oncogen,* or *blastomogen. Carcinogenesis* is the cancer development process. *Carcinoma* is generic term for a cancerous growth in epithelial tissue. *Sarcoma* is cancerous growth in supporting tissue. A specific cancerous growth is named by using the medical name for the organ involved with the suffix *-oma.*

Oncogenes. Cellular genes that bring about cancer during the normal growth and development are *protooncogenes.* Transduced oncogenes of acute transforming retroviruses were protooncogenes. Oncogenes can also be formed from protooncogenes by mechanisms other than retroviral involvement, however. Point mutations or DNA rearrangements, such as translocations or gene amplifications, also lead to oncogenes. Certain types of oncogenes are activated by specific chemical dosages. Others are activated during neoplastic progression. The loss of specific regulatory functions, such as tumor suppressor genes, is believed to be a step in neoplastic transformations. The activation of ras protooncogenes by point mutation in a tumor induced by a specific chemical may occur early in the tumorigenesis and may even be the initiation step. DNA damaging potential of chemicals to the H-ras and K-ras genes has been implicated in carcinogenesis, both in the induction of tumors and the progression from benign to malignant tumors.

Chemicals. Chemically induced cancer involves a complex, poorly understood process. Carcinogens induce malignant neoplasms or tumors following a reasonable exposure. Two stages, *initiation* and *promotion*, are believed to lead to the development of the cancer.

Initiation. When a chemical reacts directly or indirectly with DNA to produce a heritable mutation, which will lead to unrestricted growth, initiation has occurred. The mutation is caused when an electrophilic chemical or metabolite binds to DNA. The alteration in DNA structure may cause a misreading of the DNA sequence during cell replication, and that process may ultimately develop a tumor. Researchers have found a direct correlation between substances that are mutagenic *in vitro* and carcinogenic *in vivo*.

Promotion. Promotion is the second stage, in which certain events facilitate the unrestricted multiplication of initiated cells. Promotional events are peroxisome proliferation, immunosuppression, and hormonal alterations. Chemical promoters either increase the number of tumors, increase the growth rate of tumors, or decrease the latency period during which the tumor in noncancerous, at the end of which the tumor becomes cancerous. Unlike initiators, promoters do not bind to DNA and are called epigenetic carcinogens, but do not cause cancer themselves. Promoters may act by producing activated oxygen radicals including superoxides, peroxides, and hydroxyls that indirectly affect the DNA molecule. Cocarcinogens act with or precede carcinogens to increase tumor yield. Some, but not all, promoters are cocarcinogens.

When a COC or its metabolite possesses both initiation and promotion capabilities, the COC is a carcinogen. Other COCs may be either initiators or promoters, because, by themselves, they cannot cause cancer. Certain chemicals, generically called keratin stimulants, stimulate the skin to grow, leaving open the possibility of tumor formation.

Progression. Once a tumor is underway, it enters a phase called progression. If it will remain benign, the tumor, during its progression, encapsulates itself, grows slowly, and remains noninvasive. Benign tumors are controlled by excision, surgery. On the other hand, a malignant tumor does not encapsulate, grows rapidly, is invasive, distributes cells to other parts of the body, and resists treatment.

Metastasis. When other, previously healthy, tissue has been invaded by cancer cells, metastasis has occurred. The progression and metastasis phases are the concern of medical science, while the exposure assessor and related colleagues, who are interested in prevention, are concerned with the first two phases, initiation, and promotion.

Carcinogens are presumed to follow a linear, nonthreshold dose-response relationship. The so-called multistage model is used to predict risk associated with low doses of carcinogens. The model assumes that cancer has three stages and was fitted to experimental data. However, from the upper confidence level to zero, the model is linear.

The lifetime risk of cancer at dose d is P(d). The extra risk A(d) over the background rate is

$$P(d) = 1 - \exp\left[-\left(q_1 d + q_2 d^2 + \ldots + q_k d^k\right)\right] \tag{4.1}$$

where

$$q_1 \geq 0$$
$$i = 1,2,3,\ldots,k$$

and

$$A(d) = [P(d) - P(0)]/[1 - P(0)]$$

Experimentally derived data points define a unique set of q_i. Standard goodness-of-fit tests can be used to determine how well the model describes the data. A maximum likelihood estimate (MLE) and ninety-five percent upper confidence level (95-UCL) of A(d) using the 95-UCL of q_i, q_i^*. MLE, the point estimate of A(d), is the best estimate of extra risk at dose d.

The one-hit model assumes only a single stage for cancer. This model also assumes that one molecular interaction induces malignant change, therefore it is very conservative. Least conservative among models in use is the multihit model, which assumes several interactions are necessary to transform a cell. The probit model assumes a probit, or log-normal, distribution for tolerances of an exposed population and is most appropriate for acute toxicity, but questionable for cancer modeling.

Little is known beyond the effects of measurable doses external to the body. For instance, very little research has been conducted on internal dose and biologically effective dose of carcinogens. Internal dose is the amount of a carcinogen or its metabolite in body tissues and fluids as opposed to the dose received at the mouth or through skin absorption or inhaled. The biologically effective dose is the amount of the carcinogen that interacts with cellular macromolecules, such as DNA, RNA, or protein, in the target tissue or some surrogate tissue. DNA adducts are lesions that, if they go unrepaired, can produce gene mutations, which in somatic cells initiates the stages of carcinogenesis. DNA adducts can be measured in white blood cells with immunoassay techniques, a molecular technique called postlabeling, and by fluorescence spectrometry. Protein, such as hemoglobin, can often be used as a surrogate for DNA and hemoglobin adducts can be measured in red blood cells using mass spectrometry, ion-exchange amino acid analysis, high-pressure liquid chromatography (HPLC), or gas chromatography. Excised adducts in urine can be measured by HPLC or fluorescence techniques. Unscheduled DNA synthesis can be determined by cell cultures and thymidine incorporation in white blood cells. Cytogenetic techniques determining sister chromatid exchange, micronuclei, and chromosomal aberrations are performed on white blood cells, and also bone marrow in the case of micronuclei. Somatic cell mutations can be determined by measuring hypoxanthine guanine phosphoribosyl transferase (HGPT) enzyme in white blood cells using autoradiography or light microscopy.

Generally, the risk to an individual, exposed to a mixture of carcinogens, is determined by adding the constituent-specific risks, except where synergistic or antagonistic interactions are known to occur for the specific mixture. In determining the incidence of cancer in a population, the basic assumption is that the concentration estimated at a given receptor location from a given source, group of sources, facility,

TABLE 4.1
Unit Risk Factors

Pollutant	Unit Risk
Acrylonitrile	6.8×10^{-5}
Arsenic	4.3×10^{-3}
Chloroform	1.0×10^{-5}
Ethylene	2.7×10^{-6}
Methylene chloride	1.4×10^{-7}
Perchloroethylene	1.7×10^{-6}
Styrene	2.9×10^{-7}

or group of facilities, on an annual basis, is a good estimate of the concentration to which the population within a thirty mile radius (fifty km) will be exposed over a lifetime of seventy years. A lifetime unit risk factor is used to relate the exposures to risk in a surrounding population. These factors are published by EPA's Carcinogen Assessment Group. A factor is applied to each material expected to be emitted, so that the full scope of effects can be estimated. The assumption is made that the exposure is 1 mg/m³, and that the average daily inhalation rate is 20 m³, which produces a dose of 20 mg/day. Exposure is assumed to occur both indoors and outdoors, at levels predicted by the dispersion model. Table 4.1 lists some unit risk factors.

Carcinogenic potency factors ($q_1{}^*$), sometimes called potency slopes, are calculated:

$$C = \frac{R(BW)}{q_1'I} \qquad (4.2)$$

where: C = air concentration, mg/m³
R = incremental carcinogenic risk due to lifetime exposure to the chemical at concentration C, dimensionless
BW = body weight, kg
$q_1{}^*$ = carcinogenic potency factor, kg day/mg
I = inhalation rate, m³/day

Table 4.2 lists acceptable values for these factors and others. Carcinogenic potency factors for some chemicals are given in Table 4.3. In Table 4.4, the various carcinogenic designations are summarized.

Lifetime risk for cancer from a certain exposure is:

$$R = q_1' \times LADD_2 \qquad (4.3)$$

where the lifetime average daily dose is given in mg/kg day. For carcinogens, EPA requires that the average time of exposure be assumed to be 70 years at 365 days per year. For noncarcinogenic chemicals, the number of days in any assumed exposure

TABLE 4.2
Acceptable Values in Cancer Equation

Acceptable incremental risk	1×10^{-6}
Usual body weight	
Adult	70 kg
Child (1–6 years)	16 kg
Inhalation rate	
Adult	20 m³/day
Child	5 m³/day
Water consumption	
Adult	2 l/day
Child	1 l/day
Food consumption	
Eggs	0.064 kg/meal
Beef	0.112 kg/meal
Fin fish	54 g/day
Soil ingestion	
Child (1–6 years)	200 mg/day
Child (>6 years)	100 mg/day
Water exposure (nonconsumed)	
Showering	12 min./day
Swimming	2.6 hr/day
Working lifetime	80,000 hr
Biological lifetime, female	6.8×10^5 hr

After EPA.

TABLE 4.3
Carcinogenic Potency Factors

Substance	Potency Factor, q_1^* (mg/kg day)
Arsenic	14.0
Benzene	0.052
Cadmium	6.65
Carbon tetrachloride	0.13
Chromium (VI)	41.0
Polychlorinated biphenyls	4.34
Tetrachloroethylene	0.0531
1,1,2-Trichloroethane	0.0573
Trichloroethylene	0.0126

duration may be used, if so specified in the assessment report. Exposure frequency is pathway specific. Typically, the frequency is assumed to be 350 days per year for the consumption of contaminated food and water and for inhalation. The exposure

TABLE 4.4
Characteristics of Exposed Persons

Factor	Occupational	Public
Duration of exposure	8 hours/day 5 days/week	24 hours/day 7 days/week
Span of exposure (lifetime)	40 years	70 years
Exposed population	Healthy workers	Most sensitive individuals
Age	Adult	Most sensitive (child, adult, or elderly)
Risk compensation	Yes	No

After Schatz and Conway.

duration is also pathway specific and 30 years is the typical assumption for residential exposure of adults. The LADD is estimated as:

$$LADD = \frac{C \times I \times E}{BW \times L} \tag{4.4}$$

where I is inhalation rate, E is accumulated exposure, hours, and L is lifetime, days. The typical adult inhalation rate is 0.33 m^3/hr, or 20 m^3/day. The assumed body weight (BW) is 70 kg. A lifetime is considered 25,550 days, or 70 years. A year of occupational exposure is 8 hr/day, 5 days/week, for 50 weeks per year, or E = 2,000 hours.

Table 4.5 shows the carcinogen designations of various agencies and gives an abbreviated meaning.

DEVELOPMENTAL TOXICITY

Spontaneous abortions and fetal death, low birth weight, and other developmental disorders may be caused by exposure to chemicals such as certain heavy metals, herbicides, pesticides, and others. Developmental disorders can be low birth weight, prematurity, infant death, cerebral palsy, mental retardation, deafness, and blindness.

The fertilized human ovum is a *blastocyst* or *conceptus* from conception to about two weeks. From two to nine weeks, the developing human is called an *embryo*. After nine weeks, until birth, it is a *fetus*. Toxic insult during the blastocyst stage typically leads to spontaneous abortion and only rarely to teratogenic effects. The period most susceptible to teratogenesis is the embryonic phase when *organogenesis* takes place. Embryonic cells are undergoing countless divisions, migrating, and associating themselves to form rudimentary organs and skeletal features. *Anencephaly* (lack of a brain beyond the rudimentary brain stem at the top of the spine), *spina bifida* (a spinal malformation in which the spinal cord and nerves herniate and appear near the skin), and *occulta* (a spinal malformation similar to spina bifida, but without the hernia) are some of the potential developmental problems due to embryonic toxic attack.

TABLE 4.5
Summary of Carcinogen Designations

Organization	Symbol	Meaning
IARC	1	Sufficient human evidence
	2A	Limited human evidence; sufficient animals
	2B	Limited human; lack sufficient evidence in animals
	3	Not classifiable as to humans
	4	Probably not carcinogenic to humans
MAK	A1	Induces malignant tumors in humans
	A2	Unmistakably carcinogenic In animals
	B	Justifiably suspected
NIOSH	X	Carcinogen
NTP	1	Known; evidence from human studies
	2	Reasonably anticipated; limited evidence in humans; sufficient in animals
OSHA	X	Carcinogen
ACGIH TLV	A1	Confirmed human carcinogen
	A2	Suspected human carcinogen
EPA	A	*Human carc.* — sufficient evidence
	B	*Probable h.c.* — limited evidence
	C	*Possible h.c.* — limited animal evidence
	D	*Not classifiable* — inadequate evidence
	E	Evidence for non-carcinogen for humans

After ACGIH.

Developmental toxicity includes adverse effects in the developing fetus, or test organism in the case of a test. Developmental effects, in general, may be due to an exposure that occurred to either parent prior to conception, or to the mother during prenatal development, or postnatally to the new person, right up to the time of sexual maturation. Major manifestations of developmental effects include death of the developing organism, malformation, altered growth, and functional deficiency. A subclass of developmental toxicity, teratogenicity, refers primarily to malformations of the fetus, during the period characterized by cellular differentiation, functional maturation, and growth. The literal translation of teratogen is monster maker. Teratogenesis produces abnormal offspring following the exposure. Teratogenesis is the interference with normal embryonic development to produce a congenital malformation or defect. The effects of teratogenesis are not hereditary. Other congenital malformations, caused by mutations in genetic material, are hereditary. A mutagen is an agent that affects the genetic system of the exposed person, causing cancer in the subsequent cell replications of the exposed individual, or undesirable mutations in the offspring of that individual, conceived after the exposure. The genetic damage done by a mutagen may not show up until several generations later.

Ultimately, teratogenesis leads to functional disorders, malformation, growth retardation, or death. The classical definition of teratogenesis only dealt with abnormal growth, but, being as emotionally charged an issue as cancer is, the latter-day practice is to include functionally as well as structural abnormalities in the definition. Susceptibility of the developing life to a teratogen depends on the stage of develop at the time of the exposure. In the first and second week, for instance, the neural tube (spine and central nervous system) develops. The heritable traits are already locked in at conception, but teratogen attack on the developing embryo can alter the genetic code of soma cells, causing malformation of the neural tube. The two most common manifestations of neural tube malformation are anencephaly and spina bifida, mentioned above. However, each teratogen acts in a specific way. So, if a pregnant woman is exposed to a teratogen that causes malformation of the neural tube after the second week of gestation, the risk to her child is low with respect to the teratogenicity of that chemical. Substances that are carcinogenic or mutagenic are more likely to be teratogenic also, but that must be shown in laboratory testing. However, a common mechanism for both teratogenesis and carcinogenesis is the altering of DNA to produce somatic mutations.

Spontaneous abortion is a great risk in the earliest weeks of gestation, but even after that some risk remains until the birth. Therefore, the possibility of chemical interference with the pregnancy is an emotional and newsworthy topic. Determination of whether a spontaneous abortion was caused by a chemical can be impractical, if not impossible, because demographic and behavioral traits affect spontaneous abortion rates too. Maternal habits such as smoking and substance abuse increase the risk of spontaneous abortion. The age of the mother, the gestational stage at which the pregnancy was identified, and the number of previous pregnancies also affect the risk. The direct causes of spontaneous abortion are lethal embryo or fetal toxicity (chemical poisoning), chromosomal alterations, single gene effects, structural abnormalities, maternal-fetal incompatibility, or maternal abnormality. Talbot and Craun differentiate four major categories of spontaneous abortions for environmental epidemiology: chromosomally abnormal, chromosomally normal — structurally abnormal, chromosomally normal — structurally normal, and maternal abnormality.

The causes of congenital malformations are not well known. Teratogens cannot be blamed for many of the birth defects experienced, yet, the reason we must continue to test for teratogenicity is as basic as preservation of the race. A possible link between congenital malformations and chemical exposure is under study at John Hopkins University where a link is sought between the potential chemical (and biological) exposure of the Iraqi-Allied Forces Gulf War veterans and malformed hearts, anencephaly, and malformed skulls. These effects are teratogenic if chemically induced, and if pregnant mothers were exposed during the first or second week of gestation, but mutagenic if the exposure occurred (to either parent) prior to conception. If the critical exposure occurred to the pregnant mother beyond the second week of gestation, teratogenicity is questionable.

Low birth weight is another potential manifestation of teratogenicity. However, the epidemiology of low birth weight has many confounders, which can lead to misinterpretation of data, albeit innocently enough, so treat such data carefully and read reports of studies on low birth weights carefully to understand the significance of the data, or lack thereof.

TABLE 4.6
Categories of Mutagenicity Data

1. Positive data derived from human germ-cell mutagenicity studies.
2. Positive results from studies on heritable mutation events in mammalian germ cells.
3. Positive results from mammalian germ-cell chromo some aberration studies (non-intergenerational).
4. Evidence for chemical interaction with mammalian germ cells with positive mutagenicity test results from two assay systems, at least one of which is mammalian (*in vitro* or *in vivo*). Both positive results may be for gene mutations or both for chromosome aberrations but if one is for gene mutations and one is for chromosome aberrations both must be from mammalian systems.
5. Suggestive evidence for a chemical interaction with mammalian germ cells with positive mutagenicity evidence from two assay systems as described under 4. Alternatively, positive mutagenicity evidence of less strength than defined under 4, when combined with sufficient evidence for a chemical interaction with mammalian germ cells.
6. Positive mutagenicity test results of less strength than defined under 4 combined with suggestive evidence for a chemical interaction with mammalian germ cells.
7. Definite proof of mutagenicity is not possible. The chemical could be classified operationally as a mutagen for human germ cells if it gives valid negative test results for all end points of concern.
8. Inadequate evidence for either mutagenicity or chemical interaction with human germ cells.

Talbott and Craun reviewed studies on several specific teratogens. Those that are significant in the environment are methylmercury (a fungicide), heavy metals, agricultural chemicals and pesticides, nitrates in drinking water, polychlorinated biphenyls (PCBs), polybrominated biphenyls (PBBs), and organic solvents, such as trichloroethylene (TCE).

MUTAGENIC EFFECTS

Gene mutations and sperm and chromosomal abnormalities are included as mutagenic effects. It becomes difficult to separate mutagenic effects from developmental disorders. The goal is to assess the likelihood that an exposure will induce heritable changes in DNA or that the dose received will interact with human germ cells.

Epidemiological data that indicates a strong statistical association between exposure to a COC and heritable effects is evidence that a heritable mutation has been induced. The problem is obtaining such data because any particular mutation is a rare event. Among the thousands of potential human genes and genetic conditions only a very few can be used as markers in studying mutation rates. Since human data is rare, we must use experimental data on surrogate animals. Fortunately, DNA is universal in the animal kingdom and animal data derived from testing both *in vitro* and *in vivo* can be used to estimate mutation rates among human cells. Scientists have developed eight categories of data as shown in Table 4.6.

Chlorinated drinking water, while protecting consumers from potential bacterial infections, increases levels of mutagenicity. Chlorine apparently reacts with natural humic substances in water to form mutagens. One such agent produced is 3-chloro-4-(dichloromethyl)-5-hydroxy-2(5H)-furanone, called MX for short. The isomer of

MX, 2-chloro-3-(dichloromethyl)-4-oxobutenoic acid, or E-MX, is also produced by the reaction of chlorine and humic acids. According to Talbott and Craun, MX is a potent mutagen, while E-MX is somewhat less potent. Water disinfectants in decreasing order of potency for manufacturing mutagens when reacted with humic acid, are: chlorine, chloramine, chlorine dioxide, and ozone. While all water potentially contains some humic material, some areas are more prone than others to have measurable concentrations of humic acid. For instance, in the Mississippi delta, several municipalities have red to reddish-brown water due to humic acid. Municipalities and other drinking water providers do not have to submit to the mercy of nature on this issue. MX and E-MX can be easily be removed from drinking water by coagulation and flocculation, or by carbon adsorption, so the chief risk issue is identification of the problem.

Early pregnancy loss, or spontaneous abortion, is a potential early sign of exposure to mutagens or toxins that damage human reproduction and human chorionic gonadotropin (HCG), which is produced by the conceptus as early as the seventh day after fertilization, can be used to determine pregnancy in the earliest stage. This makes it easier to track potential effects of exposures. In the meantime, a paucity of data exists for mutagenic and reproductive effects of chemicals.

REPRODUCTIVE EFFECTS

Adverse reproductive effects are associated with exposure to some herbicides, pesticides and metals. Reduced fertility, impotence, and menstrual disorders are among the problems caused by exposure to certain chemicals. Also the gestation period can be affected in females. More importantly, toxic substances affect oocyte production. If secondary oocytes are damaged, temporary infertility is the result, but this is reversed as new secondary oocytes are produced. Damage to primary oocytes results in the onset of menopause, because the reserve of primary oocytes on the ovaries fell below a certain level. Total loss of primary oocytes causes immediate infertility and onset of menopause. *Embryogenesis* and fetal development consists of millions upon millions of cell divisions, any of which are susceptible to toxic insult. See teratogenesis above.

For males, the time required to restore the sperm count is a significant reproductive effect. Any toxic substance that interferes with the control mechanism for the male pituitary hormones, *Luteinizing Hormone* (LH) and *Follicle Stimulating Hormone* (FSH), can lead to male infertility. LH and FSH are regulated by hypothalamic releasing factors, which are controlled by the pituitary and hypothalamus. LH stimulates *Leydig cells*, which synthesize steroid hormones, particularly testosterone, which controls the activity of accessory sex organs and the development of secondary sexual characteristics. FSH stimulates the *Sertoli* cells to initiate spermatogenesis and maintain optimal testicular function. Male germ cells are susceptible to toxic insult during the stages of spermatogenesis, in particular, by either being killed or undergoing heritable alterations or suffering non-heritable alterations that affect morphology, motility, and viability. Reduction of sperm count leads to male infertility, while mutated sperm may lead to adverse effects in a fertilized egg.

Abnormal sperm raises the probability of spontaneous abortion, still birth, or any number of birth defects.

Successful reproduction is more than a measure of physiological health in humans; it also is important to mental health and relationships. That is not to say, however, that childless couples cannot be mentally healthy or that they cannot maintain a loving relationship, but childlessness can certainly be a stressor. Infertility can even be a stressor with couples who already have children. Not only that, but toxic insult to the reproductive system of male or female may result in decreased libido, increasing the emotional stress on the couple.

Women with occupational lead exposures have been tracked for about a century now, and the reproductive effects of lead are increased probability of spontaneous abortion or still birth, not to mention the child's increased probability of post-natal mortality, or early childhood mortality. The incidence of preterm (thirty-seven week) deliveries is strongly related to maternal blood level of lead at delivery. The risk of preterm delivery with a blood level of fourteen g lead/dl is nine times the risk at a blood level of eight. The gestation length is inversely proportional to blood level of lead. Birth weights decrease an average of three-fourths of a pound for each ten g/dl lead in the mother's blood.

ORGAN EFFECTS (NON-CANCER)

For systemic toxicants in a mixture, a hazard index is estimated only if the constituents induce the same effect by different modes of action. Since different effects occur for the same chemical at different dosages, and since biochemical mechanisms frequently are unknown or not well understood, a hazard index for a mixture should only be estimated if the RfDs on individual components of the mixture are based on the same target organ. Since many carcinogens also exhibit systemic effects, carcinogens should be considered when non-cancer individual risk from a chemical mixture is being evaluated.

NEUROPATHIC SUBSTANCES

One system affected by many common chemicals is the nervous system, which is subdivided as the Central Nervous System (CNS) and the Peripheral Nervous System (PNS). Neurotoxic chemicals act on diversified targets within the nervous system via a range of pathways. The brain and spinal cord make up the CNS, while nerve networks throughout the rest of the body make up the PNS. The nervous system has only two kinds of cells. The *neuroglial* cells are nonconducting and essential for the protection, structural support, and metabolism of nerves. The *neurons* provide the structural support and are the functional unit of each nerve.

Different types of neuroglia are found in the brain and spinal cord. The *oligodendroglia* form a fatty-like insulating sheath around the fibers of CNS neurons. *Microglia* protect neurons by engulfing invading microorganisms. *Astrocytes* aid in transporting substances from the blood to the brain, while providing a barrier to protect neurons from harm.

Outside the brain and spinal cord, the fibers of neurons are covered with *Schwann* cells, which in some instances enclose the neuron fiber with a thin layer of fatty-like material. More often, Schwann cells entwine the neuron fiber with a dense insulating cover. Schwann cells are found in the PNS.

Peripheral nerves communicate with the brain by means of their connections to the spinal cord. Nutrients are supplied to the nervous system by cerebrospinal fluid, which bathes the brain and spinal cord. Cerebrospinal fluid also serves as a barrier to protect the CNS from foreign substances, especially those which are water-soluble. Fat-soluble substances easily diffuse across the cerebrospinal fluid, however. A chemical that affects the CNS can also affect the PNS. The gastrointestinal system is loaded with as many nerve cells as the spinal cord, which allows poisons and pollutants to quickly affect the CNS. The CNS is nearly defenseless with respect to lipid-soluble toxic agents.

Part of the PNS is the *autonomic nervous system* (ANS). The ANS controls involuntary muscles, such as the heart and the breathing muscles, and has two branches: the *sympathetic* and *parasympathetic* branches, which control many of the same muscles and glands. The effects of each branch on their targets are different, allowing these organs to perform differently under different circumstances. The sympathetic branch controls the *fight or flight* responses: tachycardia (to pump more blood and therefore oxygen to the vital organs), dilation of bronchioles (to bring more air into the lungs), dilation of pupils (to allow more light to enter in order to better detect threatening movement), constriction of peripheral blood vessels (to limit blood loss if wounded), and decrease in digestive activity (to limit this drain on the body's resources during the critical moment). Parasympathetic responses are opposed to those just mentioned: bradycardia (to accommodate rest), constriction of bronchioles (to use less air more efficiently), dilation of pupil (to limit light in order to better discern background differences), increase in peristalsis (to digest food more efficiently), and an increase in secretions.

Axonopathy is damage to axons, most commonly in the PNS, that result in a particular sensory and motor dysfunction generically called a *neuropathy. Proximal axonopathies* are noted for swelling of the proximal axon, called a *giant axonal swelling.* An aminonitrile compound called IDPN causes a proximal axonopathy by blocking slow transport, but allowing the axon to produce proteins and accumulate neurofilaments, leading to giant axonal swelling, which is associated with *amyotrophic lateral sclerosis* (ALS), which has been implicated, but not proven, to be caused by environmental factors. Distal axonopathies are damage in the distal portion of the axons that vary with the offending toxicant to cause swelling, mitochondria damage, accumulation of neurofilaments, and disintegration of myelin. Chronic exposure to carbon disulfide and ethanol cause distal axonopathy.

Acetylcholine is an important neurotransmitter and the inhibition of choline by substances such as organophosphates (see Chapter 3) prevents the firing of adjacent neurons. The release of acetylcholine into the synapse fires the receiving neuron. Normally, the acetylcholine is rapidly degraded and the receiving neuron restores itself and awaits a new signal to fire. If the acetylcholine is not broken down rapidly enough, the receiving neuron keeps firing and symptoms such as muscle incoordination, nausea, and dizziness, which lead to seizures and unconsciousness, are observed. The serine enzyme acetylcholinesterase provides for the expedient breakdown of acetylcholine. Organophosphate molecules enter at the enzyme's active site and a proton donation binds the serine to the phosphate. Symptoms of the inhibition result from acetylcholinesterase being about 60-70 percent phosphorylated. Subsequently, an excess of acetylcholine creates overstimulation at synapses between nerves and voluntary muscles in the CNS and in the parasympathetic branch of the ANS. In human patients, this inhibition means constriction of the pupil, slowing of heart rate, excessive salivation, and muscle contraction. Death by asphyxiation because the diaphragm and related breathing muscles fail to function properly, or because the respiratory center in the brain malfunctions, is the ultimate symptom of this situation.

Rea, at the Environmental Health Clinic in Dallas, reports that he has observed a change in brain function in 40–77 percent of patients exposed to pesticides such as BHC, Dieldrin, Chlordane, DDT, Heptachlor epoxide, and hexachlorobenzene. For instance, sixty-one percent of patients exposed to Dieldrin presented a change in brain function, while forty percent of patients exposed to Heptachlor epoxide presented change in brain function. Exposure to pesticides apparently raises risk of brain damage. Heptachlor epoxide exposure most often leads to physical fatigue and headaches, but four percent of patients at the Dallas clinic also presented seizures and another fourteen percent presented memory loss. The halogenated hydrocarbon insecticides degrade axonal transmission by interrupting sodium and potassium transport across the axonal membrane, preventing polarization. These insecticides also inhibit enzyme controlled processes and impair the synthesis of energy. Rea cites cases of patients who resided in the Environmental Control Unit to detoxify and who were given a standard battery of tests to measure brain function after unloading chlorinated pesticides such as Heptachlor epoxide. These patients present an eighty percent decrease in signs and symptoms after treatment with a corresponding decrease in blood chlorine levels, so the prognosis is good with intervention. About 0.1 percent do not respond, however, to the Dallas ECU treatment. In the improved patients, a change for the better was observed in cerebral dysfunction as measured by the Bender–Gestalt test, in personality as measured by the MMPI, and in intelligence as measured by the WAIS-R. The point is that intervention produces good results, generally.

Typically, when a substance is neuropathological, it will also have neurobehavioral effects. Lead is a common example. The neurobehavioral effects of lead are indicated with blood levels of lead around 20–50 g/dl, but neurological changes that introduce behavior changes are induced much earlier, often during the critical developmental period of the child. A direct relationship has been demonstrated between low-level lead exposures during pre-natal and early development of the child and later deficiencies in behavior. Blood levels of 10 g/dl merits concern. Lead, one of the *ototoxins*, also reduces hearing acuity, which complicates behavior changes and the problems associated with these changes.

European studies have implicated long-term exposure to solvents in neurobehavioral disorders, such as loss of concentration, memory impairment, mood changes, personality changes, and, in the worst cases, severe and irreversible dementia. Cognitive impairment is an early sign of solvent neurotoxicity. Continuing advances in the neurosciences mean that, when a COC causes a neurobehavioral effect, scientists have a pretty good idea where to look for neurochemical and neuroanatomical effects and for mechanisms. Monoclonal antibodies can also be used to detect and characterize local cellular responses to neurotoxic exposure.

Neural catecholamine exhaustion has been identified as a cause of depression. Catecholamines are the neurotransmitters that are released by the sympathetic nervous system. Chlorinated solvents have been found to make organs more sensitive to the catecholamines. Therefore, the exposed person would have a tendency to remain at fight or flight readiness for longer than normal periods of time. However, catecholamine depletion, although it can lead to depression, does not explain major depression according to Gardner. Cerebral imbalances of serotonin, dopamine, and other biogenic amine neurotransmitters such as norepinephrine, epinephrine, and histamine have been implicated in the etiology of depression that are more likely to support long-term bouts of depression than chlorinated solvents. Norepinephrine is released by the postglanglionic neurons of the sympathetic nervous system and some neurons of the CNS. Many neurons that release norepinephrine, epinephrine, dopamine, or serotonin are found in the brain stem: medulla, pons, and midbrain. Many histamine releasing neurons are located in the hypothalamus.

The acute neurotoxic effects of solvents include narcosis, anesthesia, depression of the CNS, respiratory arrest, unconsciousness, and death. Observed neurobehavioral effects of solvents are fatigue, irritability, memory impairment, emotional instability, lack of control over impulsive behaviors, lack of motivation, lack of concentration, and decreased ability to learn. Irreversible dementia, as mentioned above, is the ultimate symptom. Dioxin, or TCDD, exposure causes impaired vision, impaired hearing, impaired smell, depression, sleep disturbance, and other neurological symptoms.

Aluminum nuclei bind with brain cells resulting in senile dementia of the Alzheimer's type. Tests performed on cats and rabbits have been criticized because the aluminum was administered intracranially, yet extracranial administered aluminum presents brain damage that is indistinguishable from damage induced by intracranial doses. Developing children in Minamata Bay, Japan exhibited neuropathologies due to exposure to methyl mercury in shellfish caught in the bay and eaten.

Arsenic compounds are absorbed into the blood stream and transported to the other organs. Arsenic encephalopathy is manifested by speech and behavioral disorder. Peripheral neuropathy of arsenic is seen in both acute and chronic forms.

HEMATOPOIESIS

Agents that act on the blood, or the *hematopoietic* system, decrease hemoglobin function, depriving the body tissues of oxygen. Symptoms of hematopoietic problems include *cyanosis*, or blueness of the skin, and loss of consciousness when the problem is severe enough. The notable hematopoietic toxins are carbon monoxide, recently notable in euthanasia exterminations in Michigan, and cyanide, often used in criminal exterminations. Lead poisoning causes anemia (hematocrit value below thirty-five percent). The lead interferes with the synthesis of heme and the formation of hemoglobin in several metabolic steps, inhibiting *aminolevulinic acid dehydratase,* increasing levels of *erythrocyte protoporphyrin* in children with blood levels as low as 15 g/dl. Impaired heme synthesis also results in general mitochondrial injury, meaning, ultimately, impairment of energy metabolism, homeostasis, and other subcellular processes. The health implications of impaired heme synthesis are reduced oxygen transport, damaged energy production in cells, disordered calcium immunoregulation, interrupted hematogenesis, disturbed calcium metabolism, impaired detoxification of xenobiotics, and disturbed metabolism of tryptophan and other endogenous agonists. Arsenic also causes damage to bone marrow. Chlordane has been reported to cause monocytic leukemia or megaloblastic anemia. Aplastic anemia has been observed after chronic lindane exposure.

Naphthalene, anticoagulant rodenticide, and inorganic arsenical exposure results in *anemia*, the insufficiency of red blood cells. *Leukopenia*, a lower than normal number of white blood cells in the circulating blood, is caused by exposure to nitrophenols and inorganic arsenicals. The anticoagulant rodenticides also cause *hypoprothrombinemia*. Prothrombin is a protein that reacts to form thrombin, an essential substance in blood clotting. Therefore, with hypoprothrombinemia, a person could bleed to death, if wounded. This is how the anticoagulant rodenticides work, anyway. The chlorates lead to methemoglobinemia, a hemoglobin that is incapable of combining with oxygen and, therefore, useless for maintaining life. Naphthalene, arsine, sodium chlorate, and copper compounds lead to the presence of free hemoglobin in plasma. Elevated alkaline phosphatase, GOT, and LDH are found with exposure to halocarbon fumigants, aminopyridine, diquat, paraquat, inorganic arsenicals, and Endrin. Methyl bromide exposure results in elevated blood bromine, as one might expect. The halocarbon fumigants, diquat, paraquat, phosphine, phosphorus, phosphides, copper compounds, and inorganic arsenicals yield *hyperbilirubinemia*, which is an excess of bile pigment in the blood, jaundice. Elevated BUN (blood urea nitrogen) and creatinine are caused by exposure to diquat, paraquat, fumigants, chlorates, inorganic arsenicals, sulfuryl fluoride, nitrophenols, chlorophenols, and phenols. Hyperkalemia, or excess potassium, is caused by naphthalene, chlorates, and copper compounds. Benzene has a metabolite produced in bone marrow that is toxic to hematological processes. Benzene is also recognized as a myelotoxic agent (destroys bone marrow) that can produce both fatal aplastic anemia (defective erythrocyte development in the bone marrow) and leukemia.

Catecholamines connect with α-adrenergic receptacles on vascular smooth muscles to produce vasoconstriction. Restriction of blood vessels increases blood pressure by increasing resistance to blood flow. Compounds that stimulate catecholamines, then, such as nitrates, nitroglycerin, and cadmium increase blood pressure.

IMMUNOSUPPRESSORS

The body's immunity system provides nonspecific barriers and specific defense mechanisms to invasion by viral and bacteria infections. Cells that form and mature in bone marrow and the lymphatic system (thymus, spleen, and lymph nodes) include polymorphonuclear (PMN) leukocytes, lymphocytes, and monocytes, collectively called the white blood cells. Using the lymphatic and circulatory system, these cells exchange chemical messengers, called lymphokines.

The foremost nonspecific protective barrier is the skin, which prevents entry of invasive substances to the body within its limitations, which, though many, generally allow the skin to be an effective barrier anyway. Epithelial cells lining the respiratory, gastrointestinal, urinary, and reproductive tracts secrete mucus to engulf and trap foreign matter, and the hair-like cilia work to sweep the contaminated mucus to the nearest bodily exit. Phagocytes are cells that engulf and digest foreign matter and cellular debris. Macrophages, phagocytes that develop from monocytes, are found in the various tissues of the body. Neutrophils and eosinophils are phagocytes that develop from PMN leukocytes and remain in the bloodstream. Immobile macrophages, such as the Kupffer cells in the liver and the microglia in the central nervous system, maintain another defense barrier for their specific systems. Free macrophages circulate throughout the body as a general line of defense. Natural killer (NK) cells are lymphocytes that seek out and destroy abnormal cells, such as cancer cells or cells damaged by viral attack, by releasing proteins called perforins that penetrate cell walls.

The complement system is a whole set of proteins that destroy cell membranes, attract phagocytes, and stimulate the immune system. Cells undergoing viral attack produce interferon, which stimulates healthy cells to produce antiviral proteins. The latter interferes with the process of viral replication, slowing down the viral attack, allowing the body's immune system a chance to catch up with defenses. Macrophages release proteins called pyrogens that produce the fever accompanying viral infections. The inflammatory response is the last line of defense. Connective tissue cells called mast cells, upon sensing the presence of cell damage, produce histamine, which causes vasodilation, increases blood vessel permeability, and attracts all defenders, such as macrophages, neutrophils, eosinophils, complement proteins, and other protein defenders to aid in the removal of debris. Vasodilation increases the blood flow to the area, making the arrival of defense forces easier, but also producing the classical signs that an inflammation is present: heat, redness, pain, and swelling.

Antigens are substances that activate specific defenses. Antigens are large molecules of protein, or a protein component called glycoprotein or lipoprotein, whole cells, or viruses containing proteins. One of these proteins produced by a specific immune response is called an antibody for that response, and antibodies may or may not stimulate production of additional antibodies. Antigens called haptens are incomplete antibodies that have only one antigenic determinant site and cannot stimulate a complete immune response, unless they bind with molecules that have second antigenic determinant sites. Activated lymphocytes mount a direct attack on the antigens. Lymphocyte-produced antibodies then attack the antigens. These attacks are called cellular immunity and humoral immunity respectively.

The lymphocytes that produce cellular immunity are called T-cells, which attack only those antigens that have been processed by other cells. Phagocytes engulf antigens and break them down. The cell fragments and proteins from the infection agent are displayed on the surface of the phagocyte, bound by cell surface proteins called human leukocyte antigen proteins, produced by genes called the major histocompatibility complex, which are unique markers that identify self from nonself. T-cells respond to the HLA protein and foreign antigen combination in a molecular lever lock and key fashion. Killer T-cells secrete a cytotoxic substance to destroy antigens. Memory T-cells are dormant until the same antigen reappears, then they attack swiftly. Helper T-cells promote T-cell activation, stimulate phagocytic activity, and enhance the humoral immunity process. Suppressor T-cells produce delayed inhibition of cellular and humoric responses.

Lymphocytes that produce humoral immunity are called B-cells. Antibodies (Ab) and immunoglobulins (Ig) are proteins on the surface of B-cells. The most abundant immunoglobulin, IgG, from about seven weeks of age onward, produces a secondary response to antigens. Before that age, the infant relies on IgG transferred from its mother transplacentally. IgM primary response antibodies active the complement pathway early in the immune response. IgA is the primary immunoglobulin found in body secretions, such as saliva, tears, and colostrum, and which are the first line of defense in respiratory and gastrointestinal infections. IgD is an important part of B-cell differentiation. IgE attaches to mast cells and other leukocytes.

An anaphylactic, or Type I, response to a chemical produces IgE antibodies that bind to mast cells on vessel walls, releasing histamine and mediators, resulting in skin irritation, rhinitis, asthma, or rapid systemic vasodilation, which causes shock if the antigen was previously exposed to the body and memory cells were produced. This response causes *angioedema* (fluid in the vessel), *urticaria* (hives), and *anaphylaxis* (shock or hypersensitivity). Another response, Type II, causes IgM or IgG molecules to bind to and destroy blood cells, and other cells of the body. Type III immune complex syndromes include *lupus vasculitis*. Type IV responses involve the slow activation and proliferation of T-cells over a period of several days.

Agents that prevent the major histocompatibility complex from operating properly lead to an autoimmune response in which the body, or part of it, begins to attack itself, and defends itself from itself. Other agents may cause immunosuppression in which case the susceptibility to infection and cancer is increased. *Immunosuppression* is part of a larger type of response called *immunomodulation*, in which the other possibility in *immunoenhancement* (which is the goal of treatment by antibiotics). *Hypersensitivity*, another immunological response, is more commonly known as allergy.

The immune system is one of the more sensitive of the body's systems for toxic attack. The effects of immunotoxicity by petroleum aromatic hydrocarbons (PAHs), for instance, are thymic atrophy, as well as severe, persistent suppression of T-cell mediated immunity. PAH targets the thymic epithelium cells, which are crucial for T-cell maturation. These effects are well documented in Japanese exposed to polychlorinated biphenyl and dibenzofuran-contaminated rice oil, and in Michigan residents who were exposed to polybrominated biphenyl-contaminated dairy products. Positive findings of immunosuppression have been reported for PAHs, aromatic solvents (benzene, toluene), aromatic amines, pesticides, heavy metals, organotin, nitrogen dioxide, ozone, sulfur dioxide, and asbestos.

Individuals exposed to asbestos and presenting either pleural thickening or parenchymal asbestos also have reduced levels of IgA, IgG, IgM, IgE, and T-lymphocytes. Viral infections such as Epstein–Barr, cytomegalic, influenza, herpes, measles, mumps, chicken pox, and others, and bacteria infections such as tuberculosis, *Pneumocystis*, *Proteus*, *Eschirichia coli*, *Staphylococcus*, *Streptococcus*, and *Pseudomonas*, among others, work against enzymatic detoxification and may exacerbate the original dysfunction. Long-term exposure to tetrachloroethylene (or perchloroethylene) can suppress the immune system. This organic chloride is the dry cleaning solvent used for many years. About twenty percent of persons exposed to terpenes suffer from immune system deregulation. PCBs have also been linked to immunosuppression.

HAZARDOUS BY DEFINITION

When assessing exposure, many chemicals have already been declared hazardous by either OSHA, EPA or both. These hazardous chemicals are so listed for one or more of the reasons discussed above. One could take the position that, if the COC is listed, merely determine exposure and take action. Certainly it saves time and money and reduces staffing. However, it is important to understand all the facts and issues of exposure. If the 1990s is the decade of exposure litigation as someone predicted in the late 1980s, it will be clear enough to defendants that all facts and issues are important.

In the meantime, we will discuss the relationship of lists to the preceding discussion of types of hazards. The Comprehensive Environmental Response, Compensation, and Liability Act of 1980 (CERCLA or Superfund) defines *hazardous substance* at Section 101(14) as:

(A) any substance designated pursuant to Section 311(b)(2)(A) of the Federal Water Pollution Control Act,

(B) any element, compound, mixture, solution, or substance designated pursuant to Section 102 of this Act,

(C) any hazardous waste having the characteristics identified under or listed pursuant to Section 3001 of the Solid Waste Disposal Act...,

(D) any toxic pollutant listed under Section 307(a) of the Federal Water Pollution Control Act,

(E) any hazardous air pollutant listed under Section 112 of the Clean Air Act

(F) any imminently hazardous chemical substance or mixture with respect to which the Administrator has taken action pursuant to Section 7 of the Toxic Substances Control Act.

The term hazardous substance does not include petroleum, crude oil, or any crude oil fraction unless it is specifically listed or designated by the definition above. It also does not include natural gas, natural gas liquids, liquefied natural gas, or synthetic gas usable as fuel. It does however include just about everything else.

Consider Section 311(b)(2)(A) of the Clean Water Act. This section requires EPA to develop a list of designated hazardous substances which

> when discharged in any quantity into or upon the navigable waters of the United States or adjoining shorelines or the waters of the contiguous zone or ... which may affect natural resources belonging to, appertaining to, or under the exclusive management authority of the United States ... present an imminent and substantial danger to the public health or welfare, including, but not limited to, fish, shellfish, wildlife, shorelines and beaches.

This passage, found at *33 U.S.C. 1321*, was last amended by Public Law 96-561 passed on Dec. 22, 1980. EPA originally provided the list required by congress on March 13, 1978, (cf. *43 FR 10479*) and it is found at *40 CFR 116*, "Designation of Hazardous Substances." This table was last amended by EPA on Aug. 14, 1989 (cf. *54 FR 33488*). Of particular note is the regulation immediately following at *40 CFR 117*, which lists these same chemicals and assigns reportable quantities (RQ) for them. Do not confuse RQ with exposure assessment terms such as ADI, PEL, TLV, SC, etc. The RQ has nothing to do with acceptable exposures; it is merely an emergency response term used to determine whether the release needs greater and presumably more sophisticated levels of response. When the emergency is over, then you get to prove to OSHA or EPA that the actual exposure did not affect workers or the environment in a harmful way, respectively.

CERCLA was enacted into law on Dec. 11, 1980 as PL 96-510 to clean up uncontrolled releases of hazardous substances that threaten the common good. Section 102 of the Act requires EPA to promulgate and revise a list of hazardous substances similar to and including the Clean Water Act RQ list, but which also would include:

> such elements, compounds, mixtures, solutions, and substances, which when released into the environment may present substantial danger to the public health or welfare or the environment...

This list may be found at *40 CFR 302*, "Designation, Reportable Quantities and Notification." Hazardous substances that are listed in this part have designated reportable quantities. Hazardous substances that are not listed, for any reason, have an RQ of 100 pounds, which was assigned by Congress, unless the substance is an unlisted hazardous waste that exhibits the leaching procedure (LP) toxicity and would then have the RQ of the contaminant for which the characteristic is based. The RQ applies to the entire waste in this latter case and not just to the toxic contaminant. If more than one toxic contaminant is present to cause a hazardous characteristic, the lowest RQ is applied. The RQ table was last amended by EPA on April 17, 1995 (cf. *60 FR 19165*).

The Resource Conservation and Recovery Act of 1976, commonly referred to as RCRA, was first enacted by congress on Oct. 31, 1976 as PL 94-580. Congress last amended the law on Nov. 2, 1994 (cf. PL 103-437). The official title is the Solid Waste Disposal Act, which was amended by RCRA. Section 3001 of the Act, set forth at *42 U.S.C. 6921*, requires EPA to establish criteria for identification or listing of hazardous waste, which Congress defined in Section 1004(5) at *42 U.S.C. 6903* to mean:

> a solid waste, or combination of solid wastes, which because of its quantity, concentration, or physical, chemical, or infectious characteristics may —
>
> (A) cause, or significantly contribute to an increase in mortality or an increase in serious irreversible, or incapacitating reversible, illness; or
>
> (B) pose a substantial present or potential hazard to human health or the environment when improperly treated, stored, transported, or disposed of, or otherwise managed.

A solid waste as defined in Section 1004(27) means:

> any garbage, refuse, sludge from a waste treatment plant, water supply treatment plant, or air pollution control facility and other discarded material, including solid, liquid, semisolid, or contained gaseous material resulting from industrial, commercial, mining, and agricultural operations, and from community activities, but does not include solid or dissolved material in domestic sewage, or solid or dissolved materials in irrigation return flows or industrial discharges which are point sources subject to permits under Section 402 of the Federal Water Pollution Control Act... or source, special nuclear, or byproduct material as defined by the Atomic Energy Act of 1954...

EPA has promulgated the required regulations at *40 CFR 261*, "Identification and Listing of Hazardous Waste," on May 19, 1980 (cf. *45 FR 33119*) last amended on June 29, 1995 (cf. *60 FR 33913*). All the listed hazardous wastes, wastes exhibiting characteristics as explained under this regulation, and materials listed in Appendix VIII (although not hazardous waste by definition) are hazardous substances under CERCLA.

Once again, we visit the Federal Water Pollution Control Act, this time to Section 307(a) (cf. *33 U.S.C. 1317*) where toxic pollutants or combinations of pollutants are defined as the toxic pollutants listed in Table 1 of Committee Print Numbered 95-30 of the Committee on Public Works and Transportation of the House of Representatives. EPA listed these 65 pollutants at *40 CFR 401.15*, "Toxic Pollutants" at *44 FR 44502* on July 30, 1979 and last revised the list on Feb. 4, 1981 (cf. *46 FR 10724*).

Section 112(b) of the Clean Air Act, set forth at *42 U.S.C. 7412*, establishes an initial list of hazardous air pollutants and charges EPA to periodically review and revise the list as necessary by adding pollutants which

> present, or may present, through inhalation or other routes of exposure, a threat of adverse human health effects (including, but not limited to, substances which are known to be, or may reasonably be anticipated to be, carcinogenic, mutagenic, teratogenic, neurotoxic, which cause reproductive dysfunction, or which are acutely or chronically toxic) or adverse environmental effects whether through ambient concentrations, bioaccumulation, deposition, or otherwise...

Finally, with regard to our CERCLA listing requirements, the *imminent hazards* of Section 7 of the Toxic Substances Control Act (TSCA) are found at *15 U.S.C. 2606*. TSCA was PL 94-469 passed on Oct. 11, 1976 and last amended by congress on Oct. 28, 1992 as PL 102-550. For the purposes of TSCA, an imminently hazardous chemical substances or mixture means:

> a chemical substance or mixture which presents an imminent and unreasonable risk of serious or widespread injury to health or the environment. Such a risk to health or the environment shall be considered imminent if it is shown that the manufacture, processing, distribution in commerce, use, or disposal of the chemical substance or mixture, or that any combination of such activities, is likely to result in such injury to health or the environment...

What we have then are: (1) the CWA list of chemicals having a designated RQ if spilled into water, (2) the CERCLA list of chemicals having a designated RQ if spilled, period, (3) all RCRA hazardous wastes whether listed or determined by characteristics, (4) the CWA Priority Pollutant list, (5) the CAA Hazardous Air Pollutant list, and (6) TSCA imminently hazardous chemical substances or mixtures. All of these compounds are potential COCs (chemicals of concern) with respect to conducting a chemical exposure assessment in the environment.

Before leaving this discussion of substances defined as hazardous, whether waste-like or not, we must also look inside the plant at the environment of our employees. The Occupational Safety and Health Act of 1970 (OSHAct) was enacted by congress as PL 91-596 on Dec. 29, 1970 and was last amended by PL 102-550 on Oct. 28, 1992. Though it does not mention hazardous chemicals or material, it does state in Section 2(b) that the purpose of OSHAct is "to provide for the general welfare, to assure so far as possible every working man and woman...safe and healthful working conditions and to preserve our human resources"

> by exploring ways to discover latent diseases, establishing causal connections between diseases and working environmental conditions, and conducting other research relating to health problems...

OSHA did promulgate regulations (or standards) regarding chemical hazards. Initially it promulgated *29 CFR 1910 Subpart Z*, "Toxic and Hazardous Substances" on June 27, 1974 (cf. *39 FR 23502*), which contains standards written for specific chemicals. This subpart was last corrected on June 29, 1995 in *60 FR 33984*. Many considered this effort by OSHA to be too little, too late, so on Nov. 25, 1983, in *48 FR 53280*, 29 CFR 1910.1200 entitled "Hazard Communication" was issued. This standard defines hazardous chemical as: any chemical that is a physical hazard or a health hazard. A physical hazard is:

> a chemical for which there is scientifically valid evidence that it is a combustible liquid, a compressed gas, explosive, flammable, an organic peroxide, an oxidizer, pyrophoric, unstable (reactive) or water-reactive.

A health hazard is:

> a chemical for which there is statistically significant evidence ... that acute or chronic health effects may occur in exposed employees. The term "health hazard" includes chemicals which are carcinogens, toxic or highly toxic agents, reproductive toxins, irritants, corrosives, sensitizers, hepatotoxins, nephrotoxins, neurotoxins, agents which act on the hematopoietic system, and agents which damage the lungs, skin, eyes, or mucous membranes.

So, to the universe of chemicals defined by EPA as discussed above, we now have added nearly every other chemical to be concerned with, for what chemical is not, in some respect, a physical or health hazard?

TABLE 4.7
Site Information for Hazard Identification

Abutting land use
Air temperature
Air turbulence
Groundwater analyses
Hydraulic gradients of groundwater
Land use
Location
Monitoring well data
Plans for future development
Sediment analyses
Soil analyses
Stream turbulence
Surface water analyses
Wind speed

In the hazard identification phase of exposure assessment, the information in Table 4.7 is collected.

CONCLUSION

The undeniable truth is that, after two decades of federal protection of the environment and the workplace, little is known about the health effects of most anthropogenic chemicals. While asbestos, lead, radon, and a few other chemicals have received a great deal of attention and research effort, the health effects of the vast majority of chemicals, including foods and medicines, are routinely approved for commercial use with little or no toxicity testing. Physicians who consider environmental and workplace exposures in diagnosis are rare. The Environmental Health Center at Dallas is one notable exception to this rarity. Neither environmental nor occupational medicine are taught in our medical schools, beyond three or four clock hours of training. [See Upton.] Our mobile society is exposed to multiple chemicals in various locations in several different situations, making an understanding of the exposure difficult as best, impossible in the worst case. This complexity, plus the severe shortage of research dollars, means that the engineer or scientist involved in exposure assessment must be able to fill in the data gap by relying on chemical and physical models to predict the behavior of the COC in the environment — inside or outside our factories.

Studies have established that nine percent of all children in the U.S. (twenty-five percent of minority children) are suffering irreversible neurologic, intellectual, and behavioral impairment resulting from chronic, low-level lead exposure. Keep that mental picture before you as the goal of the risk assessor, which is typically the user and direct customer of the exposure assessment — to eradicate harmful exposures to our children and ourselves, the ultimate customers, to the extent feasible.

QUESTIONS (ANSWERS BEGIN ON PAGE 401)

1. Explain how a neoplasm becomes a cancer.
2. What is the current risk of cancer from any cause?
3. Name four terms for cancer causing agents.
4. What is a cancerous growth in epithelial tissue called?
5. What is a cancerous growth in supporting tissue called?
6. How are oncogenes formed?
7. What is the basic involvement of chemicals in carcinogenicity?
8. What cancer model is most conservative?
9. What cancer model is used most?
10. Looking at Table 4.1, which compounds have a unit risk factor greater than one in a million?
11. If you were deciding between the following three chlorinated solvents for a new cleaning process, which would you recommend based on carcinogenicity potency? Tetrachloroethylene, 1,1,2-trichloroethane, or trichloroethylene. Why?
12. Explain the possibility of the concept that exposure to a toxic goo caused several small turtles to become crime-fighting Ninja artists who walk and talk besides.
13. How might a toxic chemical lead to male infertility?
14. What is the principle action of a cholinesterase inhibitor (besides inhibiting cholinesterase)?
15. When dealing with chemical induced neuropathy, what is the critical factor?
16. What chemical is capable of inducing both aplastic anemia and leukemia?
17. An anaphylactic or Type I response causes what to happen?

REFERENCES

Anderson, Marshall. "Activation of Proto-oncogenes by Chemicals." Section 1 of Chapter 3, "Newer Basic/Long-term Research with Application to Environmental Health Problems." In *Report of the Subcommittee on Health Effects, Appendix D: Strategies for Health Effects Research*. Research Strategies Committee, U.S. EPA, Report SAB-EC-88-040D, Revised Oct. 24, 1988.

American Conference of Government Industrial Hygienists. *Guide to Occupational Exposure Values — 1991*. Cincinnati: ACGIH, 1991.

Brauer, Roger L. *Safety and Health for Engineers*. New York: Van Nostrand Reinhold, 1990.

Burmaster, David E. and Jeanne W. Appling. "Introduction to Human Risk Assessment, with an Emphasis on Contaminated Properties." In *Environment Reporter*. April 7, 1995. pp. 2431–40.

Burrell, Robert, Dennis K. Flaherty, and Leonard J. Sauers. *Toxicology of the Immune System: A Human Approach*. New York: Van Nostrand Reinhold, 1992.

"California High Court Sets New Standards for Cancer Fear, Medical Monitoring Damages." *Environment Reporter*. Washington, D.C.: Bureau of National Affairs. Jan. 14, 1994.

Campbell, Reginald L. and Roland E. Langford. *Fundamentals of Hazardous Materials Incidents*. Chelsea, MI: Lewis Publishers, 1991.

Cockerham, Lorris G. and Barbara S. Shane, eds. *Basic Environmental Toxicology*. Boca Raton, FL: CRC Press/Lewis Publishers, 1994.

Cogliano, Vincent James. "The U.S. EPA's Methodology for Adjusting The Reportable Quantities of Potential Carcinogens." *Hazardous Materials Control Monograph Series: Health Assessment*. Silver Spring, MD: Hazardous Materials Control Research Institute, pp. 114–7.

"Company Argues for Fear-of-Cancer Rule in Case before California Supreme Court." *Environment Reporter*. Washington, D.C.: Bureau of National Affairs. Nov. 12, 1993.

Cornett, Frederick D. and Pauline Gratz. *Modern Human Physiology*. New York: Holt, Rinehart and Winston, Publishers, 1982.

"Data Growing to Link Chlorine in Chemicals to Breast Cancer, Greenpeace Says in Report." *Environment Reporter*. Washington, D.C.: Bureau of National Affairs. Oct. 22, 1993.

De Serres, Frederick J. and Arthur J. Bloom. *Ecotoxicity and Human Health: A Biological Approach to Environmental Remediation*. Boca Raton, FL: CRC Press/Lewis Publishers, 1996.

Federal Register, Vol. 57, No. 114, Friday June 12, 1992. "OSHA Preamble and Proposed Rule to Revise Air Contaminant Standards for Construction, Maritime, and Agriculture." pp. 26002–26601.

Gardner, Robert W. *Chemical Intolerance: Physiological Causes and Effects and Treatment Modalities*. Boca Raton, FL: CRC Press/Lewis Publishers, 1994.

Hamilton, Helen Klusek, *et al. Nursing Reference Library: Diagnostics*. Sprinhouse, PA: Springhouse Corp., 1984.

Hansen, Doan J. ed. *The Work Environment, Volume III: Indoor Health Hazards*. Boca Raton, FL: Lewis Publishers, 1994.

Hartman, Catherine E., George J. Schewe, Leslie J. Ungers and Michael J. Petruska. "Waste Oil Risk Assessment Study." *Hazardous Materials Control Monograph Series: Health Assessment*. Silver Spring, MD: Hazardous Materials Control Research Institute, pp. 35–8.

Hathaway, Gloria J., Nick H. Proctor, James P. Hughes, and Michael L. Fischman. Eds. *Proctor and Hughes' Chemical Hazards of the Workplace*. 3d. ed. New York: Van Nostrand Reinhold, 1991.

Kay, Robert L., Jr. and Chester L. Tate, Jr. "Public Health Significance of Hazardous Waste Sites." *Hazardous Materials Control Monograph Series: Health Assessment*. Silver Spring, MD: Hazardous Materials Control Research Institute, pp. 65–71.

Koren, Herman and Michael Bisesi. *Handbook of Environmental Health and Safety, Volume I: Principles and Practices*. 3d. ed. Boca Raton, FL: CRC Press/Lewis Publishers, 1996.

Landis, Wayne G. and Ming-Ho Yu. *Introduction to Environmental Toxicology: Impacts of Chemicals upon Ecological Systems*. Boca Raton, FL: Lewis Publishers, 1995.

Levesque, Joe. "Fire Protection." In *The Work Environment: Volume I, Occupational Health Hazards*. Doan J. Hansen, ed. Chelsea, MI: Lewis Publishers, 1991.

Luster, Michael. "Immunotoxicology." Section 5 of Chapter 3, "Newer Basic/Long-term Research with Application to Environmental Health Problems." In *Report of the Subcommittee on Health Effects, Appendix D: Strategies for Health Effects Research* Research Strategies Committee, U.S. EPA, Report SAB-EC-88-040D, Revised Oct. 24, 1988.

Mahaffey, Kathryn. "Research Advances in the Toxicology of Lead." Chapter 3 in: *Report of the Subcommittee on Health Effects, Appendix D: Strategies for Health Effects Research* Research Strategies Committee, U.S. EPA, Report SAB-EC-88-040D, Revised Oct. 24, 1988.

Manahan, Stanley E. *Environmental Chemistry.* 6th. Ed. Boca Raton, FL: CRC Press/Lewis Publishers, 1994.

Mattison, Donald, and Alan Wilcox. "Human Chorionic Gonadotropin (HCG)." Section 6 of Chapter 3, "Newer Basic/Long-term Research with Application to Environmental Health Problems." In *Report of the Subcommittee on Health Effects, Appendix D: Strategies for Health Effects Research.* Research Strategies Committee, U.S. EPA, Report SAB-EC-88-040D, Revised Oct. 24, 1988.

Neely W. Brock. *Introduction to Chemical Exposure and Risk Assessment.* Boca Raton, FL: CRC Press/Lewis Publishers, 1994.

Office of Solid Waste and Emergency Response. U.S. EPA. Draft Memorandum. "Implementation of Exposure Assessment Guidance for RCRA." Washington, D.C. Sept. 24, 1993.

Overview: A Total Management System for Loss Prevention and Control. Hartford, CT: Industrial Risk Insurers. 1986.

Partridge, Lawrence J. and Arthur D. Schatz. "Application of Quantitative Risk Assessment to Remedial Measures Evaluation at Abandoned Sites." *Hazardous Materials Control Monograph Series: Risk Assessment Volume I.* Silver Spring, MD: Hazardous Materials Control Research Institute, pp. 1–5.

Perera, Frederica P. "Carcinogen-DNA and Protein Adducts: Research Perspectives." Section 2 of Chapter 3, "Newer Basic/Long-term Research with Application to Environmental Health Problems." In *Report of the Subcommittee on Health Effects, Appendix D: Strategies for Health Effects Research.* Research Strategies Committee, U.S. EPA, Report SAB-EC-88-040D, Revised Oct. 24, 1988.

Philp, Richard B. *Environmental Hazards and Human Health.* Boca Raton, FL: CRC Press/Lewis Publishers, 1995.

Plog, Barbara A. ed. *Fundamentals of Industrial Hygiene.* 3d. ed. Chicago: National Safety Council, 1988.

"Polymerization." In *Chemical and Process Technology Encyclopedia.* Considine, ed. New York: McGraw–Hill Book Company, 1974.

Preuss, Peter W., Alan M. Ehrlich and Kevin G. Garrahan. "The U.S. Environmental Protection Agency's Guidelines for Risk Assessment." *Hazardous Materials Control Monograph Series: Risk Assessment Volume I.* Silver Spring, MD: Hazardous Materials Control Research Institute, pp. 6–13.

Rea, William J. *Chemical Sensitivity: Volume I, Principles and Mechanisms.* Boca Raton, FL: CRC Press/Lewis Publishers, 1992.

Rea. *Chemical Sensitivity: Volume II, Sources of Total Body Load.* Boca Raton, FL: CRC Press/Lewis Publishers, 1994.

Rea. *Chemical Sensitivity: Volume III, Clinical Manifestations of Pollutant Overload.* Boca Raton, FL: CRC Press/Lewis Publishers, 1996.

Reiter, Lawrence. "Neurotoxicology." Section 3 of Chapter 3, "Newer Basic/Long-term Research with Application to Environmental Health Problems." In *Report of the Subcommittee on Health Effects, Appendix D: Strategies for Health Effects Research.* Research Strategies Committee, U.S. EPA, Report SAB-EC-88-040D, Revised Oct. 24, 1988.

Rochow, Eugene G. *The Metalloids.* Boston: D.C. Heath and Company, 1966.

Roychowdhury, Mahendra. "Reproductive Hazards in the Work Environment." *Professional Safety.* May 1990, PP. 17–22.

Sax, N. Irving and Richard J. Lewis, Sr. *Hazardous Chemical Desk Reference.* New York: Van Nostrand Reinhold, 1987.

Schatz, Arthur D. and Michael F. Conway. "OSHA Standards or Risk Analysis: Which Applies?" *Hazardous Materials Control Monograph Series: Health Assessment.* Silver Spring, MD: Hazardous Materials Control Research Institute, pp. 14–7.

Schewe, George J., Joseph Carvitti and Joseph Velten. "Human Exposure Estimates Using U.S. EPA Guideline Models: An Integrated Approach." *Hazardous Materials Control Monograph Series: Health Assessment.* Silver Spring, MD: Hazardous Materials Control Research Institute, pp. 9–13.

Smith, S.L. "Nervous about Neurotoxins." *Occupational Hazards.* Dec. 1990, pp. 37–40.

Solomon, T. W. *Organic Chemistry.* 5th ed. New York: John Wiley & Sons, 1992.

Stine, Karen E. and Thomas M. Brown. *Principles of Toxicology.* Boca Raton, FL: CRC Press/Lewis Publishers, 1996.

Talbott, Evelyn O. and Gunther F. Craun. *Environmental Epidemiology.* Boca Raton, FL: CRC Press/Lewis Publishers, 1995.

Talty, John T. *Industrial Hygiene Engineering: Recognition, Measurement, Evaluation and Control,* 2d. Ed. Park Ridge, NJ: Noyes Data Corporation, 1988.

"Toxicology Study Links Ozone Exposure to Cancer in Male Mice, But Not in Rats." *Environment Reporter.* Washington, D.C.: Bureau of National Affairs. Nov. 19, 1993.

U.S. Department of Transportation, Research and Special Programs Administration. *Emergency Response Guidebook.* DOT P 5800.5. The latest is the 1996 version.

Upton, Arthur. "Environmental Factors and Human Health." Chapter 1 in: *Report of the Subcommittee on Health Effects, Appendix D: Strategies for Health Effects Research* Research Strategies Committee, U.S. EPA, Report SAB-EC-88-040D, Revised Oct. 24, 1988.

Wagner, Travis. *In Our Backyard: A Guide to Understanding Pollution and Its Effects.* New York: Van Nostrand Reinhold, 1994.

Wiesner, Paul J. and Stephen Margolis. "The Epidemiologic Approach to Hazardous Waste Problems." *Hazardous Materials Control Monograph Series: Health Assessment.* Silver Spring, MD: Hazardous Materials Control Research Institute, pp.18–20.

5 Basic Chemical Hazards to Wildlife

Improving the environment for wildlife is a critical challenge for the world in the next several decades. Two important issues are (1) improvement of the aquatic health of surface water systems and (2) better management of biological resources. Existing U.S. law has been inadequate to deal with these issues. For instance, the Clean Water Act (CWA) has not met any of its goals proposed by Congress. Pollution from point sources has more or less been reduced, but eighty percent of pollution in streams is from agriculture, which is unregulated. Biological resource management is vaguely the task of the Endangered Species Act (ESA), which has not been very effective, and is prone to crisis management and vociferous criticism. For one thing, ESA does not define the *population* of species. The problem is exacerbated by the fact that the issues are scientifically and technically complicated, place great demands on cash, and are local in nature, but sometimes global in effect. Therefore, fiscal conservatives want to give local control for endangered species management, where little, if any, resources are available for solving the problems.

Chemicals that do not have major effect directly on humans can be devastating to wildlife. Take oil, for example. A cleanup project taking place near Glenrock, Wyoming, involves an abandoned oil reclamation facility where oily pits are killing birds. Sludge pits, tank leaks, tank loading spills, and soil contaminated with oily liquids attract and trap birds and small mammals, which die from suffocation or exposure to weather. Oil transferred to eggs through oily feathers kill incubating embryos. Perhaps the death of a few birds does not seem to be a major deal in the grand scheme of things human, but spread the damage to an endangered species and the loss is invaluable. This is not necessarily a minor deal to duck hunters or other bird hunters or bird watchers, either. In Wyoming, EPA issued orders under Section 7003 of the Resource Conservation and Recovery Act to force the cleanup, thus confirming the government's contention that RCRA could be used to manage oily wastes that do not come under other federal or state laws.

Sometimes, the contamination does not have much, if any, effect on wildlife, but accumulates until consumed by humans or higher animals, which ultimately are consumed in human food products. For instance, oysters bioaccumulate several contaminants that affect human health, such as mercury, copper, and zinc.

In wastewater management, wastewater discharge permits often require a series of bioassay tests, which requirement is confusing and frustrating to industrial managers and others who do not understand the significance and importance of these tests. EPA chose bioassay organisms that represent life at three trophic levels, shown in Table 5.1.

Somewhere between five and thirty million species of living things exist on Earth, but only a million or so have been cataloged and described. The count is well

TABLE 5.1
Trophic Levels of Bioassay Organisms

Selenastrum capricornutum	Green algae, a primary producer that fixes photosynthetic energy and provides food for higher trophic species
Daphnia magna	A small freshwater crustacean grazes on plankton and algae
Pimephales promelas	Fathead minnow, a predator high in the aquatic food chain

known for larger organisms, such as the approximately four thousand species of mammals. However, the total number of all species can only be estimated, because most species of insects, deep sea invertebrates, and microorganisms have yet to be identified and counted. Most of the species on earth are insects. The abundance of species increases from few at the polar regions to the teeming life at the equator. The significance of wild species to humans, other than curiosity, and an affinity for the cuter and more cuddly species, is that they have made, and do make, substantial contributions to human agriculture, medicine, and industry, in addition to direct benefits to the environment itself. We cannot establish a total value to humanity for the goods and services provided by any one species, much less for the entire biologically diverse animal and plant kingdoms. Nor does it pay to ignore the nuisance of certain species, such as German cockroaches in residences.

So, what is the loss of a few species of animals and plants worth? Experts have concluded that twenty-five percent of the Earth's total biological diversity is at risk due to human activity in the environment over the next thirty years, and that fifty percent of species are at risk over the next century. On Feb. 25, 1992, President George Bush announced that he had enhanced wildlife protection on federal lands and developed a strategy to preserve biological diversity. Bush led the international ban on the trade in African elephant ivory, persuaded Japan to stop importing sea turtle shells, and also persuaded Japan to ban driftnet fishing on the high seas, paving the way for an international ban on driftnet fishing. Too little, too late, some experts on biological diversity say. We are taking food away from developing nations, critics of biological diversity preservation say.

What can the U.S. continue to do? Many of these biological diversity problems are caused by and are wholly within other countries. Of course, the ultimate effects of loss of species is not confined to one country, but is a human problem. The policy of the United States, established by Congress, is to:

> Recognize the worldwide and long-range character of foreign environmental problems and ... lend appropriate support to initiatives, resolutions, and programs designed to maximize international cooperation in anticipating and preventing a decline in the quality of mankind's world environment ...

> — *National Environmental Policy Act, Section 102(f)*

Due mostly to the uncontrolled loss of tropical forests, but also to habitat conversions caused by human activity in other parts of the globe, the worldwide rate of extinction of species currently exceeds the natural rate of extinction of species by one thousand times. Species extinction and destruction of entire ecosystems are human problems. As the activity of humans spreads to the farthest corners of the globe, natural areas are changed and modified with the extinction of species as one result. Deforestation, ocean degradation, desertification, and surface water and soil contamination contribute most of the driving force behind loss of species. The loss rate is now the most rapid since the extinction of the dinosaurs, a natural phenomenon, as far as we know, which occurred many millions of years ago.

Even if we do not fully appreciate that humanity is destroying species at a thousand times the natural loss of species, and whether or not we agree on which particular species are important and, therefore, worthy of the cost of protection, we should acknowledge the value of biological diversity in general. For instance, crops for human consumption are derived from wild species, and the high-yielding hybrids of modern agriculture rely on revitalization from wild genetic stock. Loss of wild genes means that eventually some predator, parasite, or infecting organism will develop a resistance to the current crop species, and no wild genes will be available to cross-breed healthy yields back into the crop. Moreover, future crops that could have been used directly, or modified by biological diversity, may be lost entirely when ecosystems are altered. Plants are currently the basis for about twenty-five percent of the prescription medicines in the U.S. Many of these plants were discovered in tropical rain forests or other wild habitats in foreign countries. Their significant contribution to the treatment of serious diseases is at risk. According to the reference booklet prepared for the United Nations conference on environment and development, in Brazil, called Earth Summit '92, each year tropical forests shrink by as much as forty-two million acres, twenty-six billion tons of topsoil are lost, new deserts appear at the rate of fifteen million acres, and tropical developing countries grossly degrade over twelve million acres of upland water shed. While the engineer preparing an exposure assessment may not be directly concerned about what is happening in foreign nations, it matters in the big picture, and should motivate one to protect as much of our own natural resources as economically feasible.

Biodiversity, or biological diversity, is simply the variety of living organisms in a particular habitat or ecosystem, measured as the numbers of species or subspecies of plants, animals, and microorganisms present in the system, as an indication of the health of an environment. Ecological systems obviously support higher biodiversity than agricultural or urban environments.

Disputes over the use of the earth's remaining natural habitats are increasing due to the understandable desire of developing countries to improve their lifestyles. While developed countries want to measure, evaluate, and preserve biodiversity, developing countries want to achieve the same level of development that the wealthy western countries have attained.

To date, each species receives more or less equal value in the distribution of protection funds and priorities. Dissimilar organisms need to be assigned greater value, though. For example, the tuatara of New Zealand is an endangered lizard-like reptile with no close relatives. Because the tuatara is genetically distinct, it contributes more to the diversity of the world's fauna than any species of true lizard. *Keystone* species, such as the tuatara, maintain the diversity of particular habitats. Another example of a keystone species is the sea otter off the Pacific Coast of North

America, which preserve diversity in subtidal waters by preying on sea urchins that feast upon kelp forests to near extinction, thus ruining the habitat for invertebrates and fishes.

ECOSYSTEMS IN GENERAL

Living, or *biotic*, and nonliving, or *abiotic*, components of the environment interrelate to form one physical system, an ecosystem, in dynamic equilibrium with its inputs and surroundings. The Earth has three spheres of life: the atmosphere, the lithosphere, and the hydrosphere. Land masses make up the *lithosphere* and the world's ocean is the *hydrosphere*. Taken together, these comprise the *biosphere*, though some authors have given the term a more narrow definition. Since life is pretty well ubiquitous in all three spheres, let us consider their totality as being the biosphere, except that the biosphere is limited as a thin mantle of life on the Earth.

The relationship of plants and animals to their physical environment includes light and heat, as solar radiation, moisture, wind, oxygen, carbon dioxide, nutrients in soil, water, and atmosphere. The biological environment includes organisms of the same kind, as well as other plants and animals. Animal species are distributed according to climatic region and dominant type of vegetation present, the combination of which is called a *biome*, forming the basic framework for the surrounding ecosystem. Terrestrial biomes are influenced by latitude, elevation, and associated moisture and temperature regimes. They vary geographically from the Tropics through the Arctic and include various types of forest, grassland, shrub land, and desert, including associated freshwater communities such as streams, lakes, ponds, and wetlands. Marine biomes include the open ocean, littoral or shallow water, benthic or ocean bottom zone, rocky shores, sandy shores, estuaries, and tidal marshes. Major biomes discussed below are tundra, taiga, desert, chaparral, wetlands, forest, open ocean, littoral, benthic, and tidal marshes, as part of the wetlands discussion.

A useful way of looking at terrestrial and oceanic biomes is to consider them as *ecosystems,* a term coined by Sir A. G. Tansley of Great Britain, in 1935, to stress habitat as an integrated whole. *Systems* are collections of interdependent parts that function as a unit, receive inputs, and yield outputs. The major parts of an ecosystem are the *producers, consumers, decomposers*, and the *nonliving,* or abiotic, component, consisting of dead organic matter and nutrients in the soil and water.

Biomes are systems of life in equilibrium. If an input to the biome is changed, the ecosystem adjusts to a new equilibrium. Hydrodynamic, biological, chemical, and physical processes in ecosystems cannot be separated, though humanity frequently considers and manages these processes separately in the name of development and progress. Obviously, multidisciplinary approaches to understanding an ecosystem is superior to continuing to attack the environment as a monolith. However, since that is the way things have always been done, you may find yourself alone, having to prepare a report to management as best you can.

The primary input to any ecosystem is the energy received from the sun. Lesser inputs are waste heat, chemical energy, and pollutants. Waste heat comes from factories, homes, and creatures, including humanity. Chemical energy comes in the form of fertilizers. Pollutants can also add energy to the environment. Other inputs are water, oxygen, carbon dioxide, nitrogen, and other elements and compounds. Outputs from an ecosystem include heat of respiration, water, oxygen, carbon dioxide, and nutrient losses.

Energy Cycle. The biosphere has three sources of energy. The sun supplies 99.98 percent of the energy available to Earth. Another 0.02 percent, roughly, is supplied by the molten core of the planet and vented through volcanoes and tidal energy is the remaining 0.002 percent. The health of an ecological system depends on environmental conditions that influence the system, but the major driving force for any ecosystem is the solar energy and all ecosystems function with energy flowing in one direction from the sun, through nutrients, which are continuously recycled. The Earth's antipodal plane, capable of intercepting solar energy, is on the order of 1.3×10^{14} square meters. The average solar energy rate received is 1.4 kilowatts per square meter per minute or 2 calories per square centimeter per minute. About thirty-six percent of solar energy is reradiated into space by deflection from the atmosphere, reflection from clouds, or back-scattering from the Earth's surface as an imperfect black-body, the total of which is termed *albedo*, the fraction of light that is simply reflected.

The atmosphere consumes about nineteen percent of solar energy for precipitation and evaporation. The remaining forty-seven percent is converted to heat at the Earth's surface and reradiated to warm our atmosphere. About 0.2 percent of the solar energy received drives convection currents in the atmosphere and ocean, dissipating as frictional heat. Although amounting to only 0.02 percent of total solar energy incident on the Earth and its atmosphere, plenty of light is available to be converted by plants to chemical energy in the form of carbohydrates and other carbon compounds, by the process of photosynthesis. Energy is transferred through the ecosystem by a series of steps called the food web that involve eating and being eaten. Each step involves multiple trophic, or feeding, levels: plants, herbivores, two or three levels of carnivores, and decomposers. A fraction of the solar energy fixed by plants follows this pathway, the *grazing food web*. Over millions of years, some of the energy fixed in plants and animals has been compressed and is now exploited as natural gas, coal, shale, oil, and less concentrated fossil fuels. At any given trophic level, as defined below, about three quarters of the energy available to the level is lost in respiration, leakage, and other losses such as friction. Only about twenty-five percent of energy coming into the trophic level is available for growth and maintenance.

Carbon Cycle. Carbon is cycled between the lithosphere, hydrosphere, and atmosphere in a never-ending process. Most carbon in the biosphere is found in sedimentary rock as carbonates (97.75%), while about 0.05 percent is fossil fuel, but only about 0.006 percent is found in land-based living, or previously living but not yet fossilized, organisms. A reservoir of carbon is maintain in the ocean as carbonate and bicarbonate (0.17%). Oceanic biomass accounts for 0.015% of carbon. Carbon dioxide in the atmosphere accounts for about 0.004 percent of carbon. This balance is slightly off due to rounding, but that is essentially where the carbon is. Plants and some algae fix carbon in the form of carbon dioxide through the process of photosynthesis. The point is that although carbon changes its compound formula through combustion, respiration, photosynthesis, and deforestation, what we have is what we get. No more carbon is to be had and water and carbon are the two essential components for life. How it is circulated in the carbon cycle is vitally important to future generations of humans, yet no economic penalty is assigned for degrading carbon into useless, nonrecoverable forms. In fact, human society currently rewards those who spend carbon into uselessness or place it in unusable storage for obtaining a transient benefit (energy).

Carbon dioxide and water react to form carbohydrates:

$$mCO_2 + nH_2O \xrightarrow{\text{SolarEnergy}} C_m(H_2O)_n + mO_2 \qquad (5.1)$$

where $C_m(H_2O)_n$ is a carbohydrate. A more complete picture is:

$$aCO_2 + bH_2O + cPO_4 + dNH_3 + eSO_4 \xrightarrow{\text{SolarEnergy}}$$
$$C_aH_{2b+3d}O_bP_cN_dS_e + \left(a + \frac{c+e}{2}\right)O_2 \qquad (5.2)$$

The CHOPNS compound is protoplasm, or cellular life. Some carbohydrate is converted to fat, proteins, and other organic compounds needed for cellular growth and nourishment.

Nitrogen Cycle. Another necessary component of life is nitrogen for protein, life's building-block. The crust of the Earth contains 77.82 percent of all nitrogen in the biosphere. Sediments account for another 22.06 percent. Nitrogen dissolved in the ocean is 0.11 percent. Oceanic biomass contains 0.005 percent of nitrogen and land-based biomass contains another 0.004 percent. Inorganic runoff from soil transports about 0.0008 percent of all nitrogen in the cycle. Inorganic nitrogen compounds in the ocean store another 0.0006 percent. Surprisingly, the remaining 0.000001 percent is found in the atmosphere. Surprising because we tend to think of the atmosphere as some infinitely large mantle around the Earth, when, in fact, it is not so large. In the atmosphere, 43.4 of the nitrogen percentage originated from biological processes on the Earth, and 14.3 percent comes from biological processes taking place in the ocean. Industrial processes account for 38.6 percent of atmospheric nitrogen in the late twentieth century nitrogen balance. Volcanoes account for 0.08 percent. Other atmospheric processes supply the other 3.62 percent of atmospheric nitrogen.

Inert nitrogen is *fixed* to ammonia, which is precipitated and absorbed by soil. Soil-ammonia is taken up by the roots of plants and converted to amino acids, then to proteins. Plants are eaten by grazing animals in which the metabolic process produces new proteins, some of which are recycled to be reused by plants and other animals. Finally, proteins are returned to the soil as animal waste and decomposing tissue. Some soil-ammonia is converted to nitrite (NO_2) by specialized aerobic organisms. Other microorganisms convert ammonia to nitrate (NO_3), which is also taken up by plants to form amino acids and proteins to become part of the ammonia sub-cycle.

Nitrogen activation involves splitting the inert bimolecular nitrogen gas molecule into two free nitrogen radicals. Fixation occurs when these free nitrogen radicals are combined with hydrogen or oxygen to form a nitrogen species (ammonia or nitrate) that plants can use. Most nitrogen is fixed by nitrogen-fixing bacteria associated with plant roots. Some nitrogen is fixed by other natural means such as lightning, meteor trails in the atmosphere, and cosmic ray activity in the atmosphere. Volcanic eruptions and weathering of igneous rocks account for a small amount of nitrogen fixation. When microorganisms oxidize nitrogen to nitrite and nitrate the process is called *nitrification*. The remainder of fixed nitrogen, nearly as much as naturally fixed, is provided as anthropogenic fertilizer at a significant energy investment. *Denitrification* occurs when microorganisms reduce nitrite and nitrate to free nitrogen for release to the atmosphere. *Ammonification* is the oxidation of amino acids and proteins to ammonia by microorganisms.

Oxygen Cycle. Oxygen composes twenty-one percent by volume of the atmosphere, mainly as the bimolecular gas, O_2. Almost all oxygen comes from the photosynthesis process, though a small amount comes from *photodissociation*, the splitting of a water molecule by solar ultraviolet energy. Oxygen is a participant in all the major cycles of life on Earth: carbon, nitrogen, sulfur, phosphorus, and water. The reservoir of oxygen in our atmosphere is due to the fact that some photosynthesis products remain unoxidized: coal, oil, and natural gas. Oxygen is highly reactive, trading electrons to oxidize carbon, nitrogen, sulfur, phosphorus, iron, and other elements. That is why oxygen is a major participant in all these cycles.

PRODUCERS

Producers are the vital members of any ecosystem. Humans tend to focus on those producers that are food crops or else esthetically pleasing (such as flowers) and consider all other producers as weeds or brush. Producers receive inorganic materials and solar energy and manufacture new organic matter by *photosynthesis*. Biota that participate in the ecosystem as producers by synthesizing their own organic materials from inorganic sources are called *autotrophs*. The amount of energy produced by an autotroph by photosynthesis per unit time is *gross primary productivity*. That amount of energy not used by the autotroph itself and remains available as food energy for higher trophic levels is called *net primary productivity*. Microorganisms that derive and use energy from the oxidation of reduced chemical compounds are called *chemotrophs*. *Heterotrophs* metabolize existing organic materials in order to live. In each ecosystem, the nutrient in shortest supply is called the *limiting nutrient*. *Homeostasis* is the tendency for an ecosystem to maintain its equilibrium. Regulatory mechanisms oppose system disruptions in order to maintain the existence of the system. Part of this regulation of equilibrium is due to *inertia* and another part is due to *resilience*.

Nutrient cycles are called *biogeochemical* cycles and begin with the release of nutrients from organic matter by weathering and decomposition in a form available to plants in soil and water for storage in their tissues. Nutrients are then transferred from one trophic level to another through the food web. Most plants and animals go uneaten by higher trophic levels, so nutrients contained in their tissues pass through the decomposer food web to be ultimately released by bacterial and fungal decomposition, which reduces complex organic compounds into simple inorganic compounds available for reuse by plants. The amount of stored energy present in plant material is called the *biomass* and is expressed as dry weight per unit area.

CONSUMERS

Biota that use organic carbon as a source of energy are the *consumers* of the ecosystem. All consumers are heterotrophs. Consumers do not directly capture energy from the sun, for nourishment. *Primary consumers* use producers for nourishment. These consumers are the *herbivores* that feed on green plants. *Secondary consumers* feed on primary consumers. Those that feed only on primary consumers are *carnivores*. Some secondary consumers nourish themselves not only on primary consumers, but on a combination of producers, primary consumers, and smaller secondary consumers. This latter group includes humanity and comprises the *omnivores*.

The number of trophic levels is limited in food webs, because a large amount of energy is lost with each transfer. For instance, the heat of respiration is lost by each level of consumers. This energy not usable for, nor transferable to, the next trophic level, and each level contains less energy than the trophic level that supported it. That is why herbivores such as deer or elk are more abundant than carnivores such as wolves.

Consumers derive energy through respiration, an *exogenic* process that releases energy:

$$C_6H_{12}O_6 + 6O_2 \rightarrow 6CO_2 + 6H_2O + (\text{energy}) \qquad (5.3)$$

Consumers that have deceased become food for other consumers known as *detritus feeders* or *scavengers*. Some are large, such as hyenas and vultures, and some are small, such as termites or earthworms, but most are microorganisms. Even detritus feeders have several trophic levels. Detritus organisms such as bacteria, fungi, or protozoa are called *decomposers*.

FOOD CHAIN

Primary and secondary consumers and producers in an ecosystem form a food chain, or food web, in which organisms eat smaller organisms and are generally eaten in turn by larger organisms. Sometimes, rather than being larger in size, the higher organisms are simply more developed, or else more efficient at hunting or at eating than the eaten organism. Regardless, lower trophic levels are *prey* to *predators* of higher trophic levels. Several trophic levels may be identified in any ecosystem. Through these levels, *bioaccumulation* or *biomagnification* can occur. When increasingly higher concentrations of a toxic substance are passed up the food chain it is biomagnification. Bioaccumulation refers to the uptake of dissolved as well as ingested phases of a COC. The tissue of the receptor has a greater concentration of the COC than the environment. According to Landis and Yu, this occurs when the

COC is more lipophilic than hydrophilic. *Bioconcentration* pertains to the accumulation of dissolved chemicals, nutrients or pollutants, in plants and animals creating potentially harmful exposures to higher trophic levels. When metabolism breaks down a COC into materials with reduced or altered toxicity, that is *biotransformation*. *Biodegradation* is any process that breaks down COCs into simpler, often less hazardous forms.

Within an ecosystem, nutrients are cycled internally, but not without losses or outputs, which must be balanced by inputs, or else the ecosystem will not function. Nutrient inputs come from weathering of rocks, from windblown dust, and from precipitation, which carries material over great distances. Nutrients are lost from terrestrial ecosystems by the flow of water and deposited in aquatic ecosystems and associated wetlands. Erosion and the harvesting of timber and crops also remove considerable quantities of nutrients. The failure to replace these nutrients results in impoverished ecosystems; hence, agricultural lands require fertilizer.

If the input of any nutrient greatly exceed output, the nutrient cycle is stressed or overloaded, resulting in pollution, which is merely an input of nutrients exceeding the capability of the ecosystem to process them. Nutrients are eroded and leached from agricultural lands and, along with sewage and industrial wastes accumulated from urban areas, drains into streams, rivers, lakes, and estuaries, destroying plants and animals that cannot tolerate their presence or else the changed environmental conditions caused by them. Biota that are tolerant to the new conditions will be favored, however. Precipitation contaminated with sulfur dioxide and oxides of nitrogen from industrial areas is reacted to form weak sulfuric and nitric acids respectively, known as acid rain, falling on terrestrial and aquatic ecosystems, upsetting acid-base relationships in the exposed ecosystems. Fish and aquatic invertebrates are killed by the lowering of pH. Soil acidity is also increased, which does not favor most food crops. Forest growth is reduced in ecosystems that lack limestone to neutralize the acid.

COMMUNITY

The functional units of an ecosystem are the *populations* of organisms, through which energy and nutrients move. A population is a group of interbreeding organisms of the same kind, living in the same place, at the same time. Groups of populations within an ecosystem interact. Interdependent populations of plants and animals make up the *community*, which encompasses the biotic portion of the ecosystem.

The community has certain attributes, including *dominance* and *species diversity*. Dominance results when one or several species control the environmental conditions that influence associated species. In a forest, for example, the dominant species may be one or more species of trees, such as oak or spruce. In a marine community, the dominant organisms frequently are mussels or oysters. Dominance influences diversity of species in a community, because diversity involves, not only the number of species in a community, but also how numbers of individual species are apportioned.

A community has layers or *stratification*. In terrestrial communities, stratification is influenced by the growth form of the plants. Grasslands are simple communities with little vertical stratification, consisting of two layers, the ground and the herbaceous

TABLE 5.2
Population Rates

Birth rate   = young produced per unit population per unit time
Death rate   = deaths per unit population per unit time
Growth rate = birth rate − death rate

TABLE 5.3
Species that Experience Exponential Growth

Produce numerous young
Provide little, if any, parental care
Produce an abundance of seeds having little food reserves
Short-lived
Disperse rapidly
Able to colonize harsh or disturbed environments

layer. The six layers of a forest are ground, herbaceous, low shrub, low tree and high shrub, lower canopy, and upper canopy. Strata influence the physical environment and diversity of habitats for wildlife. Vertical stratification of life in aquatic communities is influenced almost entirely by the physical conditions of depth, light, temperature, pressure, salinity, oxygen, and carbon dioxide.

The community provides the habitat where plants and animals live. Within a community, each organism occupies a different *niche*. That is, each species has a unique functional role within the community. For example, the woodpecker lives in a forest habitat, and gleaning insects from the trees is its niche. The more stratified a community is, the more niches it can support.

Populations have birth, death, and growth rates as shown in Table 5.2. The major agent of population growth is births, and the major agent of population loss is deaths. When births exceed deaths, population increases. When deaths exceed additions to a population, it decreases. When births equal deaths in a given population, its size remains the same and it has zero population growth.

With a favorable environment and an abundance of resources, small populations experience *geometric* or *exponential* growth in the early stages of colonizing a habitat, because they either appropriate an underexploited niche or drive other populations out of a profitable one. If the population continues to grow exponentially, the community resources are eventually stressed and the population declines sharply because of starvation, disease, and/or competition. Table 5.3 lists the characteristics typical of populations that experience cyclic exponential growth. Such organisms are called *opportunistic* species.

Some populations grow exponentially at first, but later grow *logistically*. In other words, growth slows as the population increases, then levels off as the limits of their environment or their community's carrying capacity are reached. *Regulatory mechanisms* act on such populations to maintain an equilibrium between their numbers and available resources. Animals exhibiting such population growth produce fewer young, but provide parental care. Plants in this category produce large seeds with considerable food reserves. These organisms are long-lived, have low dispersal rates, and are poor colonizers of disturbed habitats, tending to respond to changes in population density (the number of organisms per unit area) through changes in birth and death rates, rather than through dispersal. As the population approaches the limit of its resources, birth rates decline and mortality increases. Major influences on population growth involve various population interactions tying the community together, including *competition* within and among species; *predation,* including parasitism; and *coevolution,* or adaptation.

When the supply of shared resources becomes short, organisms compete and the successful survive. Within some populations, all individuals may share the resources in such a way that none obtains sufficient quantities to survive as adults or else to reproduce. Among other populations, dominant individuals claim access to the scarce resources, while weaker individuals are excluded. Individual plants claim and hold a site until they lose vigor or die, preventing other plants from surviving by controlling light, moisture, and nutrients in their immediate areas.

Adaptability is the ability of a species or system to cope with climate and food and energy sources. Interrelationships with other species is part of species adaptability. Prey must be able to escape, statistically, from predators, and predators must be able to capture, statistically, sufficient prey for nourishment. Resistance to disease and parasites is another aspect of adaptability as well as the ability to reproduce in the environmental circumstances of the community.

The sum of genes in a population is called the gene pool and we are just beginning to appreciate the value of a gene pool even though we have selectively bred animals and plants for agriculture, sports, and aesthetical purposes for thousands of years. Each individual being carries two copies of each gene, one set from each parent. Variations of these sets of genes are called *alleles.* Individuals having two identical alleles are homozygous for the gene and those having different alleles are heterozygous for the gene. The individual's genotype is the specific assortment of alleles he or she possesses. According to the Hardy–Weinberg law, even though the alleles for each gene are passed from parents to offspring in various combinations, the overall frequency of appearance for a given allele remains constant in a population. In other words, genetic equilibrium resists change in the population. Selection pressures can skew this equilibrium though and that is where the hazard assessor fits in, because COCs as toxins are one among other pressures working against genetic equilibrium. Other anthropogenic pressures are deforestation, desertification, acid rain, global warming, and ozone depletion.

Animals develop social organizations through which resources such as space, food, and mates are apportioned among dominant members. Competitive interactions involve *social dominance,* where dominant individuals exclude submissive individuals from the resource. They involve *territoriality,* where dominant individuals divide space into exclusive areas that they defend. Submissive or excluded individuals either live in poorer habitats, live without the resource, or else leave the area. Some of these animals will succumb to starvation, exposure, and predation.

Predation, the consumption of one living organism by another, moves energy and nutrients through the ecosystem, regulates population, and promotes natural selection by removing unfit genes from a population. Predation on plants includes defoliation by grazers and the consumption of seeds and fruits. The abundance of plant predators, or herbivores, directly influences the growth and survival of the carnivores. Predator–prey interactions at one feeding level influence the predator–prey relationships at the next feeding level. Predators may so reduce a population of prey species that several competing species coexist in the same area, because none is abundant enough to control the resource. When predators are reduced or removed, dominant species crowd out competitors, reducing species diversity.

In parasitism, two organisms live together, one drawing its nourishment at the expense of the other. Parasites are smaller species than their hosts, such as viruses and bacteria are smaller than their human hosts. Parasites normally do not kill their hosts, precisely because they depend on the relationship with their host for survival. This is not a hard and fast rule, however, because the great epidemics of the Middle Ages, such as the Black Plague, killed millions. Even so, host and parasite generally *coevolve* with a mutual tolerance for each other. Nevertheless, parasites may regulate their host population, as did the plague virus, by lowering the reproductive success or modifying the behavior of the host population.

Coevolution is the joint evolution of two unrelated species with a close ecological relationship. The evolution of one species depends in part on the evolution of the other. Coevolution is involved in many predator–prey relations. As predators evolve more efficient ways of capturing or consuming their prey, the prey evolves more efficient ways to escape predation. Plants acquire defensive mechanisms such as thorns, spines, hard seed-coats, and poisonous or ill-tasting sap to deter consumers. Some herbivores breach these defenses and attack the plant anyway.

Insects, such as the monarch butterfly, combine poisonous substances from food plants into their own tissues as a defense against predators. Through natural selection, other organisms acquire colors, patterns, or shapes that mimic unpalatable species. Looking like the distasteful model, mimics avoid predation. Some animals avoid predators by camouflage that blends into the background, making them appear part of their surroundings. Some animals possess obnoxious odors or poisons as a defense, and, besides, have warning colorations that act as further warning signals to potential predators.

Mutualism is a relationship where two or more species depend on one another in order to survive. *Mycorrhizae* is a relationship between fungi and certain plant roots, for instance. *Ectomycorrhizae* fungi form a cap, or mantle, about rootlets, which does a service by providing protection. Fungal *hyphae*, or threads, invade rootlets, growing between cell walls, extending outward into the soil from rootlets. The fungi, including common woodland mushrooms, depend on the tree for their energy source. In return, the fungi aid the tree in obtaining nutrients from the soil and protect the rootlets from certain diseases. Without the mycorrhizae, some trees, such as conifers and oaks, cannot survive and grow. Conversely, the fungi cannot exist without the trees.

Ecosystems are dynamic. The populations forming them do not remain the same. The vegetational community, for instance, gradually changes over time. This change is known as *succession*, beginning with the colonization of a disturbed area, such as an abandoned crop field or a newly exposed lava flow, by species able to reach and tolerate the environmental conditions. *Opportunistic* species hold on to the site temporarily. Being short-lived, poor competitors, they are eventually replaced by more competitive, longer-lived species such as shrubs, and ultimately trees. In aquatic habitats, succession is usually caused by changes in the physical environment, such as the buildup of silt at the bottom of a pond. As the pond becomes more shallow, floating plants, such as pond lilies, and emergent plants, such as cattails, invade. The rate of succession depends on the competitiveness of the species. Succession rate also depends on tolerance to the environmental conditions brought about by changes in vegetation. Interaction with animals, particularly the grazing herbivores, also affects succession. Finally, fire influences succession. When ground cover and underbrush are cleared out by fire, new species get a chance to take hold, at least temporarily. Eventually, the ecosystem arrives at a *climax*, where further change takes place very slowly, and the site is dominated by long-lived, highly competitive species. Succession still proceeds, however, and the community becomes more stratified, enabling more species of animals to occupy the area. Animals representing later stages of succession replace those found in earlier stages.

Competition among species results in the division of resources. Plant species have root systems that grow to unique depths in the soil. Shallow roots permit plants to use moisture and nutrients near the surface. Other species growing in the same place have deep roots that exploit moisture and nutrients that are not available to surface-rooted plants.

Xenobiotics (foreign substances such as COCs) typically do not have deleterious effects unless they are bound in dividing cells. Otherwise, they are metabolized to more water soluble forms and excreted. Storage of a xenobiotic results in bioaccumulation, or bioconcentration, if the concentration is increasing by diffusion from the environment through gills, lungs, or skin. If the xenobiotic is stored by consuming lower trophic levels, the process of biomagnification is occurring.

If only one physiological system, or one compartment, is involved with concentrating the xenobiotic, bioconcentration is a balance between two kinetic processes, uptake and depuration, or excretion, quantified by first-order rate constants, K_1 and K_2.

The rate of bioconcentration is

$$\frac{dC_b}{dt} = K_1 C_m - K_2 C_b \tag{5.4}$$

where: C_b = concentration of xenobiotic in the biota, $\mu g/g$,
C_m = concentration in the environment, mg/l or mg/Kg,
K_1 = uptake rate constant, mL/g-hr, or g/g-hr, and
K_2 = depuration rate constant, hr^{-1}. The assumed initial conditions are: $C_b = 0$, when $t = 0$. At any time t, then:

$$C_b = \frac{K_1}{K_2} C_m \left(1 - e^{-K_2 t}\right) \tag{5.5}$$

When uptake equals depuration,

$$\frac{dC_b}{dt} = 0 \qquad (5.6)$$

so,

$$K_1 C_m = K_2 C_b \qquad (5.7)$$

and the bioconcentration factor, K_B, referred to in EPA literature as BCF, is

$$BCF = K_B = \frac{C_b}{C_m} = \frac{K_1}{K_2}. \qquad (5.8)$$

When the exposure ceases, uptake ceases, and

$$K_1 C_m = 0 \qquad (5.9)$$

then, for depuration only,

$$\frac{dC_b}{dt} = -K_2 C_b \qquad (5.10)$$

and

$$C_b = C_{b0} e^{-K_2 t} \qquad (5.11)$$

where C_{b0} is the concentration of the xenobiotic, b, at time zero. Taking the logarithm and making a semilog plot of C_b against time, the biological half-life is calculated as

$$t_{1/2} = \frac{0.693}{K_2}. \qquad (5.12)$$

The pH of water (actually nearly all water is a dilute aqueous solution) affects the uptake of COCs as well as nutrients into producers by changing the degree of dissociation into ions. Hardness, in the form of carbonates, also affects the uptake of nutrients and COCs. Carbonates bind metal ions in solution, such as cadmium (Cd^{2+}), chromium (Cr^{3+}; Cr^{6+}), and zinc (Zn^{2+}), to make them unavailable as trace nutrients. At the same time, each metal has a specific LC_{50} that increases at certain pH and hardness concentrations. Most marine species are *poikilotherms*, meaning that water temperature greatly affects their metabolic rate, hence their circulation, transport rates, and uptake rates also. The concentration of dissolved organic carbon provides a sink for lipophilic toxicants. Oxygen concentration is another critical factor in uptake process. So is light stress, the amount of ultraviolet radiation that reaches producers.

Hydrophobic chemicals partition from the water column and bioconcentrate in aquatic organisms. Besides the method given above, from Cockerham and Shane, Howard suggests that regression equations can be used to correlate BCF with physical data, such as the octanol–water partition coefficient and water solubility, when the receiving biota do not metabolize the xenobiotic efficiently. If water solubility is greater than 1,000 ppm, the xenobiotic is highly mobile and the half-life will probably be low enough to minimize bioaccumulation. If water solubility is less than 10 ppm, bioaccumulation is expected. For a water solubility between 10 and 1,000 ppm, anyone can guess at bioaccumulation. Appendix 1 gives the solubility in water for several chemicals. The octanol–water partition coefficient, K_{ow}, indicates the affinity for bioaccumulation in fatty tissue. When K_{ow} is greater than 1,000, expect the xenobiotic to accumulate in the food chain in fatty tissue. If less than 500, the chemical will probably not bioaccumulate.

DECOMPOSERS

Some plant and animal matter is not used in the grazing food chain. Waste matter, including fallen leaves, twigs, roots, tree trunks, and the dead bodies of animals, supports the *decomposer food web*. Bacteria and fungi are the chief organisms that convert organic matter to basic minerals and organic residues that become available to be used by producers again. Bacteria, fungi, and animals that feed on dead material become the energy source for higher trophic levels that tie into the grazing food web, making maximum use of energy originally fixed by plants.

Bacteria are ubiquitous in the biosphere. Humanity must go to great lengths to achieve bacteria-free environments and these spaces are limited in size and few in number. However, most bacteria are harmless and even perform useful services to humanity, such as fermentation of alcoholic beverages, causing dough to rise, and turning milk curds into yogurt. Yet, we are only too aware that other bacteria are *pathogenic*; that is, they cause disease. Molds, mushrooms, and puffballs are common fungi.

Breaking down and recycling organic matter is the chief function of the decomposer group, which feeds on matter from dead bodies and cellular debris from all organisms and from excreta and organic byproducts of living organisms. Simply put, decomposers are heterotrophs that convert organic matter into simpler inorganic matter by the respiration process, which reciprocates the photosynthesis process with the exception that the energy released is not returned to the sun but used by other trophic levels or lost. *Aerobic* or *anaerobic* decomposers, or both, may decompose any given mass of organic substance, depending on the availability of oxygen. Aerobic bacteria require oxygen. *Facultative* bacteria can live with or without oxygen. Anaerobic bacteria thrive on an absence of oxygen.

Aerobic and facultative bacteria start decomposing organic matter until the oxygen is used up. When oxygen is depleted, facultative organisms begin to use the bound oxygen in nitrates to survive. When all the nitrate is reduced, the facultative bacteria start using oxygen from sulfates, giving off the characteristic rotten egg smell, H_2S. When sulfates are depleted and more organic matter remains, methane forming bacteria start decomposing matter, releasing CH_4 and CO_2.

Essential elements for decomposing microorganisms are carbon, hydrogen, oxygen, nitrogen, phosphorus, sulfur, manganese, copper, zinc, and cobalt. Required micronutrients include calcium, iron, magnesium, and potassium. Autotrophic bacteria use CO_2 as their sole source of carbon, while heterotrophs, or *saprophytes*, obtain carbon from organic sources.

Novotny and Olem give the aerobic decomposition as

$$C_aH_bO_cN_dP_e + \left(a + \frac{1}{4}b + \frac{3}{2}c - \frac{1}{2}d + 2e\right)O_2 \rightarrow$$
$$aCO_2 + \frac{b}{2}H_2O + dNO_3^- + ePO_4^{3-} \qquad (5.13)$$

This is a slightly different version of the equation we examined earlier and of which we said contained all the ingredients for protoplasm. Notice that oxygen is a reactant in the aerobic equation.

In the absence of oxygen, either anoxic or anaerobic decomposition takes place. *Anoxic* decomposition supplies oxygen by reducing compounds, such as nitrates (NO_3^-) and sulfates ($SO_4^=$), with the products of reaction as in *aerobic* decomposition. Bacteria break down matter in anaerobic decomposition to form methane, carbon dioxide, and ammonia in a process called *diagenesis*. Anaerobic decomposition takes place in stages. The stoichiometry of anaerobic decomposition is

$$C_{106}H_{180}O_{45}N_{16}P + 53\frac{1}{4}H_2O \rightarrow 47\frac{1}{8}CO_2 + 58\frac{7}{8}CH_4 + 16NH_3 + H_3PO_4 \quad (5.14)$$

However, as Novotny and Olem warn, this is only an approximation, because the process takes place in stages, with production of an organic acid, followed by fermentation. During the lag phase, bacterial growth is not obvious. When the bacteria start reproducing by binary fission, the exponential or log growth phase is observed as a steady increase of population. This phase is represented mathematically as:

$$\frac{dX}{dt} = \mu X \qquad (5.15)$$

where: dX/dt = growth rate of bacterial biomass, mg/l-sec,
        X   = concentration of biomass, mg/l, and
        μ = specific growth rate, l/sec.

As the decomposition matter becomes limited, the population increase is balanced by death and reproduction in the stationary phase. The matter available as food is called the *substrate* and as it decreases at a faster rate the endogenous phase begins when the bacteria die off faster than they reproduce. The decay of biomass in this phase is:

$$\frac{dX}{dt} = \mu X - k_d X \tag{5.16}$$

where: k_d = specific decay rate or endogenous decay coefficient.

The biomass production of bacteria is directly proportional to the rate of substrate utilization:

$$\frac{dS}{dt} = -\frac{1}{Y}\frac{dx}{dt} \tag{5.17}$$

where: dS/dt = rate of substrate utilization, mg/l-sec,
Y = yield coefficient (fraction of substrate converted to biomass).

Specific growth rate, μ, is dependent on substrate concentration.

$$\ln\frac{X_1}{X_0} = (\mu - k_d)(t_1 - t_0) \tag{5.18}$$

This relationship is useful when determining the growth rate of bacteria for biodegradation of pollutant-nutrients in soil or wastewater. The best treatment is achieved when the bacteria are starved for other food and they process only the pollutant–nutrient. In nature, several microorganisms are always present and compete with one another for nutrients. The most competitive species are bacteria, which dominate both aerobic and anaerobic systems. Protozoa are predators of bacteria and their presence induces vigorous development for both species.

OCEANS

The vast body of salt water that is the ocean and seas is the ocean biome, covering seventy-one percent of the earth's surface, or about one hundred forty million square miles, at an average depth of sixteen thousand feet, and a total volume of over three hundred million cubic miles. The Atlantic Ocean, the Pacific Ocean, and the Indian Ocean are major subdivisions of the world ocean, and are suitably bounded by continental masses, or by ocean ridges, or by currents. These oceans merge below latitude 40° South in the Antarctic Circumpolar Current, or West Wind Drift, where the combined ocean is called the Antarctic Ocean. The ring-shaped Arctic Ocean is landlocked, except between Greenland and Europe.

The ocean is divided into zones for ecological purposes. The *pelagic* zone is the whole ocean, while the benthic zone is the bottom. Submerged parts of each continent, called the *continental shelf*, stretch out seaward from the shoreline for an average of forty miles, out to a maximum of over nine hundred miles. At a depth of about six hundred feet, the continental shelf drops away abruptly to a steeper zone called the *continental slope*, which continues descending to twelve thousand feet. The *continental rise* is a gentler sloping zone of sediment on the ocean bottom, which extends almost four hundred miles from the base of the continental slope to flat *abyssal plains* of the deep-ocean floor. Mid-ocean *ridges* are formed by extensive mountain chains with deep *troughs* between the ridges, often isolated by *fracture zones* (cracks) in the earth's surface. The ridge system winds continuously for forty thousand miles through all the oceans, merging into the continents in several areas, where volcano activity, earthquakes, and fault lines are found.

The floor of the ocean is covered by an average of fifteen hundred feet of sediment, but the maximum sediment depth is four miles in some areas. In the center of the mid-ocean ridges, where new crust is formed, little, if any, sediment is found. Sediment consists of rock particles and organic residue with varying compositions, depending on depth, distance from continents, and proximity to submarine volcanoes or areas of high biological productivity. Detrital clay minerals, formed by weathering continental rocks carried out to sea by rivers and wind, are abundant in the deep sea. Thick deposits of detrital material are made near mouths of rivers and on continental shelves. Fine clay particles accumulate slowly on the deep-ocean floor, but are stirred up and periodically redistributed by fierce current-generated disturbances called benthic storms because they occur in the sparsely populated deep-sea habitat known as the benthic zone. The calcium carbonate shells of small organisms, such as *foraminifera*, and siliceous shells of marine *protozoans* also accumulate as sediment in the benthic zone.

In the ocean, the *littoral* zone is the horizontal reach between the inland edges of high and low tides. The shallow water over the continental shelf, beyond the littoral zone, is called the *neuritic* zone. The *oceanic* zone lies beyond the continental shelf. From the surface to a depth of 200 meters, the approximate depth of penetration of sunlight, is the *photic* zone. The *mesopelagic* zone ranges from 200 to one thousand meters. The *bathypelagic* zone lies under that.

Seawater is a dilute aqueous solution of salts from weathering of rocks and eroding of soils. Salinity is expressed in terms of total dissolved salts, in parts per thousand parts of water (ppt), reaching about forty ppt in the Red Sea, a region of high evaporation, and more than one hundred fifty ppt in the Great Salt Lake. In the ocean, salinity varies between thirty-four and thirty-six ppt. The major ions present are given in Table 5.4. These ions constitute a significant portion of the dissolved salts in seawater, with bromide, bicarbonate, silica, trace elements, and nutrients being the remainder. Ratios of the ions vary little, only total concentration changes. Major nutrients are abundant, compared with the major ions, but are extremely important for biological productivity. Trace metals are important for certain organisms, but carbon, nitrogen, phosphorus, and oxygen are universally important to marine life. Forms taken by these important elements are given in Table 5.5.

The surface temperature of the ocean ranges from 79°F in tropical waters to 29.5°F, the freezing point of seawater, in polar regions. The surface temperature generally decreases with increasing latitude, with less extreme seasonal variations than land surfaces. In the upper fifty fathoms (300 ft) of the sea, the water is almost as warm as at the surface. From fifty to five hundred fathoms, a region called the *thermocline*, temperature drops rapidly to about 41°F. Below the thermocline, temperature gradually drops to just above freezing.

TABLE 5.4
Ions in Seawater

Cation	ppt
Calcium	0.4
Magnesium	1.3
Potassium	0.4
Sodium	10.5

Anion	ppt
Chloride	19
Sulfate	2.6

TABLE 5.5
Oceanic Forms of Elements Important to Life

Element	Oceanic Form
Carbon	Bicarbonate, HCO_3^-
Nitrogen	Nitrate, NO_3^-
Phosphorus	Phosphate, PO_4^{3-}

Gyres, or surface currents, are kept in motion by prevailing winds, but direction is altered by the rotation of the earth due to the *Coriolis Force*. The Gulf Stream in the North Atlantic and the Kuroshio in the North Pacific are major gyres, both of which warm the climates of the eastern edges of their respective oceans. Where prevailing winds blow offshore, such as the west coasts of Mexico, Peru, and Chile, surface waters move away from the continents until upwelling colder water from the deep, which is rich in nutrients, mixes and cools the original current. The regions where mixing occurs have high biological productivity and provide excellent fishing. Deep water is richer in nutrients than shallow water because decomposition of organic matter exceeds production in the deeper water. Plant growth is more productive where photosynthetic organisms have access to light. Nutrients are recycled when organism remains are oxidized and consumed in the deeper water. High productivity is enhanced by strong vertical mixing in the upper regions of the ocean. The western edges of the continents, plus the entire region around Antarctica, are regions of high productivity, where surface water sinks after being chilled, being replaced at the surface by deeper water.

The deeper circulation in the ocean is a function of density differences between adjacent water masses, known as *thermohaline* circulation. Salinity and temperature determine density, and processes that change salinity or temperature affect density. Evaporation increases salinity. Cooling of seawater increases its density. Ice forms at higher temperatures in less saline water, so partial freezing increases the salinity of the remaining cold water, making it very dense. Below the surface, water cannot exchange gases with the atmosphere. Dissolved oxygen is used up in the oxidization of dead organic matter, and slowly depletes as the water remains beneath the surface for as long as fifteen hundred years.

Oceans are potentially a major food source for the future. Production is measured by the amount of organic matter fixed, or changed into stable compounds, by photosynthetic organisms in a given unit of time. The potential annual ocean production amounts to about two hundred ninety billion tons of fish, shellfish, and edible plants. Rooted plants grow in shallow coastal waters, while plankton consists of plants and animals that float in deeper water. Phytoplankton are producers that turn carbon into organic matter, with the aid of sunlight. Zooplankton are floating animals that are primary consumers. Zooplankton and fish graze on phytoplankton, and each grazer, in turn, is fed upon by its own predator. This organic matter is mostly recycled and reused, so that the standing crop of material is only a small fraction of the annual total. The amount that is harvestable is a function of technology, tastes, needs, and the ability of the system to sustain this harvest. Benthic organisms such as clams and snails live in the bottom sediment. Nekton are fish and insects that swim underwater. Fish, having gills, are excellent indicators of COC insult. Gills are very sensitive. Irritants easily leave evidence on gills, for instance. Also, neoplasms in flatfish are often used as an indicator of oncogenesis in marine life. Neuston are organisms that swim or rest on the surface. Saprotrophs are aquatic bacteria, fungi, and flagellates. Macroinvertebrates, including bivalves and *cnideria*, which are somewhat sessile, also provide good indication of COC attack.

Chemicals are introduced to marine environments in several ways. Individual ships dump garbage and trash overboard, although an international agreement exists against the dumping of floating or harmful materials. Many of the world's population centers are located within a short distance from an ocean or major sea, and a certain amount of municipal sewage, sewage sludge, trash, and garbage is, all too frequently, disposed of in the seas. Chemicals also are introduced to the ocean through incineration at sea, and incineration and other forms of air pollution within about thirty miles of an ocean, a small transport distance compared to the one hundred six mile transport distance of pollutants in the ocean.

Contaminating substances can yield harmful concentrations in a water column, though, at some remote point from the source, the concentration becomes so diluted as to be undetectable. The problem is similar to a contaminant plume in the atmosphere. Eventually, most contaminants end up in the residual materials on the ocean floor, or in the animal and plant life that came into contact with it, as it settled and dispersed. Transport in the ocean depends on the decay rate of the species in water and the adsorbed–dissolved partitioning coefficient.

Background concentration of most pollutants is considered to be zero for the ocean.

In order to asses damage to a water column ecosystem the following must be determined. The mass quantity of organisms including, pelagic and demersal fish, living in a water column, M_w lb., is

$$M_w = \rho_w A_w h \qquad (5.19)$$

where the density of living organisms in the water column, ρ_w lb./ft^3, the cross-sectional area of the water column as a cylinder, A_w ft.^{2}, and the depth of the water column, h ft are the parameters needed. M_w is sometimes referred to as the weighted volume of water affected.

The extent of damage, D, to an oceanic ecosystem due to the contamination of a known chemical is a function of the ambient chemical concentration, C, to which the ecosystem is exposed in the water column. The damage function is determined with the threshold, C_T, and catastrophic concentration, C_P, of the chemical in the water column. If C is less than C_T, no damage to the ecosystem has occurred, and D = 0. When C exceeds C_P, the ecosystem is damaged, and D = 1. Between C_T and C_P the damage is assumed to be a linear function of the logarithm of concentration. Damage is assumed to be the same for the water column and the sediments.

Some fraction of organisms living in a water column (DC_w) is severely damaged due to ambient chemical concentrations and a quantity of living organisms is severely affected in the water column: $DC_w MA_w$, where C_w is the middle point of the concentration range between ambient and contaminated. The total quantity of affected organisms in the water column is

$$M = \Sigma_{C_w} \rho_w h A_w C_w DC_w \qquad (5.20)$$

where $A_w C_w$ is the area of the water column associated with the concentration range, with the middle point C_w. This term is the physical transport model, while DC_w is the saltwater ecosystem damage function.

If the density of living organisms in the water column is uniform, and not dependent on location,

$$M = \rho_w \Sigma_{C_w} h A C_w DC_w \qquad (5.21)$$

Assessing damage to a sediments ecosystem requires the following information: weighted area of ocean, river, or lake floor affected, and damage to the most exposed sediments ecosystem, called the sediments MEE. The mass quantity of organisms, including benthic fish living in a given sediments area, is M_s lb.

$$M_s = \rho_s A_s \qquad (5.22)$$

where ρ_s is the density of living organisms on the floor of the water body, and A_s is the area of the sediments. As with the water column, DC_s is the fraction of severely damaged organisms in the sediment due to ambient concentration. The quantity of living organisms severely affected is calculated as

$$M = DC_s MA_s \qquad (5.23)$$

The total mass quantity of living organisms affected in the sediments are summed over all concentrations ranges as before:

$$M = \sum_{C_s=0}^{C_s} \rho_w A_s C_s DC_s \qquad (5.24)$$

If the density of organisms is constant with respect to location, the density term can be factored out of the summation operand.

WETLANDS

According to the U.S. Fish and Wildlife Service, wetlands are transition areas between terrestrial and aquatic systems, where the water table is usually at, or near, the surface, or else the land is covered by shallow water. Wetlands also have one or more of the following attributes: (1) at least periodically, the land supports predominantly hydrophytes, (2) the substrate is predominantly undrained hydric soils, or (3) the substrate is a nonsoil (organic matter) mixture with water or covered by shallow water at some time during the growing season each year.

Coastal wetlands extend from the outer edge of the continental shelf to the reach of *spring tides*, the highest tides. These wetlands are called *tidal flats*. Subtidal areas are continuously submerged, whereas intertidal areas experience alternate exposure and flooding of the substrate, soil, and organic bottom matter. Estaurine systems have deepwater tidal habitats that are partly surrounded by land, but where open, partly blocked, or infrequent access to the open sea or ocean dilutes the salt water deposited by the tide with stormwater runoff from the land side. The estaurine subtidal wetland is continuously submerged, while the estaurine intertidal wetland is exposed and flooded alternatively by the tides.

Palustrine wetlands are nontidal, or inland, wetlands with the exception of wetlands adjacent to river channels and littoral zones of larger lakes. *Forested* and *scrub* land wetlands are dominated by woody vegetation. *Emergent* wetlands are dominated by erect, rooted herbaceous plants that are distinctly found in these environments, such as cattail, reeds, rush, bulrush, and sedge. *Nonvegetated* wetlands are beds of standing water.

Riparian wetlands is a term used for riverine and lacustrine wetlands. *Riverine* wetlands join deepwater stream channels. *Lacustrine* wetlands are littoral wetlands in topographic depressions, surrounding lakes, or impounded rivers. This definition is reserved for lakes or impoundments that are less than twenty acres in area, are more than six feet deep, or have active wave-formed or bedrock shoreline features. *Mangroves* are saltwater marshes in tropical and subtropical areas that are dominated by salt tolerant trees and bushes. Other marshes or coastal wetlands are dominated by grasses, reeds, sedges, and cattails, which typify emergent vegetation, that is, plants having roots in soil that is covered or saturated with water and leaves that are held above water.

Swamps are wetlands that are dominated by trees and shrubs. *Marshes* are frequently, some are continually, inundated and have emergent vegetation adapted to saturated soil. Tidal salt marshes are covered by salt or brackish water. Inland salt marshes are found along the edges of saline lakes. Tidal freshwater marshes are covered by freshwater, mostly, and are necessarily located further from the coast.

Freshwater marshes develop along the shallow margins of lakes, and slow-moving rivers, when these shallow areas fill with sediment. The plant composition, species richness, and productivity of a marsh are strongly influenced by its relationship to surrounding ecosystems, which affects the supply of nutrients, the movement of water, and the type and deposition of sediment.

Freshwater marshes in the prairie pothole country of glaciated central North America undergo a renewal cycle induced by periodic drought. The renewal of these marshes also depends on the feeding habits of muskrats. The cycle begins with a nearly dry marsh in which seeds of aquatic plants germinate in the mud. When the marsh is water filled, aquatic plants grow densely and muskrats eat large areas of the emergent vegetation, creating patches of open water. The shallow-water emergents decline, but submerged and floating species persist. When the next drought comes, the cycle begins anew.

The most developed salt marshes occur on the Atlantic coasts of North America and Europe. In eastern North America, low marshes are dominated by salt-marsh cordgrass. High marshes, on the other hand, consist of short cordgrass, spike grass, and glasswort. The latter is dominant in Pacific Coast salt marshes. The saw-grass wetlands of the Everglades and tidal salt marshes, which are swept by water flowing in sheets across the surface, are typically dominated by one or two species of emergents. In marshes where the water flows in channels, instead of in sheets, either irregular snow melt or heavy precipitation brings in nutrients and sediment varying the water depth, and creating conditions favorable for a variety of wetland species. Deep marsh water is colonized by submerged pond weeds and floating plants, such as pond lilies. Reeds and wild rice prefer shallow water. Very shallow water supports sedges, bulrushes, and cattails. Sediments and organic deposits eventually raise the bottom of a marsh above the water table, so aquatic vegetation is gradually replaced by shrubs and, ultimately, by a terrestrial ecosystem of upland grasses or forest trees.

Freshwater marshes provide nesting and wintering habitats for waterfowl, shorebirds, muskrats, frogs, and aquatic insects. Salt marshes are wintering grounds for snow geese and ducks, a nesting habitat for herons and rails, and a source of nutrients for estaurine waters. Marshes are important in flood control, in sustaining high-water tables, and as settling basins to reduce pollution downstream. Despite their great environmental value, marshes are continually being destroyed by drainage and filling. A slough is a swamp, shallow lake system, a slowly flowing swamp, or marsh.

Peatland, also called *mire,* is an ecosystem that produces organic matter faster than it decomposes, thus accumulating partially decomposed vegetative material, called peat. A *bog* is a peat-accumulating wetland having no significant inflows or outflows. The peat becomes so thick that the surface vegetation is insulated from mineral soil, forcing the plants to depend on precipitation for both water and nutrients. Bogs are dominated by acid-forming sphagnum moss and other *acidophilic* mosses. *Fens* are partly drained, so peat never accumulates to the point where plants lose contact with water moving through mineral soil. Fens are dominated by glass-like sedges.

Peatlands are found mostly in northern regions, where drainage of water is blocked, precipitation is retained, and decomposition of organic matter is slow. *Raised bogs* are peatlands formed when accumulated organic matter and sediments fill in a pond or basin above the level of the water table. These sphagnum-dominated basins are so high in acidity and low in groundwater minerals that sedges and grasses retreat to the edges. Occasionally, sphagnum moss forms a floating mat over water, thickens and supports associated vegetation, mostly heaths, and eventually reaches the bottom to create a *quaking bog*. Sphagnum invades higher ground under certain conditions, such as deforestation, where the peat is compressed, blocking the drainage of water to create *blanket mires* or *moors*.

FRESHWATER BODIES

Lentic ecosystems include lakes and ponds. As explained elsewhere, lakes are water bodies having no link with the Earth's ocean. Though the distinction is somewhat arbitrary, ponds are generally considered as small, shallow lakes where the rooted plants of the top layer reach the bottom. Light and temperature in lakes vary with depth and season, but are more uniform in ponds, except for seasonal variations. Lakes, therefore, have three distinct zones: littoral, limnetic, and profundal. Aquatic life and rooted plants are plentiful in the *littoral* zone where the sunlight reaches to the bottom. Recall that a pond, as we have defined it, has only one zone, the littoral. The *limnetic* zone runs from the surface, beyond the limits of the littoral zone, down to the depth of light penetration, allowing photosynthesis to take place here as well. Plankton, nekton, and neuston inhabit the limnetic zone. The *profundal* zone is the deep water where light does not penetrate, where life depends on oxygen concentration, organic concentration, and temperature. This zone supports littoral and limnetic organisms that feed on the profundal zone and the benthic organisms that inhabit the bottom sediments, such as anaerobic bacteria, clams, shell fish, snails, and worms.

Young, vibrant lakes are *oligotrophic*, or nutrient poor, and life in them competes vigorously for nutrients. When lakes become nutrient rich, called *eutrophic*, they fill up with algae, and become oxygen deprived. Life slows down in eutrophic lakes, competing sluggishly for oxygen. Humans exacerbate the problem by contributing sediment and unwanted nutrients, making the lake die a slow death.

Lotic water consists of moving bodies of water. Streams and rivers are running water bodies that eventually end up in the ocean. Water velocity is the chief characterizing factor of these bodies of water. Up to two feet per second is considered slow flow, while a fast stream has a greater velocity. The *riffle* zone of fast streams are shallow, have a fast velocity, and are popularly called rapids. Only organisms that can cling to a substrate against the current reside here. Algae, water moss, and strong swimmers live in the riffle zone. Under rubble and river gravel, where they are protected from the current, other organisms such as crayfish, salamanders, and mussels reside. Mayfly nymphs, alderflies, caddisflies, stoneflies, and true flies are typical riffle zone insects. Downstream of each riffle zone, usually in the inside of a bend in the stream, is a *pool* zone where the water runs deeper and slower, almost standing still sometimes. Silt drops out of suspension here due to the low velocity of the water. The community of the pool is similar to those of lakes and ponds.

TERRESTRIAL ENVIRONMENTS

Terrestrial environments include tundra, taiga, deserts and related environments, chaparral and related environments, and forests.

TUNDRA

Tundra, or arctic plain, covers most of the earth's surface north of the coniferous forest belt. Tundra is dominated by sedge, heath, willow, moss, and lichen. Alpine tundra are high plains that occur above the timberline on mountains. Tundra have severe winters, low average temperatures, little snow or rainfall, and short summers. The arctic tundra is influenced by permafrost, a layer of permanently frozen subsoil in the ground. The surface soil of arctic tundra tends to be rocky, thawing in the Summer to varying depths. The combination of frozen ground and flat terrain impedes the drainage of water, which forms ponds and bogs that provide moisture for plants. Recurrent freezing and thawing of the soil, at drained locations, forms cracks in the ground in regularly patterned polygons. Poorly drained areas produce irregular land forms such as hummocks, knolls, frost boils, and earth stripes. Thawing slopes move soil down slope in warm months to produce *solifluction*, or flowing soil terraces. Alpine tundra commonly has bare rock-covered ground, called *fell-fields*, which support lichen growth.

The tundra has few plant species, and with low growth, most biomass is concentrated in the roots. With a short growing season, tundra plants typically reproduce vegetatively by division and budding, rather than sexually by flower pollination. Typically, arctic vegetation includes cotton grass, sedge, dwarf heath, mosses, and lichens, all adapted to sweeping winds and frost heaves, two kinds of soil upsets. These plants adapted to the environment by supporting photosynthesis at low temperatures and light intensities over long periods of time. Alpine plant communities consist of mat- and cushion-forming plants, which are adapted to gusting winds, heavy snows, and widely fluctuating temperatures. They photosynthesize under brilliant light during short days.

Arctic wildlife is *circumpolar*, that is, the same or closely related species are found in other places around the world. However, the variety is limited in the challenging environment. Dominant large grazers include muskox, caribou, and reindeer, which feed on grass, sedge, lichen, and willow. Small prey mammals, such as the Arctic hare, snowshoe rabbit, and lemming, feed on grass and sedge. Predators are the wolf, arctic fox, and snowy owl. Polar bears, and sometimes brown bears, are also found in the Arctic. Several bird species nest in the tundra shrubbery in summer, but migrate to milder climates before winter sets in. Invertebrate life is scarce, but insects such as black flies and mosquitoes are abundant. Alpine animal life includes the mountain goat, big-horned sheep, pika, marmot, and the ptarmigan. Flies are scarce, but butterflies, beetles, and grasshoppers are abundant.

The equilibrium of the tundra is fragile. Generally, the tundra is unable to recover from environmental insult without loss of species. Disruption of the vegetative cover causes deep melting of permafrost, collapsing of the ground, and destruction of soil.

Vehicle tracks make deep gullies that persist for decades. Tundra wildlife is susceptible to habitat destruction.

TAIGA

The taiga is a vast belt of coniferous forest in northern Eurasia, which extends from the Baltic to the Okhotsk Sea, covering most of Siberia, as a transitional zone between the tundra to the north and the steppes and mountains to the south. Canada has a similar region. Pine, larch, spruce, fir, and cedar populate the taiga, where numerous swamps and peat bogs are found. Taiga soils are typically *podsolic*. The wildlife population includes elk, deer, reindeer, lynx, brown bear, other fur-bearing mammals, and numerous birds and insects.

DESERT

A *desert* is a region that receives less than ten inches of annual rainfall, has an evaporation rate that exceeds precipitation, and, usually, has a high average temperature. Lack of soil moisture and low atmospheric humidity causes most of the incident solar energy to penetrate to the ground, meaning daytime temperatures can reach 130°F plus. The desert floor radiates heat back to the atmosphere at night, causing the temperature to drop to near freezing.

Deserts are formed by dry air masses moving over the surface of the earth. As the earth rotates, enormous eddies are created in the atmosphere. Hot air rising over the equator flows northward and southward towards the poles. As the air currents cool in the upper regions of the atmosphere, they descend as high pressure areas in the subtropical zones. Closer to the poles, air ascends and pressure drops. Still further from the equator are polar regions of cool descending air. As air rises, it cools and loses moisture, creating clouds. As it descends, air warms and picks up moisture, causing the land mass below to dry out.

Downward movements of warm air masses over many thousands of years have produced two belts of deserts, one along the Tropic of Cancer in the northern hemisphere, and the other along the Tropic of Capricorn in the southern hemisphere. Northern deserts include the Gobi in China, the deserts of southwestern North America, the Sahara in North Africa, and the Arabian and Iranian deserts in the Middle East. The southern desert belt includes Patagonia in Argentina, the Kalahari Desert of southern Africa, and the Great Victoria and Great Sandy deserts of Australia.

Other desert areas were formed by the influence of ocean currents on land masses. As cold currents from the polar regions flow toward the equator and strike continental coastlines, the currents are further cooled by cold water rising from the ocean depths. Air currents cool as they move across this cold water, carrying fog and mist, but little rain. Cold currents flowing along the coastal regions of southern California, Baja California, southwest Africa, and Chile often shroud these areas in mist, but make them into deserts.

Mountain ranges also develop deserts by creating rain shadows on the leeward sides. Moisture-laden winds flow upward over windward slopes, cooling, and losing moisture in the form of rain and snow. Dry air descends the leeward slopes, evaporating moisture from the soil. The Great Basin, a desert of North America, for instance, lies in the rain shadow produced by the Sierra Nevada.

Some interior deserts are formed by prevailing winds, which, being far removed from large bodies of water, have no moisture by the time they reach interior regions such as the Gobi and Turkestan.

Stark desert landscape is wind-blown, but also shaped by water. Desert soil, unprotected by vegetation, easily erodes when rain falls. *Arroyos* are canyons that form where water runs off desert hills. Erosion resistant rocks form angular peaks. Alluvial fans lead away from these peaks to deposit debris in the form of large slopes, called *bajadas*, which flatten out to form *playas*, literally meaning beaches but referring here to the beach-like situation. During infrequent rains, playas fill with water, which evaporates later, leaving behind, on the surface, a layer of salt dissolved from the ground. Salt lakes are common in deserts, therefore. The Great Salt Lake of Utah, though, is not one of the desert lakes, but the remnant of an inland sea fed by an inflow of fresh water. Evaporation, which is never complete in desert lakes, concentrates salt in the lake water. Wind erodes rocks into unusual shapes in the desert, building up dunes, which are typical in the Sahara and parts of the North American desert.

All but the most arid deserts support life. In fact, desert life is frequently lavish, being well adapted to the scarcity of water and daytime heat. Desert plants are very efficient water consumers and storekeepers. Some flowering plants are short-lived, living for a few days at most. Seeds of these plants lie dormant for years, until a soaking rain allows them to germinate and bloom. Woody plants either have long root systems to reach deep water sources, or spread shallow roots to quickly take up surface moisture from heavy dews and occasional light rains. Desert plants typically have small leaves to reduce surface area from which transpiration takes places. Other plants shed leaves during the driest period. Photosynthesis is usually conducted in the stems of desert plants, not in the leaves. Desert succulents store water in leaves, stems, and roots. Thorns, which are modified leaves, protect the precious water from animals. Thorny plants take in and store carbon dioxide only at night. During the day their *stomata,* or pores, close up to prevent evaporation. Desert plants thrive on saline soils and concentrate salt in sap, then excrete the salt through their leaves.

A few amphibian species adapted to desert life by remaining dormant for the long dry periods. When rains come, they mature rapidly, mate, and lay eggs within a few hours. Many birds and rodents reproduce only during or immediately following winter rain that stimulates the growth of vegetation. Desert rodents, such as the North American kangaroo rat and the African gerbil, feed on dry seeds. Their metabolic processes are extremely efficient at conserving and recycling water. Their urine is highly concentrated to aid the precious water balance. Several desert mammals, the camel for example, withstand considerable dehydration. Mammals and reptiles are nocturnal, remaining in cool underground burrows or in the shade by day. Reptiles, such as the horned toad, control metabolic heat by varying their heart rate and the rate of metabolism. Mammals, such as the desert oryx, store body heat by day, releasing heat at night.

Desert soil is naturally fertile because so little water is available to carry nutrients away. Crops are easily grown on the desert with just a little irrigation water. However, desert causes salt accumulation on the surface when irrigation water evaporates, eventually rendering the soil useless for further crop production. By tapping reservoirs of fossil water deep beneath the desert, humans are, in effect, mining water, which, once gone, is irreplaceable. Burning and overgrazing of semiarid lands on the periphery of deserts, irreversibly damages plants that concentrate moisture and hold the soil together.

Thus, deserts encroach on arable land. This serious world problem, called *desertification*, threatens about thirty-five percent of the earth's land surface.

CHAPARRAL

Chaparral is a shrub-land biome dominated by evergreen vegetation with small leaves, with a Mediterranean type of climate with warm, wet winters, and long, dry summers. The shrub lands of California and Baja California, dominated by scrub oak and dense chamiso and manzanita shrubs, are typical of chaparral also. The health of chaparral depends on fire, which wipes out decadent growth, disposes of accumulated litter, recycles nutrients, and stimulates new, vigorous growth from seeds and sprouts. Fire reduces complex organics to their elemental state to be recycled. Fire also reduces the nitrogen content of soil by converting vegetative litter to ash. Minerals in soil are exposed when fire burns the ground cover and this stimulates the germination of many types of seeds.

Shrub lands in the American Southwest with similar vegetation are sometimes called chaparral, but lack chamiso, and the summers are not as long and dry. Equivalent plant communities are called *tomillares* in Spain, *macchia* in the other Mediterranean countries and South Africa, *phrygana* in the Balkans, and brigalow shrub in South Australia.

FOREST

Forests, dominated by trees and other woody vegetation, occupy an extensive area of the earth's surface. Left alone, forests are self-regulated, remaining relatively fixed ecosystems over long periods of time. Climate, soil, and the topography determine the dominant trees of a forest, and associated shrubs and herbs. Forest floor vegetation is influenced by the larger and taller plants, but the low vegetation affects the organic composition of the soil. Disturbances, such as forest fires or timber harvesting, force a shift to another forest type. Left undisturbed, ecological succession eventually results in a *climax* forest community. Eight general types of forest are classified on the basis of leaf characteristics and climate: temperate deciduous, deciduous monsoon, tropical savanna, northern coniferous, tropical rain, temperate evergreen, temperate rain, and tropical scrub forests.

Temperate Deciduous Forests. The forests of the eastern United States are typical *temperate deciduous* forests, with two subtypes. Temperate forests in the northern and southern hemispheres are radically different, due to the *continental climate* of the northern hemisphere and the *oceanic climate* of the southern. Deciduous forests have four strata: two canopy layers, a shrub layer, and a field layer. The latter layer consists of ground covering herbs, ferns, and mosses.

Monsoon Deciduous Forests. *Deciduous monsoon* forests are found in Bengal, Burma, Southeast Asia, India, and the Pacific coastal regions of Mexico and Central America. The climate in these forests is characterized by heavy daily rainfall, seasonally relieved by dry periods, during which time the trees shed their leaves.

Tropical Savanna Forests. *Tropical savanna* forests are found in regions such as the *campos* of Brazil, where forest and grassland meet. African and South American savannas are dominated by grasses and sedges, with open stands of widely spaced, frequently thorny trees. Australia also has tropical savannas. Savannas are created by fire, or by grazing and browsing mammals and have prolonged dry seasons followed by a period with forty to sixty inches of rainfall and are dominated by grazing and burrowing animals.

Northern Coniferous Forests. *Northern coniferous* forests form a worldwide belt in subarctic and alpine regions. Conifers are limited to the northern hemisphere where long, cold winters are found. Gnarled scrub trees dominate tree lines and mountaintops. Spruce and fir trees populate the more northern forests. Pine, larch, and hemlock dominate farther south. A belt of conifers extends from the Atlantic coast of Canada, westward and northwest to Alaska, and includes spruce, balsam, fir, and pine, with poplar and white birch being the principal deciduous species represented. Usually, northern coniferous forests occupy regions that were glaciated long ago, leaving behind lakes, bogs, and rivers. Growing under the coniferous canopy are mostly ferns and mosses, but ground cover growth is poor at any rate.

Tropical Rain Forests. *Tropical rain* forests, found in central Africa in the Congo, Niger, and Zambesi basins, and the Amazon and Orinoco watersheds in South and Central America, and in the Indo–Malay–Borneo–New Guinea belt of Asia, are always active because plant growth is profuse, and leaves fall and regrow constantly throughout the year. Tree species are highly diverse, but usually have smooth, straight trunks and large, simple leaves. Large vines are common, but tangled jungle growth occurs only where the normal forest area has been abused, or at a river's edge.

Temperate Evergreen Forests. *Subtropical* regions of North America and the Caribbean islands have *temperate evergreen* forests, because of the warm maritime climate. The typical temperate evergreen forest is found along the U.S. Gulf Coast, and in the Florida Everglades. Characteristic trees are live oak, magnolia, palms, and bromeliads. The South Atlantic and Gulf states chiefly have softwoods such as longleaf, shortleaf, loblolly, and slash pines. Dense hardwood stands in the eastern United States, particularly around the Mississippi and Ohio river valleys, have oaks, black walnut, yellow poplar, and sugar maple. The forests of the Appalachians are dank mesophytic stands. South of the Canadian coniferous belt, the mixed forests around the Great Lakes, Saint Lawrence Seaway, and Acadian regions, produce eastern white pine, red pine, eastern hemlock, spruce, cedar, fir, yellow birch, maple, oak, and basswood.

Temperate Rain Forests. *Temperate rain* forests, with broad-leaved evergreen trees, are common on Mediterranean coasts. Rainfall may be low, but the ocean-cooled air is moisture laden, with frequent fogs. The temperate West Coast rain forests of the U.S. are dominated by hemlock, cedar, spruce, fir, and redwood. The western forests of the Rocky Mountains and the Pacific coast contain Douglas fir, ponderosa pine, western white pine, Engelmann spruce, and white fir.

Tropical Scrub Forests. *Tropical scrub* forests occur in regions of slight rainfall bordering wetter forests.

TERRESTRIAL POLLUTANTS

One group of persistent pollutants in the terrestrial environment is heavy metals. In nature, these metals appear in their oxide forms, the lowest level of energy, and, therefore, are at equilibrium in nature, and at peace with the universe. Humanity adds vast quantities of energy to make these metals more valuable, and, consequently, hazardous. Arsenic, cadmium, lead, mercury, selenium, antimony, bismuth, indium, tellurium, and thallium are the predominant metal pollutants in soils.

Metals enter plants by uptake from the soil by active transport against a concentration gradient through roots or foliage. Metal ions are absorbed through roots by passive diffusion through cell membranes. Within the plant, metals are transported via the xylem to various plant parts. The concentration of metal ions is greatest in the roots, and least in the seeds, regardless of the relative mobility of the ion. Foliar uptake occurs by absorption through the leaf cuticle. Some plants do not fair well with large concentrations of metal ions in the soil, but other species use them as nutrients.

Terrestrial invertebrates, such as earthworms, millipedes, and isopods, stimulate decomposition of plant material by increasing the surface area for microbial attack. Centipedes and spiders prey on terrestrial isopods, transferring metals from primary to secondary consumers.

ECOSYSTEM MONITORING

Though not in the mainstream of industrial investigations, more and more monitoring of ecosystems, surrogates, and indicators is being performed. Table 5.6 lists some of the testing that is being tried.

Climate, soils, watery systems, and species combine to form natural systems for which humanity is the chief steward. As we develop a better understanding for the interdependencies in these systems, we also gain a greater appreciation of our responsibilities as the caretaker of Earth.

QUESTIONS (ANSWERS BEGIN ON PAGE 401)

1. Name the spheres of life comprising Earth's biosphere.
2. What influences terrestrial biomes?
3. What are some major biomes?
4. What is the role of primary consumers in an ecological systems and how might their exposure to pollutants affect humanity?
5. When sales literature advertises a product as biodegradable, what is the message being transmitted, and what is probably the truth of the matter?

TABLE 5.6
Ecosystem Testing

Artificial stream modeling	Metabolic inhibition
Behavioral studies	Metabolic transformation rates
Bioassays	Microcosm simulation
Biochemical markers	Molecular biomarkers
Bioluminescence inhibition	Mutation rates
Biomass	Nutrient cycling
Cellular biomarkers	Photosynthesis rates
Cellular metabolism	Photosynthetic activity
Chromosomal abnormalities	Population growth rate
Competition tests	Predator–Prey tests
Detritus processing	Primary productivity
Diversity indices	Relative abundance
Enzyme activity inhibitors	Reproduction productivity
Exotic invasion	Seed germination
Heavy metal uptake	Species richness
Infrared vegetation assessment	Tumors
Mesocosm simulation (aquatic ATP assays field test)	

After de Serres and Bloom.

6. How might a human community cause the acceleration of the eutrophication of a lake?
7. What four factors affect the uptake of nutrients (and pollutants) into biota from soil or water?
8. Why can pockets of relatively high concentrations of a pollutant be found in the ocean hundreds of miles from its entry point?
9. What is the usual indicator of oncogenesis in marine life?
10. Describe the biome where you live and the potential adverse impact of humanity on it. What are some ways that humanity can sustain development in your location?

REFERENCES

"Agreement Calls for Five Companies to Clean up Oil Reclaiming Facility." *Environment Reporter.* Washington, D.C.: Bureau of National Affairs, Inc. March 11, 1994.

Boulding, J. Russell. *Practical Handbook of Soil, Vadose Zone, and Groundwater Contamination: Assessment, Prevention, and Remediation.* Boca Raton, FL: CRC Press/Lewis Publishers, 1995.

The Bush Administration and the Environment: Accomplishments and Initiatives. U.S. Executive Office of the President. Feb. 25, 1992.

Cockerham, Lorris G. and Barbara S. Shane. *Basic Environmental Toxicology.* Boca Raton, FL: CRC Press/Lewis Publishers, Inc., 1994.

Daugherty, Jack. *Industrial Environmental Management: A Practical Handbook.* Rockville, MD: Government Institutes, Inc., 1996.

De Serres, Frederick J. and Arthur D. Bloom. *Ecotoxicity and Human Health: A Biological Approach to Environmental Remediation.* Boca Raton, FL: CRC Press/Lewis Publishers, 1996.

"Health of Water Systems Among Major Items on Conservation Agenda, Administrator Says." *Environment Reporter.* Washington, D.C.: Bureau of National Affairs, Inc. Feb. 11, 1994.

In Our Hands: Earth Summit '92. A Reference Booklet about the United Nations Conference on Environment and Development in Rio de Janeiro, Brazil, June 1–12, 1992. United Nations Environmental Programme.

Howard, Philip H. *Handbook of Environmental Fate and Exposure Data for Organic Chemicals: Volume I, Large Production and Priority Pollutants.* Boca Raton, FL: CRC Press/Lewis Publishers, 1989.

Kaczmar, Swistoslav W., Edwin C. Tifft, Jr. and Cornelius B. Murphy, Jr. "Site Assessment under CERCLA: The Importance of Distinguishing Hazard from Risk." *Hazardous Materials Control Monograph Series: Health Assessment.* Silver Spring, MD: Hazardous Materials Control Research Institute.

Karam, Joseph G. and Martha J. Otto. "Ocean Disposal Risk Assessment." *Hazardous Materials Control Monograph Series: Health Assessment.* Silver Spring, MD: Hazardous Materials Control Research Institute.

Koren, Herman and Michael Bisesi. *Handbook of Environmental Health & Safety: Principles and Practices, Volume I.* Boca Raton, FL: CRC Press/Lewis Publishers, 1996.

Landis, Wayne G. and Ming-Ho Yu. *Introduction to Environmental Toxicology: Impacts of Chemicals upon Ecological Systems.* Boca Raton, FL: CRC Press/Lewis Publishers, 1995.

Ney, Ronald E., Jr. *Fate and Transport of Organic Chemicals in the Environment: A Practical Guide.* 2d. Ed. Rockville, MD: Government Institutes, Inc., 1995.

"NOAA Finds Increased Contamination in Tampa Bay." *Environment Reporter.* Washington, D.C.: Bureau of National Affairs, Inc. Jan. 14, 1994.

Novotny, Vladimir and Harry Olem. *Water Quality: Prevention, Identification, and Management of Diffuse Pollution.* New York: Van Nostrand Reinhold, 1994.

Philp, Richard B. *Environmental Hazards & Human Health.* Boca Raton, FL: CRC Press/Lewis Publishers, 1995.

Stine, Karen E. and Thomas M. Brown. *Principles of Toxicology.* Boca Raton, FL: CRC Press/Lewis Publishers, 1996.

6 Basic Chemical Hazards to Community and Landscape

Our first concern as exposure assessors is for the health and well-being of exposed humans. In the third and fourth chapters, we discussed some of the ramifications of exposure to humans. In the last chapter, we examined the impact of chemicals on the flora and fauna of the environment, the biota of biomes and eco-systems. In this chapter, we shall look at the impact of chemicals on the abiota of the environment, on natural, as well as anthropogenic, structures. We shall also examine the potential damage to human social and commercial intercourse.

COMMERCE AND ECONOMY

Economic progress is measured in terms of the beneficial aspects of productivity improvements. Adverse, or even neutral, consequences of productivity are ignored. This is why, before the advent of governmental micromanagement of the business affairs of private companies, well-meaning people poured terrible substances into holes in the ground behind the plant. This is also why those micromanaging laws and regulations are still necessary. Without those awful regulations, we would soon be forced to return to old practices in order to compete!

In his book, *Earth in the Balance,* now Vice President Al Gore points out that classical economics fail to properly account for all the costs associated with consumption. Therefore, we ignore the waste side of consumption in our economic calculations and naturally cannot justify paying money to protect the environment, not voluntarily anyway. Hence, until the accounting procedures are changed, we have to live with regulations. To believe otherwise, one has either not grasped the big picture or is not grounded in reality.

Natural resources ought to be depreciated in calculating the gross national product (GNP) of any country. Also, a gross world product (GWP: not to be confused with Global Warming Potential discussed later), which accounts for the depletion of natural resources along with productivity, needs to be devised, for measuring the health of the entire global economy.

For instance, how much soil has been washed away into the ocean in order to produce the grain crops of the world? What percentage of freshwater aquifers have been lost, due to the use of pesticides and fertilizers to increase grain yields? Where does the U.S. GNP account for the depletion of the tropical rain forests, so vital to global climate? What has that to do with us? What about U.S. interests in extracting trees for pulpwood, lumber, and the derivation of medicines?

TABLE 6.1
Al Gore's Sustainable Development Plan

1. Change GNP to include environmental costs and benefits.
2. Define productivity to include environmental improvement or decline.
3. Global agreement to eliminate use of inappropriate discount rates, adoption of better ways to quantify effects of decisions on future generations.
4. Global elimination of public expenditures that subsidize and encourage environmentally destructive activities.
5. Global communication of the environmental impacts of specific products to consumers.
6. Full disclosure by companies of responsibility for environmental damage.
7. National research into costs and benefits of environmental efficiency.
8. Include environmental harm in national antitrust laws.
9. Environmental protection clauses in international treaties and trade agreements.
10. Evaluation of environmental concerns by international finance institutions before granting development funds.
11. Environmental stewardship in return for international debt relief; debt-for-nature swaps.
12. An international treaty limiting CO_2 emissions by country and establishing a market for emission rights.

Gore's call for a Global Marshall Plan drew only derision and laughter from his congressional colleagues on Capitol Hill (he was still a Senator when he wrote the book), but, so far, no one has proposed anything better. Our legislators still act as if they believe that all natural resources are limitless and free for the taking. No one seems the least bit concerned about the effects our current decisions about the use of natural resources will have on future generations. Yes, some people are concerned about the consequences of acid precipitation, global warming, and ozone depletion, but many other abuses of humanity upon the Earth go entirely unnoticed.

Gore suggests (1) printing the environmental impact of consumer products right on the label or package, (2) accounting for depletion of natural resources or destruction of environments in calculating economic benefits, and (3) stopping wars and violence as ways of halting depletion and destruction of resources and environments. Admittedly "Pollyannaish", but it is the direction we need to head in. Table 6.1 is a summary of Gore's plan for sustainable development.

Even business leaders like Paul Hawken are calling for a redesign of such commercial institutions as the GNP. In *The Ecology of Commerce*, Hawken states: "To create an enduring society, we will need a system of commerce and production where each and every act is inherently sustainable and restorative. Business will need to integrate economic, biologic, and human systems to create a sustainable method of commerce." Table 6.2 lists Hawken's initial objectives for achieving sustainable commerce. Why are we discussing these things in a book on calculating chemical exposures? Because, the exposure assessor must have some vision of the larger picture, in order to understand the meaning of his or her task, and have some feel for the meaning and implications of his or her results. Exposure assessment, in my opinion, should not be performed blindfolded, or in a vacuum, but should be an integral part of a logical whole. Perhaps you disagree with Gore and Hawken; I

TABLE 6.2
Paul Hawken's Sustainable Commerce Objectives

1. Reduce absolute consumption of energy and natural resources in the northern hemisphere by 80% in the next five decades.
2. Provide secure, stable, and meaningful employment for people everywhere.
3. Be self-actuating as opposed to regulated or morally mandated.
4. Honor market principles.
5. Be more rewarding than our present way of life.
6. Exceed sustainability by restoring degraded habitats and ecosystems to their fullest biological capacity.
7. Rely on current income.
8. Be fun and engaging, and strive for an aesthetic outcome.

certainly do not agree with them totally, but they still, unfortunately, have the only alternative game in town, and their views are a starting point for discussions on where and how to go from here. Unless you are still convinced that nothing is wrong with the old game. Perhaps you are, but why, then, are you interested in exposure assessment, if nothing is wrong in the world?

The chief problem in our current economic system is pointed out in technical terms by Hawken. Economists measure efficiency in monetary terms, while ecologists, and other scientists and engineers, like you and me, measure efficiency in terms of thermodynamics, and conservation of resources. In managing the wealth and resources of our human communities, we technicians agree with economists on the inefficiency of pollution and waste. We can also agree that increasing production efficiency will save money, improve the economy, and reduce pollution, especially reduce emissions of global warming gases, such as carbon dioxide. Where we technicians depart from the economists is our understanding that, once expended, energy is irreplaceable. This chaos of disordered energy we call *entropy*, and the economists have no analogous term. Economists are not concerned about disordered wealth, or recovering spent wealth, or maximizing the use of wealth. In fact, wealth seems to pour from some undefined cornucopia, and it matters not whether its source was a wealth producing hole in the ground in Africa, or a wealth transferring fast-service hamburger joint in Los Angeles, it is still wealth and just as valuable to the economist. Spending, saving, or making wealth has no efficiency. Scientists and engineers are very much interested in efficiency, understanding entropy as they do. Hawken points out that the U.S. economy would have spent two trillion dollars less on energy in the last decade, if only we had been using it as efficiently as Japan and Sweden! That's about one thousand dollars per capita, annually! I could use a bonus like that in my home.

Table 6.3 lists Hawken's principles of restoration.

Despite the fact that economic modeling ignores environmental damage, the cost is staggering. In Germany, the total cost has been estimated at five to six percent of the GNP, and Georg Winter claims that number is low. This successful German industrialist joins Gore and Hawken in calling for revamping the modeling of national economics. He writes:

TABLE 6.3
Hawken's Principles of Restoration

1. Eliminate waste from industrial production.
2. Change from a carbon-based economy to a hydrogen/solar-based economy.
3. Create systems of feedback and accountability that support and strengthen restorative behavior.

> What we are so far lacking are politicians capable of altering the framework conditions of our free market economy such that economic success can only be maximized by maximizing the environmental acceptability of products and services. Environmentally acceptable business activity must be made economically profitable to enable the dynamic forces of free competition to work to the advantage of the restoration of the environment. There is plenty of job creation potential in the process.

Another way in which the community's economy is affected by pollution is the destruction of infrastructure and monuments by air pollution. Building materials are soiled by soot and other particulate matter (PM), but that problem is insignificant compared to the destruction of concrete structures by acid precipitation. Concrete is a solid mixture of alkaline materials, and acid destroys it quite handily. Also, sulfuric acid reacts with carbonate building stones to form gypsum ($CaSO_4 \cdot 2H_2O$) and calcium sulfite ($CaSO_3 \cdot 2H_2O$), both of which are soluble in water. Water transports the acids into the interior of the stone or concrete, producing soluble salts that may precipitate from the very materials that provide strength, and form encrustations, which may grow and further weaken the structure, or else the precipitate may be washed away by rain to erode the structure.

Acidic erosion of priceless, irreplaceable historical monuments and works of art has occurred in Western Europe and along the eastern seaboard of North America. The Coliseum and other Roman monuments and works of art, especially those made of marble, also have suffered greatly. In India, the Taj Mahal is in an advanced state of dissolution by acid precipitation.

Bronze and copper statues and building structures have also been corroded by acid deposition. Other nonferrous metals are corroded by basic carbonate coating that forms on exposed surfaces. Iron-based structures take on the characteristic reddish appearance of rust. SO_2 and acidic PM accelerate corrosion of metal surfaces. Connections in electrical and electronic equipment are corroded, resulting in serious operational and maintenance problems. About one and one-half billion dollars are lost annually due to the corrosion of metals by acid deposition.

The appearance of surface coatings, such as paints and varnishes, can be damaged by acid deposition. The durability of coatings is also damaged by acidic or alkaline PM, H_2S, SO_2, and O_3. Sometimes, PM serves as a wick that allows chemically reactive substances to penetrate coatings to corrode underlying metals. Deleterious effects on surface coatings include soiling, discoloration, loss of gloss, loss of scratch resistance, decreased adhesion, and decreased strength.

Ozone cracks rubber compounds. This leads to the occasional replacement of rubber tires and other rubber products exposed to the outdoors. The depth and nature of cracking are a function of O_3 concentration, the rubber formulation, and the degree of stress on the rubber item. Especially susceptible are the double bonds of unsaturated natural and synthetic rubber compounds, including butadiene-styrene and butadiene-acrylonitrile. More resistant are saturated compounds such as polysulfide, butyl, and silicon polymers, and chlorinated unsaturated compounds, such as neoprene. Nonresistant electrical wire insulation, for instance, is damaged by O_3 exposure. Damage to paints and rubber products account for five hundred million dollars lost per year.

Cellulose-based fabrics, including cotton, hemp, linen, and rayon, are sensitive to acid aerosols and acid gases. Synthetic fabrics, such as nylon, are hardly damaged, though NO_2 may oxidize nylon polymers, reducing the fabric's affinity for dye. Reactions with fabric dyes cause fading, leading to fairly substantial economic losses, especially in carpeting. Damage to textiles and fabrics by acid deposition is estimated at two billion dollars annually in the U.S. alone.

Finally, documents have suffered from acid deposition. Many important historical documents are now archived in acid-free environments, for this reason.

PUBLIC HEALTH

The immune system of every unborn child in the world will soon be irrevocably damaged. While, as exposure assessors, we are concerned with the hypothetical Most Exposed Individual (MEI), it is the public health we are ultimately protecting. We should never forget that real people, somewhere out in the community, are going to be, or perhaps already are, affected by the exposure.

JUSTICE

Gore points out in *Earth in the Balance* that "wherever people at the grass-roots level are deprived of a voice in the decisions that affect their lives, [both] they and the environment suffer."

Corruption is a major cause of environmental destruction, and, based on about seven thousand years of history, more if you count ancient scriptures and myths, something of a blight on human civilization that will always plague us. While our assessments are about exposures, not corruption, we must be aware of the impact corruption has on potential exposures, and be vigilant for it. Needless to say, let us as exposure assessors avoid being part of this moral cancer.

Does everyone have a right to a clean, safe environment? The environmental justice movement says everyone does. According to Swanson, environmental justice has also been called environmental equity, toxic colonialism, and environmental racism. On February 11, 1994, President Clinton signed Executive Order 12898, *Federal Actions to Address Environmental Justice in Minority Populations and Low-income Populations,* which was amended by Executive Order 12948 on February 1, 1995. The Executive Order requires that where environmental justice concerns, or the potential for these concerns, are identified, an appropriate analysis of the issues is to be conducted. To the extent practicable, the ecological, human health,

TABLE 6.4

Minorities around Cement Plants and Incinerators

	Cement Plants (%)	Incinerators (%)
Co. avg. + >5%	27	37
Co. avg. − >5%	36	44
Co. avg. ±5%	38	20

TABLE 6.5

Poverty vs. Cement Plants and Incinerators

	Cement Plants (%)	Incinerators (%)
Co. avg. + >5%	18	36
Co. avg. − >5%	22	37
Co. avg. ±5%	60	28

and socio-economic impacts of any proposed exposures in minority or low-income communities are to be evaluated. The human health evaluations are to take into account subsistence patterns and sensitive populations. EPA is directed to achieve environmental justice as part of its mission, by identifying and addressing dispro- portionately high and adverse health and environmental effects where minorities and low-income peoples live. EPA must compare how a new policy will impact minority or low-income communities vs. majority and affluent communities. The agency is also required to examine how subsistence farming or fishing patterns relate to its risk assessment policies. In short, EPA is to ensure that no segment of the population, regardless of race, color, national origin, or income bears a disproportionate or high burden of adverse health or environmental effects.

Does environmental injustice exist, or is it a myth? Using the 1990 census, EPA found that the populations in the vicinity of landfills and land treatment units were more likely to be minority populations than the National distribution of minorities would have predicted. Table 6.4 shows the percentage of cement plants and incin- erators that have minority percentages, within a one mile radius, that exceed the corresponding minority average for the county by more than five percent. Table 6.5 shows the same analysis for low-income populations. Excuse the blasphemy, but I do not see the emperor's new clothes. Perhaps, if EPA could break out the precise locations of the first row (exceeding county average by more than five percent), the numbers might be more meaningful. For instance, if all the cement plants and incinerators that fall into that category are located in one region of the country, such as the industrialized northeast, then we may have a problem. EPA did find that of all the TRI (Toxic Release Inventory) reporting plants, households with less than $15,000 annual income, and minority and urban populations were slightly over-represented in the communities where the plants are located. Otherwise, I do not see a general problem. However, we must be sensitive to this possibility as we complete our exposure assessments, though, frankly, I think a demographer will have to do that portion of the study. Nevertheless, many grassroots groups have sprung up in the past five years to combat so-called environmental injustice.

Perhaps minorities and poverty-level families are more endangered in the workplace than in the community at large. According to Swanson, migrant farm workers, mostly Hispanic, are suffering three hundred thousand cases of pesticide poisonings every year. Their wives are experiencing seven times the national average for miscarriages. These numbers are complicated by the fact that farm work is physically demanding, migrant workers receive spotty medical care, and, due to low incomes, attempt many deliveries at home. So, the environmental justice issue, in my thinking, is not a black and white issue, no pun intended. Many other justice issues, decent wages, availability of medical care, nutrition, and other poverty issues play heavily into the picture. At best, the situation is a shade of gray. I am not calling for scientists and engineers to bury their heads in the sand, rather to make a reasonable assessment of every situation to spot environmental injustice and report it, if found.

INNOCENT BYSTANDERS/UNSUSPECTING NEIGHBORS

In causing serious diseases and injuries to the body as discussed in Chapters 3 and 4, we see that the unchecked release of COCs into the environment has the potential to cause much grief in the community and to divide everyone into us and them. How many times have we read newspaper stories about how the siting of a new waste management facility or a new industrial plant that will be a particularly offensive neighbor is being protested by some members of the community. How many times have we laughed and joked and acted scandalized about the NIMBY factor, while being secretly glad the facility is not in our neighborhood?

Sometimes the NIMBY demonstrators are wrong, dead wrong. Sometimes they are right on, to borrow the expression. Mostly, they're not even there. The demonstrations mirror the perceived danger, which, in a vague way, reflect actual dangers, or at least potential for actual danger. The point I wish to make is that the demonstrations remind us, or should remind us, that real people stand to be exposed. Innocent bystanders and unsuspecting neighbors deserve our consideration and protection. The EPA is a people protection agency first,and an eco-systems protection agency second.

We cannot brush off the issue of innocent bystanders and unsuspecting neighbors by assigning a low probability to the danger, as if we were God! It'll never happen? That's what the chemical manufacturers were saying prior to Bhopal. It almost happened again in Institute, West Virginia, didn't it? Blaming the Bhopal accident on incompetent foreigners, by the way, is dodging the issue: things do happen, despite our best planning and precautionary efforts. It'll never happen, the railroads and chemical manufacturers were saying, before the Florida Panhandle and Miamisburg, Ohio accidents. Those two accidents were my personal wake up call. The Florida derailment affected close friends of mine, and the Miamisburg accident affected a sister and two brothers and their families. One sister and her husband stood on the levee of the Miami river, a few hundred yards from a burning railcar, taking photographs of it for nearly four hours before anyone warned them to move away. Apparently, the emergency responders thought they were a news crew and left them alone until a BLEVE scare forced an evacuation. When I heard their story, NIMBY suddenly took on a whole new meaning for me, and you will not find me cracking jokes or deceiving myself about it anymore. Not in my backyard, Buster!

I am not suggesting that no more industrial facilities, particularly chemical manufacturers, be sited. I do suggest that chemical exposure assessment teams examine every potential site on behalf of the innocent bystanders and unsuspecting neighbors. That is primarily the responsibility of the regulatory enforcement agencies, but more and more businesses are undertaking exposure assessments on their own. I see that as a sign of encouragement about future possibilities for a more just way of locating dangerous processes.

We will always need dangerous chemicals and dangerous processes, so long as we refuse to return to the cave. So, we might as well get used to learning how to better coexist with danger, since danger is to be our constant companion. We also need to learn to communicate better about danger and hazards and exposures.

THE GLOBE

Several global problems have been caused by, hitherto fore, uncontrolled releases of pollutants: global warming, acid precipitation, and stratospheric ozone depletion.

GLOBAL WARMING

Atmospheric carbon dioxide reduces the amount of solar radiation that reradiates from the earth's surface and escapes into space. Thus, carbon dioxide regulates atmospheric temperature and maintains the relatively warm temperature that we enjoy, called the *greenhouse effect.*

Only a portion of the solar visible radiation is absorbed in the stratosphere. Much of it is reflected by clouds and the earth's surface, but some is used to heat the earth's surface, which is reradiated in the infrared region, from 4 to 100 μm, at night. However, the 13 to 100 μm portion is absorbed by CO_2 and moisture in the atmosphere, while the 4 to 7 μm portion is also absorbed by atmospheric moisture, but the 7 to 13 μm portion escapes into space. CO_2 and moisture in the atmosphere, then, acts as a greenhouse, trapping moderate levels of heat to maintain the comfortable temperatures that enhance life on earth.

The CO_2 concentration of the atmosphere was constant, within a narrow range, before the Industrial Revolution. According to Valsaraj, the average concentration of CO_2 in 1850 was around 270 ppm. Each year, since 1958, the level of CO_2 in the atmosphere has increased from around 315 ppm in 1958 to around 360 ppm in the early 90s. Before human civilization began pouring tons of heat absorbing gases into the atmosphere, the thin blanket of *greenhouse gases* around the planet effectively trapped just enough solar heat to keep the surface air from plunging to extreme cold temperatures at night. The greenhouse gases are carbon dioxide, water vapor, methane, nitrous oxide, and chlorofluorocarbons. By adding more greenhouse gas to the atmosphere than needed to maintain an equilibrium, the blanket of gases has grown significantly thicker, leading to an increase in atmospheric temperature, and eventually, to increased ocean temperatures, and ultimately, to major climatic changes with serious implications for agricultural productivity. Hence, the term global warming, or greenhouse effect.

One consequence of global warming is the that the melting of polar ice caps will cause the sea level to rise and low countries, such as the Netherlands and Bangladesh, will be covered by ocean. This will not be sudden and catastrophic, but the eventual loss of their country will be grievous nonetheless for these people. In the U.S., southern Louisiana, and other parts of the Gulf Coast will be inundated, as well as localized areas along both seaboards. Where will these people go? Who will reimburse them for their property losses?

Since 1850, a mean rise in global temperature of approximately 2°F has occurred, but this could just be part of a natural fluctuation, which has been observed for tens of thousands of years in both short-term and long-term cycles. The difficulty is distinguishing between anthropogenic and natural sources of carbon dioxide emissions. For this reason, legislation regarding greenhouse gas control has been slow in coming. The potential consequences of global warming are so great that some scientists urge immediate international cooperation.

The Global Warming Potential (GWP) is a time-integrated measure of the warming effect that an instantaneous release of 1 kg of any greenhouse gas has, compared to 1 kg of CO_2. The contribution of CO_2 is great, because of the enormous amount pumped into the atmosphere, although its GWP is least. Contribution, C_{GW}, is calculated

$$C_{GW} = GWP_{GG} \times M_{GG} \tag{6.1}$$

where: M = mass emitted into atmosphere, kg, and
$_{GG}$ = greenhouse gas species.

Table 6.6 lists some characteristics of greenhouse gases. CO_2 is produced by the burning of fossil fuels and deforestation, some of which is natural, such as in forest fires caused by lightning. The source of methane is from decomposing materials, wetlands, rice paddies, and cattle. CFCs are entirely anthropogenic, and although they are the least global warming gas in number of tons emitted, they are the most potent with respect to GWP. Cleaning solvents, paint stripping solvents, refrigerants, foam-making chemicals, and aerosols (before 1976), release CFCs to the atmosphere. With recent laws and international treaties, some of these chemicals are no longer in production and use. However, notice that CFCs have the longest atmospheric life, enduring up to four hundred years, depending on the species.

Acid Precipitation

The use of tall stacks to disperse air pollutants from heavy industry, but mainly from electric power generating plants that burn fossil fuels, has resulted in the spreading of pollution over great distances from the source, even across international boundaries.

TABLE 6.6
Greenhouse Gases

	CO_2	CH_4	CFCs	No_x	Others
GWP		1	21	4500–7300	290
C_{GW} 1990	61%	15	12	4	8
$t_{1/2}$, yrs.		5–20	1	6.5–40	15

Tall stacks solve pollution by dilution, a method that is not allowed in water pollution laws and other waste management laws.

Sulfur emissions from the burning of high-sulfur fuels, especially coal, combine with moisture in the air to form sulfuric acid. Nitrogen oxides, produced by combustion, combine with atmospheric moisture to form nitric acid. Finally, atmospheric carbon dioxide forms carbonic acid, as shown in this equation, to make the pH of water $\approx$ 5.6, but the reaction is driven to the left when strong acids are present and the pH is lowered to 4.3, or less.

$$CO_2 + H_2O \Leftrightarrow H_2CO_3 \rightarrow H^+ + HCO_3^- \tag{6.2}$$

Acid rain is the result of the oxides of sulfur and nitrogen combining with atmospheric moisture to yield mineral acids, which may be transported over long distances from the source before being deposited by rain. The acidic pollution may also take the form of snow or fog or be precipitated in dry forms. Although the term acid rain is derived from atmospheric studies that were made in the region of Manchester, England during the last century, the more accurate term is *acid deposition* because the dry form of acid precipitation is just as damaging to the environment as the liquid form.

Acid rain became a noticeable problem started during the Industrial Revolution and has grown ever since. The severity of local effects has long been recognized in the acid smog common in heavily industrialized areas, though the widespread destructiveness of acid rain became evident only recently. The most studied affected area is northern Europe, where acid rain has eroded structures, damaged crops and forests, and ruined freshwater lakes. By 1984, almost half of the trees in the Black Forest of Germany had been damaged by acid rain. The northeastern United States and eastern Canada have been particularly affected by acid deposition, but damage has been detected in other areas of the world, as well.

Industrial emissions are blamed as the cause of acid rain, but industry has challenged such findings, calling for more studies. Studies by the U.S. government in the early 1980s, strongly implicate industry as the main source of acid rain in the eastern U.S. and Canada. In 1988, as part of the international Long-Range Transboundary Air Pollution Agreement, the United States and twenty-four other nations ratified a protocol freezing nitrogen oxide emissions at 1987 levels. The 1990 amendments to the U.S. Clean Air Act require holding sulfur dioxide emissions from power plants to ten million tons per year by January 1, 2000, about one-half the emissions of 1990.

OZONE DEPLETION

A thin layer of ozone molecules in the stratosphere, from twelve to thirty miles above the earth's surface, is responsible for absorbing much of the harmful ultraviolet energy bombarding earth from the sun. Ozone concentrations as high as ten ppm occur in this ozone layer. The presence of these molecules of ozone is due to diffusion from the troposphere where they are generated in abundance from several processes, including, mainly, lightning and photochemical pollution. Naturally occurring nitrogen compounds have kept the stratospheric ozone concentration at a fairly stable level for millions of years.

At about thirty miles out, solar radiation strikes the stratopause with an average wavelength of around 180 nanometers (nm). Photolysis of bimolecular oxygen in the stratosphere leads to the formation of ozone (trimolecular oxygen), some of which is transported slowly to the ozone layer, at around fifteen miles in altitude. Photodissociation of anthropogenic chlorofluorocarbons (CFCs) produces chlorine and chlorine monoxide molecules. Dissociation of N_2O produces NO, which with the Cl, and ClO, reacts with the ozone to produce bimolecular oxygen. The chlorine radicals remain available to participate in other reactions.

Finlayson–Pitts and Pitts report the Chapman cycle in the stratosphere as:

$$O_2 + hv(\lambda \lesssim 220\,nm) \rightarrow 2O$$
$$O + O_2 \xrightarrow{\quad M \quad} O_3$$
$$O + O_2 \rightarrow 2O_2$$
$$O_3 hv \rightarrow O + O_2 \qquad\qquad (6.3)$$

UV light down to 180 nm penetrates the stratosphere and is absorbed by O_2 and O_3, releasing heat, and causing an increase in the temperature of the stratosphere. In the troposphere, temperature decreases with altitude, but temperature increases with altitude in the stratosphere. The pressure in the stratosphere is also lower than in the troposphere. CFCs persist in the troposphere for many years, slowly diffusing across the tropopause into the stratosphere. The atmospheric life of the CFCs appears to be fifty to one hundred fifty years, depending on the species. All this time, the CFC molecules are undergoing a chain of photolysis reactions, including hundreds of ozone depleting reactions, in the stratosphere. CFCs do not themselves undergo photodissociation until they reach high in the stratosphere, above the ozone layer, where the UV wavelength distribution shifts to the shorter wavelengths, which the CFCs can absorb.

The chlorine reacts in a catalytic chain, which destroys O_3:

$$Cl + O_3 \rightarrow ClO + O_2$$
$$ClO + O \rightarrow Cl + O_2 \qquad\qquad (6.4)$$

The net of the reactions in Equation 6.4 is:

$$O_3 + O \rightarrow 2O_2 \qquad\qquad (6.5)$$

TABLE 6.7
Ozone Depleting Chemicals

	CFCs	Halons	CCl_4	CH_3CCl_3
ODP	1	10	1.1	0.1
$t_{1/2}$, yr.	13.9	10.1	6.7	0.8

In the atmosphere, chlorine chemicals rise until the molecules are broken down by sunlight. Chlorine molecules then react with and destroy ozone molecules, prompting the ban of CFC aerosols in the U.S. and elsewhere. Chemicals such as halocarbons and nitrous oxides may also attack the ozone layer.

In the late 1970s, scientists in Antarctica detected periodic losses of ozone high over the continent. This ozone hole develops in the Antarctic spring, continues for several months, then closes up again. The percentage of ozone over the Antarctic is declining. The average concentration has dropped by fifty percent in the Antarctic Spring, lasting from August to October. The hole is twice as large as the contiguous states of the U.S., and, in a sizable area over Antarctica, ozone has vanished completely. Though the worst depletion occurs above Antarctica during its Spring, the average year-round decline for all latitudes south of 60°S is five percent. A similar problem is developing over the Arctic, to a lesser degree. Between 30° and 60°N latitude, where most of the world's population lives, the stratospheric ozone decline now averages 4-5% annually, with 6-7% maximum depletion in the winter months. Table 6.7 shows the characteristics of somé ozone depleting gases. While the typical Ozone Depleting Potential (ODP) for CFCs is 1, they actually vary around unity, depending on the specific species of CFC. Halons are a trademark class of chemicals design to provide fire suppression, and are very good at it. Unfortunately, the Halons are the most potent ozone depleters. Their contribution is small, about five percent, due to the fact that they are only released when suppressing a fire. However, they are no longer in production in the U.S.

In 1987, thirty-six nations, including the U.S., banned the use of CFCs. In order to monitor ozone depletion, NASA launched the seven ton Upper Atmosphere Research Satellite in 1991, which orbits at an altitude of 372 miles, and measures ozone at different altitudes, providing the first complete picture of upper atmosphere chemistry.

CONCLUSION

Perhaps the best way to conclude this chapter, which is admittedly slanted towards my own opinion about the state of affairs we find ourselves in today regarding the environmental impact of human commerce, is to give you Georg Winter's six major reasons for being sensitive to environmental damage. This German industrialist was such a convinced convert that he hired consultants to go to his employees homes in order to identify and correct any sources of chemical exposures that might be found in them. He paid for the project as the chief executive officer of his family of

TABLE 6.8
Winter's Six Major Reasons for Environmentally Focused Economics

1. The human race cannot expect to survive [with] a life that's worth living [headed in the direction we are headed now].
2. [The environmentally focused economics are required to build] a public consensus with the business community [without which] there can be no free market economy.
3. Environmentalist business management [is needed to capitalize on] fast-growth market openings [and to prevent] being held liable for enormous sums of money for environmental damage, thus [protecting] the future of the company and all of the jobs dependent on it.
4. [To avoid jeopardizing jobs and careers of] boards of directors, management executives, heads of department, and other members of the staff.
5. [To capitalize on cost reduction — by pollution prevention as we call it in the U.S.]
6. [To avoid] conflict with our own consciences — and [to retain] self-respect, [without which] there can be no real sense of identification with one's job.

companies as a benefit to his employees! Radical! That is conversion! He is such a devout environmentalist (yes, he calls himself the E word) that he just might be responsible for every major environmental law in the European Union during the past decade. He is an industrialist, not a politician. Table 6.8 lists the six major reasons he gives for encouraging others to convert to his new way of thinking. The brackets are my attempt to turn some clumsy negative statements, as translated from German, into positive English statements having the same meaning. My apologies to Georg Winter, if I failed to retain his meanings.

QUESTIONS (ANSWERS BEGIN ON PAGE 401)

1. Why do current measurements of economic progress fail the environment?
2. What is our essential need to insure that human development on the planet does not damage the environment to the point of placing large numbers of people and/or large areas of the world in imminent danger?
3. What does entropy have to do with sustainable development?
4. What makes the environmental justice issue so controversial?

REFERENCES

Daugherty, Jack E. *Industrial Environmental Management: A Practical Handbook.* Rockville, MD: Government Institutes, 1996.
"Federal Actions to Address Environmental Justice in Minority Populations and Low-income Populations." *59 FR 7629.* Feb. 16, 1994.
60 FR 57787-8.
60 FR 66417.
61 FR 17479.
61 FR 33616.

Finlayson–Pitts, Barbara J. and James N. Pitts, Jr. *Atmospheric Chemistry: Fundamentals and Experimental Techniques.* New York: John Wiley & Sons, 1986.

Gore, Al. *Earth in the Balance.* New York: Houghton Mifflin Company, 1992.

Griffin, Roger D. *Principles of Air Quality Management.* Boca Raton, FL: CRC Press/Lewis Publishers, 1994.

Hawken, Paul. *The Ecology of Commerce: A Declaration of Sustainability.* New York: Harper Collins Publishers, 1993.

Lipfert, Frederick W. *Air Pollution and Community Health: A Critical Review and Data Sourcebook.* New York: Van Nostrand Reinhold, 1994.

Liu, Paul Ih-Fei. *Introduction to Energy and the Environment.* New York: Van Nostrand Reinhold, 1993.

Swanson, Sandra. "Can We Balance the Scales of Environmental Justice?" *Safety + Health.* October 1995, pp. 76-80.

Valsaraj, Kalliat T. *Elements of Environmental Engineering: Thermodynamics and Kinetics.* Boca Raton, FL: CRC Press/Lewis Publishers, 1995.

Winter, Georg. *Business and the Environment.* Hamburg, FRG: McGraw-Hill Book Company GmbH, 1988.

7 Principles of Exposure

The procedure for exposure assessment includes such activities as making field observations or taking measurements, conducting laboratory exposure studies, performing transport and fate modeling and conducting literature searches. These activities may be done alone or in combination with each other. The most appropriate approach is determined on a case-by-case basis. To assess the significance of an exposure the items in Table 7.1 must be determined and understood. Typically however detailed assessments are not possible due to the fact that some or all of this information is not available.

First, we must decide on the major contaminants, or COCs, for the site. The COCs are the subject of what EPA refers to as a source assessment, that is, a characterization of the sources of contamination. One needs to know, not only what the COCs are, but how they are being added to the environment. In other words, by what mechanisms, chemical and/or physical, do they enter the environment and get dispersed? Knowledge of the physical state of the COC is implied. These requirements hold true, whether the environment is a workplace, or the outdoors. Information may be obtained by chemical analysis, or by examination of documents, such as Material Safety Data Sheets, purchasing records, shipping/receiving records, inventory records, stock distribution records, and disposal records. Unfortunately, all the necessary information is not always readily available. Due to the frequent lack of information, it is common practice to choose surrogate COCs to represent the others.

Environmental pathways for exposure are determined by a pathways and fate analysis. Here, the workplace and outdoor environment cases begin to differ. In the first case, ventilation flow rates and patterns affect the distribution of COC through the air. Skin contact is the second most common workplace pathway. Rarely is ingestion a problem, much less a feasible pathway for any but minor exposure. In the outdoor environment, the general public can also be affected by the inhalation pathway, but weather and climate patterns govern transport. Skin contact is, usually, the least significant threat to the general public, but not always, as in the case of the dioxin release at ICMESA, in Italy; whereas ingestion of soil and plants and animals, which have bioaccumulated the COC, is a much greater factor. Ultimately, then, the pathways and fate analysis is a description of how the COC is transported from source to the receptor — the potentially exposed population — be that one person in a work station, or a community of exposed individuals. Any number of complicated, super-sophisticated models for pathways is available, but, oft-times, it is sufficient to use simple mathematical models to screen the travel of a COC by any of several pathways, to determine if any of them are insignificant relative to the others. You can save a lot of time, money, and headaches by using screening models

TABLE 7.1
Significant Items of an Exposure Assessment

1. Contaminants (or chemicals) of concern (COC).
2. The nature of the source of COCs.
3. Environmental pathways for exposure.
4. Health impacts on the receptor of exposure.

After Kay and Tate.

before delving into the more sophisticated ones. You can also have a better, general idea of the ball park your answer lies in, which gives you better quality control when you use the computer model.

Before we can assess health impacts of exposure, we must estimate the environmental concentration, which we call exposure. Dose, the amount received, which causes health effects or not, is some fraction of exposure. In the absence of hard evidence to the contrary, we assume dose = 100% exposure. Exposure can be calculated or measured. All sorts of monitoring devices and analytical schemes are available for measuring exposure, but the subject of this book is calculation of exposure. Where monitoring for a particular substance is impractical, or infeasible, determine whether an analogous or surrogate substance can be monitored. This is an effective monitoring strategy, especially where similarities exist in the physical and chemical properties of the substances. Other similarities that should be examined are the source and receiving environments, the pathway, quantity handled, and activities of the exposed persons. For whatever reason monitoring might be infeasible or inadvisable, we want to be able to estimate the potential for human contact with the COC at some location remote from the source. Even though our estimates are based on sound engineering and science, we want to examine the results for realism. Where such calculations exceed the PEL, STEL, IDLH, or other accepted exposure standard, recheck all assumptions and calculation methods for validity.

The analysis of health impacts on receptors is begun with a population analysis. What is the size of the population? Where is it located relative to the source? What habits of the potential human and biological receptors affect dose? In this phase of our analysis, we must be scientific enough to be objective, yet be sensitive to the needs and fears of humans, who may be affected by our decisions. Science is science, but sensitivity is especially prudent when explaining science to a worker who fears a specific exposure, or to the general public, which likewise may be especially fearful. Scientific wisdom is a thin veneer on human society, and our insensitive handling of risk communication can melt that thin layer of reason in an instant.

The integrated exposure assessment is the combination of all these preliminary steps, the calculation of exposure levels, and evaluation of any uncertainty. The exposure assessment is rarely an end in itself, though. Typically, it is weighed against the Big Picture, and a policy decision is made, for action or inaction. The process and information, other than the exposure assessment, needed for making this decision is the risk assessment. Table 7.2 shows the relationship of the exposure assessment to the risk assessment. The focus of this book is the exposure assessment and

TABLE 7.2
Risk Assessment, Including Exposure Assessment

1: Characterize chemical(s) of concern (COC).
    Identify chemicals present.
    Determine quantity, form, location on site.
    Select indicator(s).
2: Assess exposure.
    Identify exposure pathway(s).
    Characterize population(s)-at-risk.
    Estimate release probability/quantity.
    Evaluate environmental fate/transport.
    Model human intake/adsorption.
3: Assess toxicity.
    Identify type(s) of toxicity.
    Determine dose-response/threshold/NOAEL.
4: Characterize risk.
    Combine exposure and toxicity information.
    Develop various risk measures.

preceding tasks. Although the scientist or engineer may get involved in the risk assessment, it is primarily the business of policy makers. The purpose of this book is to be able to deliver the best exposure assessment feasible, given the time and circumstances of its preparation.

STRESSOR — SYSTEMS AT RISK — UNDESIRED EFFECTS

The focus of this book, and the science of exposure assessment, is that of stressors (chemicals in this case) acting (exposure) on systems at risk (humans or the environment) to cause undesired effects (health effects or environmental damage).

The stressors, as stated above, are chemicals of concern (COC). What exposures are we worried about? What chemicals are present? Are they contained? Stored properly? In a compatible storage environment? Obviously, the first task is to identify the COC(s). Is our concern general, or specific? That is, are we entering an unfamiliar place to search for any COC that may be present? Or, are we concerned only with one particular chemical in a workplace? Or perhaps, has one particular chemical been spilled by accident into the environment? Or leaked from a railcar? Just as obvious, we need to make haste in some cases, but proceed with deliberation in others. We cannot take long to deliberate about acute exposure, for instance, if we are concerned about an organophosphorus compound leaking from a railcar, on a derailed train, in the downtown area, at the noon hour! On the other hand, if we are concerned about chronic trichloroethylene exposure for operators of a vapor degreaser, we do not want to be too hasty.

At uncontrolled releases, such as Superfund sites or disaster sites, we may only be able to piece together a sketchy picture of the nature and extent of contamination or contaminant release. If more than one contaminant is involved, we may have to select a limited group of them as our COCs. Walker and Hagger suggest that we rank identified contaminants by the extent of their contamination and toxicity. This assessment must include major factors which influence transport through each medium: air, water, soil. Using this method we can limit our study to ten or fifteen COCs, maximum, for each medium.

About the stressor (COC), we will want to know about the occurrence of releases: quantity, concentration, reactivity, compatibility, environmental persistence, and transportability, in addition to its toxicity. In other words, we must determine the fate of the stressor in the environment, with the expressed hope that, from point of generation (cradle) to its ultimate fate (grave), the stressor has minimal impact on the environment and any humans who happen to be there. Absence of sufficient information about transport variables, or sufficient toxicological or epidemiological data, will preclude evaluation of some COCs. Only about twenty-one percent of commercial chemicals have a minimal amount of toxicity information available, and less than ten percent have available data on subchronic, chronic, and reproductive effects.

Consider the body as a system. It is composed of several systems actually, but, for the moment, consider it as one monolithic system, with a definite boundary, as discussed in the chapter on nomenclature. The process of a chemical stressor entering this system is described in two steps: contact and transfer. Contact is exposure, while actual entry, crossing the boundary, is transfer. Transfer is usually, but not always, effected by absorption, and leads to the availability of some amount of the chemical to react at biologically significant sites within the body. This availability at a site is called internal dose, which, if present in sufficient amount, causes an effect.

Contact, or exposure, is relatively simple to visualize. Consider a contaminated atmosphere (outdoor ambient air quality or indoor air quality) and visualize particulate matter (PM), or welding fumes, or an evaporating solvent, stressors in other words, in the breathing zone of some person (system). That person is exposed to the contaminant — or pollutant, chemical species, or stressor.

Or, another way to look at it, a certain amount of the chemical species in the air is available to cross the boundary into the person. That is exposure. A stressor has arrived at a point in space and time that is next to, but not inside (dose), the body.

Generically, we are interested in two types of human exposure: 1) exposure of a human population and 2) exposure of the most exposed individual (MEI). Human population exposure assessments have five factors of interest. (1) The average individual exposure is the expected number of cases, multiplied by a severity factor, divided by the number of people exposed. (2) Total exposure is the expected number of cases, multiplied by the severity factor. (3) Medium of concern is the avenue of most exposure: air, groundwater, surface water. (4) Another parameter often used is the percentage of total exposure attributed to the medium of concern. (5) Finally, the chemical of concern (COC) is the constituent in the source stream causing the most concern over all media. For the most exposed individual exposure assessment, four factors interest the assessor. The exposure to the MEI is the first factor. Medium of concern, percentage of total exposure attributed to the medium of concern, and the COC itself are the other three factors of interest.

Environmental exposures are not unlike those to humans: ecosystem population exposures and exposures to the most exposed ecosystem (MEE). When examining ecosystem populations, we are interested in the average ecosystem damage, or the average damage over points where damage occurs. Another factor of interest is the weighted volume/area of water/sediments/soils/air affected. The medium of concern is of interest, just as with humans, as is the percentage of total exposure attributed to the medium of concern. Finally, the COC is a factor we are interested in examining.

The purpose of calculating the exposure of the environment to oil or chemicals is to determine whether natural resources came into contact with the oil or chemical(s) from the incident. In disaster planning, the calculation of potential exposures allows responders to focus on those natural resources or public services that are most likely to be affected by the incident. Exposure to oil is broadly defined, by the National Oceanographic and Atmospheric Administration (NOAA), to be the direct physical or indirect exposure to oil or other substance. An example of an indirect exposure would be the injury to an organism as a result of the disruption of a food web. Documentation of exposure is required for those who would claim injury to natural resources, from a spill of oil, except for response-related injuries, and injuries from substantial threats of discharges. Evidence of exposure, by itself, does not lead to the conclusion that injury to a natural resource has occurred. For instance, an injury is not necessarily demonstrated by the presence of petroleum hydrocarbons in oysters. Selection of methods for demonstrating oil exposure depend on the type and volume of discharged oil, natural resources at risk, and nature of the receiving environment.

It takes more than analysis of the water column, or sediments, or soils to prove that organisms living in these systems have been exposed. It also takes more than analysis of the organisms to prove that contamination in the water column, or air, or sediment, or soil caused the linkage of the exposure to the discharge. Both approaches may be necessary to make such a linkage.

Transfer is more complex. A chemical crosses the boundary from outside to inside the body by one of two major processes: intake or uptake. Let's consider these terms separately for the moment.

Intake is the physical movement of a chemical through an opening in the outer boundary, such as the nose or mouth. Inhalation (or respiration) and eating or drinking (ingestion) are common activities of humans that involve intake. Air, water, and food have trace chemicals and contaminants, which are introduced to the body in this manner. The estimate of how much contaminant enters the body through intake depends on how much of the carrier medium (water, food, air) enters the body. Bulk flow is the mass transfer process in intake. The amount of chemical crossing the boundary per unit time is called the *chemical intake rate* and is the product of the exposure concentration times the ingestion, or inhalation, rate. These latter rates are the amount of carrier medium crossing the boundary per unit time. Inhalation rates are given in such units as cubic meters per hour (m^3/hr), liters per hour (l/hr), or cubic feet per hour (ft^3/hr). Ingestion rate is typically given in kilograms per day (kg/day), or pounds per day (lb./day), for food, and liters/day (l/day), or pints per day (pt/day), for water.

Ingestion and inhalation rates are not constant over time. We do not eat or drink constantly, nor is our breathing at the same rate all the time, even though we do breathe constantly (almost — not counting sleep apnea). Since the ingestion of food and water is intermittent, it is often expressed as the amount of medium per event, such as kg/meal, rather than as a rate. To convert from such terms to a rate expression, we multiply it by the frequency of contact:

$$I = M \times E \qquad (7.1)$$

where: I = intake, kg/time
         M = weight of medium/event, kg
         E = event frequency, time^{-1}

Uptake is the process whereby a chemical crosses the outer boundary of our bodies by absorption. As with intake, the chemical may be contained in a carrier medium, but the chemical is absorbed at a totally different rate than the carrier is. An example of direct uptake is the dermal absorption of a chemical species. Uptake through the lung, gastrointestinal tract, or other internal barriers following inhalation or ingestion is indirect. Chemical uptake rate is the amount of the chemical absorbed through a period of time. Transfer occurs by diffusion in this process, which means that uptake depends on the concentration gradient across the boundary, and permeability of the barrier:

$$U = f\left(C, k_p, A\right) \qquad (7.2)$$

where: U = uptake rate,
         C = exposure concentration,
         k_p = permeability coefficient, and
         A = exposed surface area.

Uptake rate is generally expressed as a flux (rate/unit area).

Contact, intake, and uptake are used to derive equations for exposure and dose for all routes of exposure. Typically, it is not possible, except under experimental conditions, to quantify dose, or even to evaluate the dose-response relationship. That is why most dose-response curves are really exposure-response curves. Keep in mind that unless exposure results in a significant dose, then, in all likelihood, no health effect can be pinpointed as caused by that exposure. Table 7.3 is the hierarchy of dose-response data acceptability.

Once a dose is absorbed in the tissue of the target organ, an effect is produced. The science of studying these effects, called toxicology, is outside the scope of this book and of an exposure assessment, too. However, a brief word about some toxicological concepts, before moving on. Fundamentally, a relationship exists between the dose of a stressor and the effect, or response, produced in a biological system. This is called the dose-response relationship, and, graphically, it is the dose-response curve. One should be aware that many so-called dose-response curves are, in reality,

TABLE 7.3
Dose-Response Hierarcy of Data

1. Human epidemiological data.
2. Data from animals that respond most like humans.
3. Biologically acceptable data set from long-term animal studies showing the greatest sensitivity.
4. Data from the exposure route of concern; if data from other exposure routes are used the considerations used to make route-to-route extrapolations must be described.
5. Count the number of animals having one or more of the significant tumors (pooling) when there are multiple tumor sites or types.
6. Combine benign tumors with malignant tumors for risk estimates.

exposure-response curves. This distinction does not always matter, but be alert, in case it does matter in your situation. For instance, if a dose-response curve was prepared using the airborne concentration levels determined by area monitoring, then the true dose-response curve may look a lot different from the supposed one.

Let's examine the assumptions that are made in order to establish a dose-response relationship. It is assumed that the response is in fact due to the chemical administered and not to some other cause. Of course, the relationship of dose to response is itself an assumption, for which a mechanism must be verified. The chemical either interacts with a molecular or receptor site, or the response is related to the concentration of the chemical at the reactive site, or the concentration at the site is related to the dose. Another assumption is that a method exists of quantifying toxicity, with a precise means of expressing it. For carcinogens, a linear multistage model is often used to extrapolate responses to low doses, consistent with some proposed mechanisms of carcinogenesis.

For mutagens, a dose-response assessment is only valid with data on germinal mutations induced in intact animals. Morphological-specific locus and biochemical-specific locus assays can provide data on the frequencies of recessive mutations. Data on heritable chromosome damage can be extracted from the heritable translocation test. For tests involving germ cells of whole mammals, few dose points will be available, so a linear extraction is used.

System toxicants and noncarcinogenic health effects require another approach to dose-response assessment. A reference dose (RfD) is calculated, which represents the dose below which we expect no significant risk of health effects. RfD is related to the ADI, but removes the elements of risk management. Literature is researched to find 1) the critical toxic effect, which is the adverse effect that first appears as dose is increased and 2) the highest dose at which the effect does not occur. This latter dose is the No-Observed-Adverse-Effect-Level (NOAEL). NOAEL is divided by an uncertainty factor ranging from 10 to 10,000, depending on several factors.

Data for mixtures are used, if available, if the mixture itself is being evaluated. Where no mixture data are available, but data on components are available, inferences are made from the available data. When threshold components are toxicologically similar, divide each estimated intake level by its RfD, and take the sum of components to compute a hazard index. If the index is less than one, no significant risk is expected from the mixture. If the hazard index is greater than one, a significant risk is probable. Sometimes, it is necessary to make inferences about effects caused by similar compounds.

Where experience is available with workplace responses, this information is studied by toxicologists and epidemiologists. To supplement that data, or to provide data where no human experience is available, tests are performed on mice and rats under precise, controlled circumstances. Using statistics, these animal tests give scientists the LD_{50}, the dose expected to be lethal in 50% of the exposed population, and the LC_{50}, the atmospheric concentration expected to be lethal to 50% of the exposed population. Autopsy of the test animals also identifies the target organs of the contaminant species. The best practice is to suspect that the same effects will occur in humans, if they are exposed to a sufficiently high atmospheric concentration for a period of time, or to an oral dose proportional to body weight. While not precisely true, this is the safest assumption. Where industrial and accidental exposures provide human data, the response curve is updated to reflect the better information.

PRIMARY AND SECONDARY STRESSORS

Stressors either work directly or indirectly on the system. Those that work directly are called *primary stressors*. Those that work indirectly are called *secondary stressors*. Exposure assessments may address single or multiple chemical, physical, or biological stressors. A *stressor* is simply any chemical, physical, or biological entity that can induce an adverse response. In this book, we are interested only in chemical stressors and the physical stresses of chemicals (such as corrosiveness due to pH or fire hazard due to a low flash point). Some things not covered here are physical stressors, such as ionizing radiation or noise or extreme temperature, and biological stressors, such as bacteria and viruses. Under certain circumstances, a stressor could produce neutral, or even beneficial, effects. We will not concern ourselves with that possibility, but the student of exposure studies should be aware of it, nevertheless.

CHARACTERISTICS OF CHEMICAL STRESSORS

Since chemical stressors are our focus, let us now turn our attention to them. We will discuss types of chemical stressors, intensity of exposure, duration of exposure, frequency of exposure, timing of exposure, and the scale of exposure.

TYPE OF CHEMICAL STRESSOR

Many chemical stressors have natural counterparts or multiple anthropogenic sources. Most metals, for example, occur naturally. So do many poisons. Another category of stressors, mentioned above, is differentiation between primary and secondary stressors. Creation of secondary stressors alters the conclusions otherwise drawn about an exposure assessment. Secondary stressors may be formed by biotic or abiotic transformation processes, and may be of greater or lesser concern than the primary stressor. The duty of the exposure assessor is to ensure that all potential secondary stressors are identified and evaluated as well as the primary stressor.

For chemicals, the evaluation of secondary stressors typically focuses on metabolites, or degradation products, or chemicals formed through abiotic reactions. Microbial action, for instance, increases the bioaccumulation of mercury by transforming it from inorganic to organic form. In another case, azo dyes, which are nontoxic due to their large molecular size, are hydrolyzed in anaerobic environments, forming more toxic, water soluble species.

Secondary stressors may also be formed through ecosystem processes. Nutrients entering an estuary, for example, decrease the dissolved oxygen content, because primary production and subsequent decomposition increases.

As for the primary chemical stressors themselves, some participate in aquatic chemistry, some in atmospheric chemistry, others are relevant in the biosphere or the geosphere.

In aquatic chemistry, we deal with COCs in groundwater, lakes, rivers, estuaries, and the ocean. The study of water movement in the geosphere is *hydrology*. The branch of hydrology that deals with freshwater chemistry, physics, and biology is called *limnology*. The saltwater counterpart is *oceanography*. The water soluble compounds are the chief interest, since they can be transported great distances in solution. Water insoluble compounds may be a hazard locally to the point of origin, however.

Besides being a good solvent, water has a high dielectric constant, meaning that it is not only a good solvent for ions, but promotes ionization in solution. The high surface tension of water controls its behavior in physiological processes and on surfaces. Being transparent to visible light and the longer wavelengths of ultraviolet light allows energy to be available for biological activity at great depths of water. Large bodies of water retain a tremendous amount of energy, as heat, because water has the highest heat of vaporization of any substance. This characteristic regulates the transfer of heat between the ocean and the atmosphere. Its large heat capacity allows for the stabilization of temperatures by geographical region.

We've talked a good bit about atmospheric chemistry already, but Table 7.4 shows the atmospheric components.

The geosphere is the land mass of the Earth's surface. Soil is a variety of material, typically ninety-five percent minerals and five percent organic matter.

The biosphere is the portion of the environment that includes all organisms and biological materials. Unlike the geosphere, hydrosphere, and atmosphere, the biosphere is not specific in location, but is the living part of the geosphere and hydrosphere. The most important chemistry in the biosphere is photosynthesis.

$$CO_2 + H_2O \xrightarrow{\text{hv}} \{CH_2O\} + O_2(g) \qquad (7.3)$$

Plants are algae are *autotrophic*, that is, they use solar or chemical energy to fix elements from simple, abiotic materials into complex molecules that build living things. Biodegradation, on the other hand, breaks down living material. In the presence of aerobic respiration:

TABLE 7.4
Atmosphere Components

Nitrogen, 78.1%
Oxygen, 21.0%
Argon, 0.9%
Carbon dioxide, 0.03%
Ammonia, carbon monoxide, helium, hydrogen, krypton,
methane, neon, nitrogen dioxide, nitrous oxide,
ozone, sulfur dioxide, xenon, 0.002% combined
plus
Water vapor, 1–3%

All components listed in percent volume.

$$\{CH_2O\} + O_2(g) \rightarrow CO_2 + H_2O \tag{7.4}$$

Anaerobic respiration yields:

$$2\{CH_2O\} \rightarrow CO_2(g) + CH_4(g) \tag{7.5}$$

Notice that either reaction returns carbon dioxide to the atmosphere, plus, the latter reaction is a major natural source of atmospheric methane. Undegraded remains provide the organic materials for soil and sediment. Fossil fuel amounts to carbon that was once photosynthetically fixed in plant matter.

In addition to categorizing stressors with respect to their sphere of influence, we can also type them as ignitables, corrosives, reactives, and toxics. These categories are discussed in Chapter 3 with respect to their impact on human health. Finally, we can categorize stressors in more traditional chemical terms, as shown in Table 7.5. In order to tie these compounds to specific toxicological responses, a search must be made of toxicological databases, such as the EPA IRIS database, or literature published by NIOSH or professional journals and professional societies.

Generic toxicological responses are another way to categorize stressors: acute, sub-acute, sub-chronic, or chronic. Also, they can be categorized according to the target organs they affect. Obviously, many ways exist by which we can categorize stressors. The safest approach to take is to search Material Safety Data Sheets, for all their limitations, that match purchasing, receiving, storage, or shipping records. More generically, chemical stressors can be industrial effluents and emissions, waste streams from manufacturing processes, or pesticides.

Neither in the workplace nor in the environment are receptors of chemical stress exposed to only a single chemical, except in rare instances. Where a single stressor is the case, the exposure assessment is as straightforward as possible. However, where more than one stressor is present, even physical or biological stressors in addition to the chemical stressor, the assessment is complicated by the fact that the extrapolation of laboratory studies involving single chemicals, and lacking matching

TABLE 7.5
Chemical Stressor Categories

Elements/forms of elements	Inorganic compounds
Halogens	Acid halides
Heavy metals	Asbestos
Ozone	Carbon monoxide
White phosphorus	Cyanide
	Halogen oxides
Organometallics	Interhalogens
Carbonyls	Nitrogen oxides
Organolead	Phosphorus compounds
Organotin	Silicon compounds
	Sulfur compounds
Alkanes	Alkenes
Alkynes	Aromatics
Polycyclic aromatic hydrocarbons (PAH)	
Epoxides	Alcohols
Phenols	Aldehydes and ketones
Carboxylic acids	Ethers
Acid anhydrides	Esters
Aliphatic amines	Carbocyclic aromatic amines
Pyridine	Nitriles
Nitro compounds	Nitro compounds
Nitrosamines	Isocyanates
Organonitrogen pesticides	Organohalides
Polychlorinated dibenzodioxins	Chlorinated phenols
Organosulfur compounds	Organophosphorus compounds

physical and biological stresses, does not match the real world very closely. Shugart, in De Serres and Bloom, suggests that a family of dose-response curves, rather than a single curve, exists for such complicated situations, with each curve representing a different combination, or permutations, of interacting chemical and other environmental stresses.

Field investigators and enforcement officers are more interested in stress indicators than exposure assessors are, but one should be aware of indicators, as they generally represent environmental or health damage. For instance, areas of dead grass or trees and shrubbery are the obvious environmental indicators. A fish kill is another that we read about in the newspapers. Weinstein lists changes in phytoplankton, cessation of fish reproduction, disappearance of benthic crustaceans, appearance of filamentous algae, decreases in primary production, decreases in rates of decomposition, and changes in nutrient concentration as others to investigate closely. Beagle dogs are frequently used for inhalation studies as their lung capacity and physiology is similar to ours. Rabbits are used for skin and eye irritation studies as their skin and eyes are similar to ours. Rats and mice are used for many tests because they are cheap and more expendable and are reproduced under careful genetic scrutiny in order to minimize the possibility that inherited traits will predispose the individual test animal to present the response under observation. Someone once suggested the use of attorneys for toxicological testing as toxicologists and their technicians often grow attached to the rats. Seriously, McBee and Lochmiller advocate the use of wild animals for toxicity testing of COCs that are potential environmental stressors, even for human health comparison, because they live in a naturally stressful habitat, where temperature extremes are seasonal stressors, competition in the food web is behavior stress, and nutritional deficiencies, or physiological stress, is a way of life. Nevertheless, a stressor poses absolutely no risk to an environment unless an exposure also exists, but the premise of this book is that exposure is a potential and we want to assess that possibility.

INTENSITY

Intensity is the concentration of exposure or the dose received. The more intense the stress, the more likely damage is to be done, eventually, to the environment or to the health of a human receptor. Intensity is either measured by sampling and analysis or calculated by modeling.

Intensity has another aspect, which is not normally examined, and, admittedly, may be more properly an aspect of characteristic. However, for the purpose of discussion, let us consider the aspect of energy state of the COC species. For instance, take heavy metals. Highly energized ions of lead, mercury, chromium, and others cause varying responses from madness to cancer. Humanity removed these metals as low-grade ore from the ground and expended vast amounts of energy to raise them to the ionic, least oxidized, but most highly energized state for use in any of many applications.

Typically, in the natural state, though, metals are in their most oxidized, least energized state. At the lowest energy level possible, most metals are not very dangerous. Do not take that too literally because lead ore and cinnabar, the ore of mercury, are notable exceptions. Humanity mines metals and elevates the energy level to a reduced state, wherein the metals are generally rendered more hazardous to humans, and often to the environment, too. That is not to say, however, that metals in the natural state have no toxicity at all. The Romans were forced to use slave labor to mine cinnabar in Almaden, Spain because this mercury-bearing ore caused madness within a few years. In fact, a citizen who volunteered to work as a cinnabar miner for a two year period was rewarded with an exemption from service in the Roman legions, an equally dangerous occupation.

DURATION

Duration, frequency, and timing are the temporal characteristics of COCs. Duration is the length of time for which the exposure exists. Sometimes the duration may be several intervals in sequence, such as when an employee goes to his or her shift, eight or ten hours out of twenty-four for four, five or six days in a row, then one or two days without exposure. This example is not particularly complicated. Try tracking a swing-shift worker's exposure duration when his or her job assignments rotates, almost, but not quite, at random.

Besides the duration compared to short-term time intervals, such as a work shift or day, which is typically the focus of occupational exposures, long-term intervals are typical of environmental exposures. Here, we are interested in lifetimes. Whether humans or wildlife, how much COC is received over the lifetime is a necessary quantity to know or assume in order to predict the consequences. Remember, we trade hard data for conservative assumptions.

FREQUENCY

Another temporal aspect of stress is frequency. A one-time dose that sends you to the hospital is one thing, but many one-time doses either go unnoticed, or present mild or confusing symptoms of exposure, if at all. For the great majority of COCs, such an exposure leads to no further consequences. A few, as discussed in Chapter 3 and 4, may cause edema in the lungs at the most unexpected times, a dangerous, life-threatening situation. If we accidentally ingested a single thimble of gasoline, we would expect no lasting effects, once we got over the immediate effects. If we ingest a thimbleful, once a month, who knows for sure, but maybe no lasting damage. Once a week? Let's send this fellow to the doctor, for surely he or she is suffering internally, even if we do not observe signs and symptoms, and he or she does not complain. Daily? We ought to be observing some signs and symptoms, as we speak. Realistically, who would ingest gasoline that often by accident? If this really occurred and you became aware of it, send the person for possible psychiatric counseling. A more realistic situation is the swallowing a minute amounts of metalworking fluids that is present in low concentrations in almost any machine shop. Some is inhaled, but some is swallowed as it sticks to mucous in the nasal passage and throat. What about a child who ingests an ounce or so of contaminated soil daily for a number of years? Even as adults, we take in a certain amount of soil from dust in the air, unwashed vegetables and fruits, handling dirty tools, and, yes, even handling our dirty little children who have been playing hard all day out in lawn and garden. Some of you fathers who have never had to bathe a filthy toddler may not appreciate this latter circumstance, but I am positive that most mothers, excepting those whose children are totally sheltered from lawn and garden, will know exactly what I mean.

TIMING

Does the release and transfer of COC occur when a dose can be received? By definition, it must. Otherwise, no exposure occurs.

SCALE

How far from the source does the stressful contamination affect the environment? The spatial aspect is what is meant by scale. Damage is proximate (near the source) or remote.

THE SYSTEM

One reason the system, a human body, an animal or plant, or an ecosystem, is of interest to us in exposure assessment is the response. While study of response is the

TABLE 7.6
Classification of Toxicity

LD$_{50}$ wt/kg (1 oral dose)	LC$_{50}$ ppm (4-hr inhaled)	Toxicity Description	Toxicity Rating
1 mg or less	<10	Extremely	1
1–50 mg	10–100	Highly	2
50–500 mg	100–1000	Moderately	3
0.5–5 g	1000–10,000	Slightly	4
5–15 g	10,000–100,000	Practically nontoxic	5
15 g or more	>100,000	Relatively harmless	6

subject of another phase of an overall risk assessment project, the person conducting the exposure assessment needs some understanding of the system, too. For this perspective, we are interested in two primary terms: toxicity and hazard. These terms are not synonymous.

Toxicity is the chemical's ability to cause damage to biological tissue. Table 7.6 shows the classification of toxicity based on lethal doses and airborne concentrations.Hazard is the likelihood that the chemical will cause damage under given conditions. Hazard is dependent on toxicity, for one thing, but also on how the chemical is absorbed, metabolized, and excreted from the body. Hazard also depends on the rate of reaction the chemical may have with tissue or other chemicals in the body and on the warning properties of the chemical: smell, sensation. Hazard also depends on the potential of the chemical to catch on fire or cause an explosion. The manner in which the chemical will be encountered will affect the hazard.

ROUTES OF EXPOSURE

After determining what can be known, in the time available, about the stressors, the next step is to evaluate each reasonable route of exposure or pathway to the receptor. The chief routes of exposure are: inhalation, ingestion, and absorption through the skin. A less common route of exposure is by parental administration (subcutaneous, intramuscular, intravenous or intratesticular injection) which is rarely, if ever, encountered in workplace or environmental exposures. The route of exposure is important, because it affects the relative toxicity of a chemical. Once the chemical enters the blood stream, by any route of exposure, it is distributed throughout the body, but especially to its target organ(s). On the other hand, if a chemical is readily detoxified in the liver, it is less toxic if administered orally, where it can reach the liver more rapidly, than if inhaled. Another factor is the absorbability of the chemical. When a certain dose of a chemical causes death, regardless of the route of exposure, and within the same period of time, then we must assume that the chemical is easily and rapidly absorbed. What we do not want to get into is spending an inordinate amount of time studying a route of exposure that is insignificant, relative to some other route. Therefore, it is prudent to do some simple screening calculations, or take some measurements, as suggested by Walker and Hagger.

Stressors can be transported along many pathways. Chemical stressors can be transported by air current, in surface water, over the soil surface, including through the soil water zone, and through groundwater. Chemical stressors can also be transported through the food web.

When estimating individual daily doses (IDD), exposures from different pathways are to be added, if the effect is the same for each route of exposure, and a reasonable expectation exists that the same individuals are exposed. When carcinogens are involved, add the exposure across pathways, if the constituent is carcinogenic, through each and every route considered, such as: oral and inhalation; ingestion and dermal. For most noncarcinogens, it is not usually appropriate to add oral and inhalation exposures, unless literature indicates that the oral reference dose and the inhalation reference concentration are based on the same effect. Generally, oral and inhalation exposures have the same effect and can be combined. Dermal and ingestion exposures also typically have the same effect and are combinable. Before combining exposures though, consider whether the same individual is likely to be exposed through each of the routes.

INHALATION

When the carrier medium is air, the COC, as a gas or aerosol, enters the body, most often by route of inhalation. However, if the aerosol is heavy, skin absorption is another possibility. Ingestion of aerosols is minor. Inhalation of particulate matter (PM or dust), generated from activities on a contaminated site, is a route of exposure we should concern ourselves with for those workers on site and for neighboring residents, if wind conditions justify concern.

The amount of gas, aerosol, or particulate matter that is inhaled depends on many circumstances, such as: the airborne concentration of the COC, the amount of time the receptor spends in the contaminated atmosphere, the breathing rate of the receptor, the nature of the receptor's activities, the physical and chemical properties of the COC, the temperature changes, seasonal changes, effectiveness of engineering controls, effectiveness of personal protective equipment, and effectiveness of preventive hygienic measures.

The important factor in determining how much inhaled material reaches the alveoli is the water solubility of the stressor. Species that are highly soluble in water readily dissolve in the mucous membranes of the nose and upper respiratory tract (URT). These sites will be irritated, but the deep lungs are spared, if the airborne exposure was a fairly low concentration. The nose and URT literally become a wet scrubber for the soluble species. For higher concentrations in air, some of the gas or vapor crossing the boundary of the epithelial tissues may escape the scrubbing process and make it to the alveoli in sufficient quantity to cause irritation and pulmonary edema. Inorganic acid gases are almost completely soluble in water. Ammonia, an alkali, also has a high solubility.

Many insoluble gases and vapors are not removed in the nose and URT and end up in the alveoli. The human respiratory system is diagrammed in Figure 7.1. Finally, some species irritate the lung independently of solubility. Among these are ozone and chlorine.

TABLE 7.7
Typical Assumption for Inhalation
of Contaminated PM by Adults

PM concentration, on site	5 mg/m³
Breath volume	1 m³/hr
Duration of exposure	8 hr
Body weight	70 Kg
Absorption into body	100%

TABLE 7.8
Typical Assumptions for Inhalation
of VOCs by Adults

Breath volume	8 m³/day
Body weight	70 Kg
Absorption through lungs	100%

Gases such as carbon monoxide are absorbed into the blood stream at the alveoli and cause systemic intoxication.

Whether aerosols of solid particles reach the alveoli depends on particle size. Larger sizes (ten microns and greater) are removed in the nasal passages by nose hair and mucous. Respirable particles are ten microns or less in size. From one to five microns, PM is removed by sedimentation in the bronchioli. Particles less than 1 micron diffuse within the alveoli. PM may deposit on the ciliary epithelium down to the terminal bronchioli.

Tables 7.7 and 7.8 give typical assumptions used for inhalation of PM and volatile organic compounds (VOCs) by adults respectively.

The worst case assumption for dust or PM, contaminated or not, is 5 mg/m³, which is the allowable respirable portion of the OSHA nuisance dust PEL.

Unless we can separate respirable from nonrespirable dust, we must assume all dust is respirable.

DERMAL ABSORPTION

Actually, the skin, with its underlying layer of lipid (fat), forms an effective barrier for many chemicals. However, some chemicals, such as acids, alkalis, and organic solvents, react with the skin to deactivate the natural barrier, and the resulting irritation may worsen to the point of inflammation and bleeding of underlying tissue. Fever or moisture on the skin can aid the absorption of chemicals. A few chemicals penetrate the skin and cause sensitization (allergic reactions in subsequent exposures). A chemical may penetrate the skin and enter the blood stream where it can have systemic effects. Fortunately, very few chemicals are absorbed through the skin in amounts that are hazardous.

TABLE 7.9
Typical Assumptions for
Ingestion of Fish by an Adult

Dilution factor	50%
Bioconcentration	Steady state
Fish consumption	6.5 g/day
Body weight	70 Kg
Absorption into body	100%

TABLE 7.10
Typical Assumptions for
Ingestion of Soil by a Child

Consumption	1.0 g/day
Body weight	20 Kg
Absorption into body	100%

The actual entry, when it occurs, is through epidermal cells, sweat glands, sebaceous glands, or hair follicles. The main pathway into the body is through epidermal cells and the overlying stratum corneum into the blood. The stratum corneum is the majority of the surface area of the skin and is critical to overall permeability. Where the skin is abraded or inflamed, absorption is quicker than through intact skin. Severe dermatitis cases, with bleeding and inflammation, are more susceptible to less toxic chemicals, due to the open areas in the stratum corneum.

Evidence supporting the use the ratio of skin surface area to extrapolate from a test species to humans is not strong. Often, the dermal exposure is over a small area anyway.

INGESTION

For workers, ingestion occurs when eating or smoking with contaminated hands. Inhaled material is ingested to a minor degree with little damage, unless the material is highly toxic. The inhaled material that gets ingested is moved to the throat by ciliary action, where it is swallowed. Ingested material is absorbed into the blood in the intestinal tract. Tables 7.9 and 7.10 list typical worst case assumptions for adults ingesting a contaminant in fish, and a child ingesting soil, respectively.

TYPES OF EXPOSURE

Generally, four different types of exposure are: acute, subacute, subchronic, and chronic, as explained in Table 7.11. One must be careful not to confuse acute and

TABLE 7.11
Types of Exposure

Exposure Type	Exposure Period	Exposure Frequency
Acute	≤ 24 hours	Single or repeated
Subacute	≤ 1 month	Repeated
Subchronic	1–3 months	Repeated
Chronic	> 3 months	Repeated

chronic exposures with acute and chronic health effects, a common mistake. We discussed health effects earlier, but, in general, acute exposures can have both immediate and delayed-onset responses. Chronic exposures also can cause both immediate reactions as well as long-term health effects.

The duration of exposure must account for both the expected operational life of the source and the time period of residence of the receptor. For many exposure pathways, the exposures may continue long after the source ceases generating the exposure. This may be due to continued cycling of contamination in and between biota, soils, and sediments. Exposure duration generally represents less-than-lifetime. Sometimes, however, it is reasonable to expect an individual to be exposed for a lifetime. Estimate duration via a given pathway whenever possible. Census data and limited site-specific surveys are used to establish likely duration of individual exposures.

Frequency of dosing significantly influences the magnitude of a toxic effect. A large single dose of an acute toxin typically has more than three times the effect of one-third the amount administered three different times. The same total dose administered in ten to fifteen fractional doses may have no effect at all. The dosing pattern is important because some of the chemical may be detoxified and/or excreted between successive doses. Also, the route of exposure, may be denied through healing, such as a lesion, or physical change or barrier occurs.

A chronic effect occurs, if one or more of three things happen. If a toxic substance accumulates because the dose absorbed is greater than the body's ability to transform or eliminate the chemical, then a toxic effect occurs. A chronic effect may occur, if the chemical produces irreversible adverse effects. Finally, if a chemical is administered in such a way that damaged organs and systems have inadequate time to repair and recover, the effect is chronic. Table 7.12 summarizes. Effects of concern are caused by the theoretical twenty-four hour exposure of an individual over a lifetime (70 years). Air concentrations in ambient air for known, or suspected, pollutants are linked to population effects (either acute or chronic), through unit cancer risk factors, or modified Threshold Limit Values (TLVs). The TLV is modified by putting it on a TWA basis, consistent with the concentration calculated by the dispersion model.

For chronic exposure, the average daily lifetime exposure (E_{ADL}) in mg/kg-day is

$$E_{ADL} = \frac{D_T}{W_B \times L} \qquad (7.6)$$

TABLE 7.12
Chronic Effects

A toxic substance accumulates faster than body can transform or eliminate it
A toxic substance produces irreversible effects
A toxic substance is dosed faster than injured system can repair or recover

where: D_T = total dose, mg
W_B = body weight, kg
L   = lifetime, 70 years expressed in days

Total dose is

$$D_T = C \times R \times T_E \times F_A \qquad (7.7)$$

where: C   = contaminant concentration (exposure), mg/kg or mg/m^3
R   = contact rate, mg/day, m^3/day or m^3/m^2
T_E = exposure duration
F_A = absorption fraction

The exposure or contaminant concentration (C) in the environment is estimated or measured in the appropriate medium (air, water, soil), which can by any means come into contact with the human body. or some aspect of the environment (biota or structural), which requires protection from over exposure. The contact rate (R) is the rate at which the medium contacts the body, biota, or structure. In the case of the human body, and most animals, we are concerned with media contact through inhalation, ingestion, or dermal contact. The units are typically mass/time for ingestion, volume/time for inhalation, and volume/surface area for dermal contact. The length of time the contaminated medium has contact with a person is called the exposure duration. The absorption factor is the portion of the COC that actually contacts and enters the body, by crossing one of three exchange membranes: skin, alveolar membrane, or gastrointestinal tract.

Bodily response to toxic insult varies greatly for several reasons. Some chemicals have immediate effects, while others have delayed effects, or, at least delayed onset of symptoms of the effect. Some carcinogens have latency periods as long as forty years. Several chemicals having adverse ocular effects may not cause overt symptoms until hours after exposure.

Health effects of chemicals may be reversible or irreversible. Reversibility depends on two things: site of action and magnitude of dose. The liver can regenerate itself to a point, so, when the liver is the site of action, the effect is usually reversible. Site of action is a factor that itself varies widely.

Local and systemic effects may also vary. Local effects are lesions produced at the site of first contact between the chemical and the organ damaged. Systemic effects involve the absorption of the chemical from the site of entry to a distant site, where the toxic response is manifested.

To prepare a list of potential human health effects, consider the organ(s) of concern, the target organ(s) for the COC. What is the adverse change to the organ that constitutes an injury? What are the potential degree and spatial/temporal extent of the injury? Is there evidence indicating injury? What is the mechanism by which injury has occurred? Is there evidence indicating exposure? What is the linkage between source and receptor? Is there a natural recovery period? What first aid and personal protection actions are feasible in the short-term? These are some of the questions an exposure assessor must ask.

SYSTEM VS. BOUNDARIES

The boundary and the receptor system are often confused. The pertinent question is, when does the contaminant actually enter the body? Where is the boundary, where on one side, the contaminant is still outside the body for the purpose of potentially causing health effects, and, on the other side, is definitely located in the body?

For the inhalation pathway, one might be tempted to envision the nostrils as the boundary. Surely, contact with nasal membranes and other membranes occur all the way down to the lungs, but this is dermal contact, not inhalation. The boundary for inhalation is the alveolar membrane. When a contaminant crosses the alveolar membrane, it is inhaled for the purposes of determining its health effects by inhalation.

Suppose the contaminant is received by ingestion, wouldn't the lips be the boundary? After all, the COC may likely start absorbing through tissues in the mouth. Again, this is dermal absorption. The boundary for ingestion is any membrane in the gastrointestinal tract: stomach, small intestine, colon.

The skin itself is the boundary for dermal absorption.

QUESTIONS (ANSWERS BEGIN ON PAGE 401)

1. What is the intake in kg/day if 0.25 kg fish is consumed per week?
2. What affects the rate at which chemicals are absorbed into the body in the digestion system?
3. Why are benign tumors counted with malignant tumors in risk assessments?
4. Most dose-response curves are in reality _____.
5. List six characteristics of chemical stressors.
6. Name four generic toxicological responses.

REFERENCES

Ayers, Kenneth W., *et al. Environmental Science and Technology Handbook.* Rockville, MD: Government Institutes, Inc. 1994.

Brauer, Roger L. *Safety and Health for Engineers.* New York: Van Nostrand Reinhold. 1990.

EPA. "EPA Guidelines for Exposure Assessment." *57 FR 22890.* May 29, 1992.

Hathaway, Gloria J., Nick H. Proctor, James P. Hughes and Michael L. Fischman. *Proctor and Hughes' Chemical Hazards of The Workplace.* 3d ed. New York: Van Nostrand Reinhold. 1991.

Jones, Baxter and Ken Kolsky. "Approaches to Computer Risk Analysis at Uncontrolled Hazardous Waste Sites." *Hazardous Materials Control Monograph Series: Risk Assessment, Volume I.* Silver Spring, MD: Hazardous Materials Control Research Institute, pp. 74–9.

Karam, Joseph G. and Martha J. Otto. "Ocean Disposal Risk Assessment Model." *Hazardous Materials Control Monograph Series: Health Assessment.* Silver Spring, MD: Hazardous Materials Control Research Institute, pp. 1–8.

Kay, Robert L. and Chester L. Tate, Jr. "Public Health Significance of Hazardous Waste Sites." *Hazardous Materials Control Monograph Series: Health Assessment.* Silver Spring, MD: Hazardous Materials Control Research Institute, pp. 65–71.

Landis, Wayne G. and Ming-Ho Yu. *Introduction to Environmental Toxicology: Impacts of Chemicals upon Ecological Systems.* Boca Raton, FL: CRC Press/Lewis Publishers, 1995.

Manahan, Stanley E. *Environmental Chemistry.* 6th. Ed. Boca Raton, FL: CRC Press/Lewis Publishers, 1994.

McBee, Karen and Robert L. Lochmiller. "Wildlife Toxicology in Biomonitoring and Bioremediation: Implications for Human Health." Chapter 6 in: *Ecotoxicity and Human Health: A Biological Approach to Environmental Remediation.* Ed. By Frederick J. de Serres and Arthur D. Bloom. Boca Raton, FL: CRC Press/Lewis Publishers, 1996.

NOAA. "NOAA's Notice of Proposed Rule on Natural Resource Damage Assessments." *60 FR 39804.* Aug. 3, 1995.

Office of Solid Waste and Emergency Response. U.S. EPA. Draft Memorandum. "Implementation of Exposure Assessment Guidelines for RCRA." Washington, D.C.: Sept. 24, 1993.

Preuss, Peter W., Alan M. Ehrlich and Kevin G. Garrahan. "The U.S. Environmental Protection Agency's Guidelines for Risk Assessment." *Hazardous Materials Control Monograph Series: Risk Assessment, Volume I.* Silver Spring, MD: Hazardous Materials Control Research Institute, pp. 6–13.

Schewe, George J., Joseph Carvitti and Joseph Velten. "Human Exposure Estimates Using U.S. EPA Guideline Models: An Integrated Approach." *Hazardous Materials Control Monograph Series: Health Assessment.* Silver Spring, MD: Hazardous Materials Control Research Institute, pp. 9–13.

Shugart, Lee. "Biomarkers of DNA Damage." Chapter 4 in: *Ecotoxicity and Human Health: A Biological Approach to Environmental Remediation.* Ed. By Frederick J. de Serres and Arthur D. Bloom. Boca Raton, FL: CRC Press/Lewis Publishers, 1996.

Walker, Katherine D. and Christopher Hagger. "Practical Use of Risk Assessment in The Selection of A Remedial Alternative."

Hazardous Materials Control Monograph Series: Risk Assessment, Volume I. Silver Spring, MD: Hazardous Materials Control Research Institute, pp. 40–4.

Weinstein, Charles E. "Ecotoxicology: Environmental Fate and Ecosystem Impact." Chapter 2 in: *Ecotoxicity and Human Health: A Biological Approach to Environmental Remediation.* Ed. By Frederick J. de Serres and Arthur D. Bloom. Boca Raton, FL: CRC Press/Lewis Publishers, 1996.

8 Fundamentals of Source Assessment

We have examined the potential effects of chemical stressors on the human body and the environment and covered some of the preliminary questions we must ask. Either we find answers in literature, take measurements, or make assumptions about certain things. Now, the fun begins. We are going to assess and model the source.

A source description begins with the origin of the stressor. Where did the stressor come from? What other stressors were generated? Are other sources of the stressor viable? Exposure analysis may begin with the source, if it is known. If not, the analysis begins with a known exposure and works backwards to link the exposure to a discovered source, or at least to the profile of a source. Still other analyses may begin with a known stressor and work in both directions to identify a viable source and to quantify a potential exposure. Needless to say, the source is the first component of the exposure pathway, and whether you work forward, backward, or from the middle, the analysis is ultimately the same. However, the source influences significantly where the stressor will be found and when. Table 8.1 lists questions for source description.

A source may be defined in several ways. For instance, source may be defined as the place where the stressor is released. In this sense, a liquid evaporating from a manufacturing process, or coming from a smokestack, or from a wastewater effluent pipe, or contaminated sediments at the bottom of a body of water are sources. The source may also be defined as a management action or practice. The dredging of a surface impoundment or a river would be a source in this sense. Sometimes, the original source no longer exists. For instance, a drum dumped into the woods illegally, or unwisely before such activity was declared illegal and we knew better, may have corroded away and its contents are now both the source and the stressor. Or contaminated sediments on the property of a plant that no longer operates could be included in this latter category.

In exposure assessments, we have three basic alternatives for evaluating exposure. We can monitor (measure) the exposure from the source. We can model the exposure. Or, we can do both. Monitoring is good and necessary. With a monitoring report in our hands we have hard data we can use. However, modeling is important too. First, modeling gives us the ability to predict a wide variety of scenarios. We can screen many scenarios if we have a good working model. Modeling also relieves us of the problems and expense of being onsite to monitor. Suppose the site does not even exist yet, for instance, and the purpose of the assessment is to evaluate a greenfields siting? Or suppose the source no longer exists and the contamination is underground, as in a CERCLA site where groundwater is contaminated? Models are good and useful, but not the end all, of course.

TABLE 8.1
Questions for Source Description

1. Where does the stressor originate?
2. What environmental medium first receives stressors?
3. Does the source generate other constituents that will influence the ultimate distribution of a stressor in the environment?
4. Do other sources of the same stressor exist?
5. What are background sources?
6. Is the source active?
7. Does the source produce a distinct signature that can be seen in the environment, organisms, individuals, or communities?

Always remember though when using a model, trade conservatism for hard data. If you are totally ignorant of exposure or toxicity, err on the side of overestimation when making assumptions.

SITE AS SOURCE

The initial step in conducting exposure assessments is, normally, the gathering of information about the source. This amounts to sampling the atmosphere, soil, or water at an existing site, or modeling the source at a proposed site. Generically, the information of concern is the level of chemicals in the water, soil, and/or air. At a Superfund site, you may be interested in all media, including fish and game, crops, wild but edible plants, as well as air, water, and soil. If you are doing a work-related exposure assessment, you may just be interested in airborne contamination. In other words, some or all of the media (air, water, soil) may be the focus of the study. Sometimes the chemicals and processes used at the site in the past are of interest. This would be the case at a Superfund site or RCRA corrective action.

Needless to say, you will need information about the physical characteristics of the site, if you do not already know that information like the back of your hand. Table 8.2 lists some of the non-chemical information you will need. Even if you know some of this information well, you should verify what you think you know. You may be surprised. Gather information about potentially exposed populations. In the workplace, this is relatively easy. Who works there? For environmental exposure assessments, you may have to get some of the demographic information from the municipality, county, state, local census office, or public health office.

Obviously, then, you start with an inventory of chemicals present now or in the past, unless you know the chemical you are interested in (such as in the case of employee exposure to a particular chemical). The problem can be defined in the terms of Table 8.3. However, it is rarely practical to include more than a few substances in your subsequent investigations. You will invariably be forced to select a few chemicals as surrogates for all others. Which ones do you choose? The choices are as many as sites to investigate. Literally, the choice must be site-specific. Table 8.4 steps you through a logical sequence of actions and decisions. In general

TABLE 8.2
Source Characteristics — Non-Chemical

Site Latitude/Longitude
Site Transverse Mercator Location
Appearance
Areas of Observed Contamination
Topography
Geology
Hydrogeology
Site History
Area Demography
Flora/Fauna Populations
Processes

TABLE 8.3
Type of Chemical Inventory Required

Abandoned site.	All past and present chemicals.
Operating site, past practices.	Chemicals used during past practice of interest.
Operating site, current or ongoing spill or leak.	Chemicals spilling or leaking.
General personal exposure.	Chemicals in use.
Specific personal exposure.	Name the chemical of concern.

TABLE 8.4
Choosing Chemicals of Concern

1. Inventory the total number of substances present, as possible, and as shown in Table 8.3.
2. Determine the quantity of each.
3. Determine the likelihood of release of each.
4. Determine each one's environmental stability.
5. Summarize each one's toxicology.

your decision to include a chemical for study, or not, boils down to this: those substances whose chemical properties are well-defined, which are present in large amounts, which are likely to be released to the environment and persist there, and which have at least a moderate toxicity. Or, you may use the selection paradigm of Burmaster and Appling, summarized in Table 8.5.

Chemicals of concern (COCs) may be substances from uncontained spills, releases, or potential releases. What is a COC, more precisely? If the quantity and concentration of a substance, at its point of release (source), can cause potential harm to humans or food chain organisms, the substance is a COC.

TABLE 8.5
Factors for Selection of COCs

High average and/or maximum concentration.
Persistent in the environment.
Toxic, relative to other candidate COCs.
Carcinogen.
Teratogen.
Detected more often or in more places than competing COCs.
Mobility in the environment.
Degree of public concern or awareness.
Relationship to human activities at the site.
Relationship to natural or anthropogenic background levels.
Essential nutrient for mammals.

Some very highly toxic substances may be COCs in very small quantities and concentration, so be conservative about screening out potential COCs too soon.

If potential COC releases are the concern, what is the condition of containment vessels and structures? What is the potential to release, in terms of quantity, concentration, rate and direction of flow, and duration of release? What is the likelihood of the release?

If a substance is being released and undergoing transport, determine its concentration at a location downstream from the point of release, and let the degree to which that much of the substance would affect anyone suggest whether the substance is a COC or not. Measurement at a downstream location is precisely the objective of personal industrial hygiene monitoring under OSHA compliance standards. The environmental concentration is compared with applicable standards, too. In the case of the workplace that would be OSHA standards. In the larger environment that may be OSHA standards as a point of reference, but may also be standards and guidance from EPA, as shown in Table 8.6, or other standards recommending bodies.

In the case of carcinogens, or noncarcinogens for which no benchmark exists, EPA has established a screening concentration (SC). The SC for cancer corresponds to that concentration which yields a 10^{-6} (one in a million) individual cancer risk for inhalation exposures via an air pathway, or for ingestion exposures via groundwater pathway, or for drinking water and human food chain threats in a surface water pathway, or for a soil exposure pathway. For hazardous substances, sampled at a discrete location, which are carcinogens, calculate an index, I, as follows:

$$I = \sum_{i=1}^{n} \frac{C_i}{SC_i} \tag{8.1}$$

where: C_i = concentration of hazardous substance i in sample
SC_i = 10^{-6} screening concentration for cancer for hazardous substance i
n   = number of hazardous substances in sample.

TABLE 8.6
Standards Used as Media-Specific Benchmarks

Source/Name/Abbreviation	Use
EPA/Maximum Contaminant Level Goals/MCLG	Groundwater migration pathway; drinking water threat in surface water migration pathway.
EPA/Maximum Contaminant Levels/MCL	Groundwater migration pathway; drinking water threat in surface water migration pathway.
FDA/Action Level/FDAAL	Fish or shellfish — human food chain threat in surface water migration pathway.
EPA/Ambient Water Quality Criteria/AWQC	Aquatic life — environmental threat in surface water migration pathway.
EPA/Ambient Aquatic Life Advisory Concentrations/AALAC	Environmental threat in surface water migration pathway.
EPA/National Ambient Air Quality Standards/NAAQS	Air migration pathway.
EPA/National Emission Standards for Hazardous Air Pollutants/NESHAP	Air migration pathway. Use only NESHAPs promulgated in ambient concentration units.
OSHA/Permissible Exposure Limits/PEL	Pathways restricted to the workplace. Use REL, TLV, WEEL, MAK, etc. only if PEL not available or alternate level is more restrictive.

SC for noncarcinogens correspond to the RfD for inhalation exposures via air pathway or for ingestion exposures via groundwater pathway, drinking water and human food chain threats in a surface water pathway, and a soil exposure pathway. The same relationships as the SC for cancer. For those hazardous substances for which an RfD is available, calculate an index, J, as follows:

$$J = \sum_{j=1}^{n} \frac{C_j}{CR_j} \qquad (8.2)$$

where: C_j = concentration of hazardous substance j in sample
CR_j = RfD screening concentration for noncancer for hazardous substance j
n = number of hazardous substances in sample

Potentially toxic air pollutants from manufacturing and industrial sites are a concern for individual communities; however, some air pollutants have an impact on the global warming phenomenon, and/or stratospheric ozone depletion, and/or acid deposition. Exposures may also come from steam and power production plants, storage facilities, and such small, local operations as dry cleaners or automobile body repair shops. EPA has established National Ambient Air Quality Standards (NAAQS) for primary air pollutants (particulate matter, ozone, carbon monoxide, sulfur dioxide, nitrogen oxides, and lead).

Also, what is the toxicity vs. the mobility of the substance? Two additional factors are important with respect to determining whether the presence of a substance makes it a COC: stability and persistence. Any number of physical, chemical and biological processes can degrade or distribute a substance in the environment. These processes are somewhat limited in the workplace, but less so in the outside environment. Fate processes are mostly predictable and help determine whether the substance is a COC.

If the fate processes transform the substance into something relatively harmless, at a reasonable rate, perhaps the substance is not a COC for this assessment. On the other hand, if the hazardous substance will persist in the environment, or be transformed into something more harmful, make it a COC. Classify highly persistent substances that bioaccumulate as COCs. Even at low initial concentrations, bioaccumulation can make a very toxic substance a definite COC. An example of the latter case is dioxin. The National Oil and Hazardous Substances Contingency Plan is one resource that gives persistence ratings for organic compounds.

Where do you get information about COCs, what their identity is, and facts about them? Their identity can be determined from operational practices, process materials, purchasing records, material safety data sheets (MSDSs), technical data sheets (TDSs), processing job orders, process instructions, standard operating procedures, inventory of storage spaces, sampling and analysis, interviews with process operators and their supervisors, interviews with process design engineers and other personnel who may have knowledge. This may take a little sleuthing. In the active workplace, the COCs should be no secret. In abandoned facilities, a bit more digging will be required. Facts about COCs may be found in any of the ways used to determine their identity, but standard references and handbooks may also be used to obtain chemical and physical characteristics.

A well-prepared exposure assessment addresses the complexity of an individual source, and does not just plug numbers into a computer model, without regard for what the model imitates. A model should allow the source emission characterization to be consistent with reality, and the selection and use of meteorological data should be consistent with observed weather patterns in the area. It should also use reasonable population statistics and health risk factors, such as the concentration that, when breathed over some period of time, will cause one cancer during the interval.

Table 8.7 shows the steps to take in modeling community exposure from one or more sources.

The physical and chemical characteristics of contaminants of interest are needed, along with source characteristics, such as quantity and quality of the discharge stream. Table 8.8 shows some information to be gathered about potential COCs. As we covered much of this material in the chapter on human health effects, let us now turn our attention to the generation of a release of a hazardous chemical. What types of sources would these be and how would we model them?

When monitoring data from occupational or environmental settings are available, as appropriate, it is to be preferred to use this data alone rather than modeling alone. Rarely is monitoring data sufficient, however. Therefore, some combination of modeling and monitoring is ideal. Since the quality of data in literature and computerized data bases varies, and the quantity of data available is always limited, an analogous

TABLE 8.7
Modeling Steps

Step	Description
1.	Review source emission characteristics.
2.	Perform screening analysis of individual sources to determine critical downwind distances.
3.	Determine rural or urban treatment (per model), flat or complex terrain, particulate or gaseous, property boundaries, downwash calculations.

TABLE 8.8
Contaminant Characteristics

Molecular Weight
Chemical Formula
Vapor Pressure
Solubility in Water
Solubility in Other Solvents
Sensory Detection Limits (Taste/Smell)
Bioconcentration Factors
Soil/Sediment Adsorption Coefficients
Octanol/Water Partition Coefficient
Melting Point
Boiling Point
Degradation Rate in Water
Degradation Rate in Soil
Degradation Rate in Biological Media
Analogous Compounds (Structure/Activity/other)

COC study may be advisable. Use caution when using analogies and check assumptions and calculation methods more than once for validity, especially when estimates indicate established exposure standards may be violated.

The analogy method for vapor concentration in air is given by Equation 8.3 when vapor generation is driven by evaporation of the liquid from an open surface or in the case of the displacement of saturated vapors from a container. The bulk liquid temperatures and mass transfer coefficient (K) of the COC and surrogate must be similar. Ventilation rates (Q) and mixing factors (k) for workplaces where the COC will be used and those where the surrogate was measured must be nearly identical. The windage and atmospheric stability must be similar in environmental settings. Quantities of the two materials must be similar. Finally, Raoult's law must be valid for both cases. The estimate is

$$C_1 = C_0 \frac{P_1 X_1}{P_0 X_0} \tag{8.3}$$

where: 0 = the surrogate substance, 1 = the COC, and
        C_1 = estimated airborne concentration of the COC, ppm
        C_0 = measured airborne concentration of the surrogate, ppm
        P_1 = vapor pressure of COC, mmHg
        P_0 = vapor pressure of surrogate, mmHg
        X_1 = mole fraction of COC in mixture,
        X_0 = mole fraction of surrogate in mixture.

The vapor pressure of a bulk liquid or solid is the pressure it exerts when it is in equilibrium with its own vapor. A liquid always has a vapor pressure and it is directly proportional to the absolute temperature. Vapor pressure increases with temperature increases and decreases with temperature decreases.

$$C_m = C_v \frac{M}{V} \tag{8.4}$$

where: C_m = airborne concentration, mg/m^3
        C_v = airborne concentration, ppm
        M = molecular weight, g/g-mole
        V = molar volume, l/mole.

At 25°C and 760 mmHg, V = 24.45 l/mole and Equation 8.4 becomes

$$C_m = C_v \frac{M}{24.45} \tag{8.5}$$

When particulate matter (PM) is involved, analogous data may be expressed as either the concentration of a specific substance or total solids. Assuming the composition of the PM is the same as the bulk material, calculate the airborne concentration of the COC by taking a ratio of the weight fractions of each.

$$C_1 = C_0 \frac{Y_1}{Y_0} \tag{8.6}$$

where: Y_1 = weight fraction of the COC in the mixture, and
        Y_0 = weight fraction of the surrogate.

If the bulk material is the surrogate, $Y_0 = 1$, and Equation 8.6 becomes:

$$C_1 = C_0 Y_1 \tag{8.7}$$

Factors such as hygroscopicity, moisture content, density, particle shape, particle size of interest, particle size distribution, and static buildup potential should be similar for the COC and its surrogate. Particle size information may be available as

TABLE 8.9
Mesh Sizes

U.S. Series Mesh Size	microns
25	707
50	297
100	149
200	74
325	44
400	37

a complete particle size distribution, or be more limited, such as average diameter, or percent plus or minus a cutoff value. The mineral industry used a shorthand for denoting particle size. For instance, ten minus 325 (written 10 – 325) means that ten percent of the particles are finer than 325 mesh in diameter. If the mesh is not clearly stated, be sure that the number is not written in microns because 325 microns is not the same as 325 mesh. In fact, Table 8.9 shows that the mesh sizes are very large compared to the micron diameter range of interest in inhalation studies.

Respirable PM, on the other hand, have an aerodynamic diameter of 3.5 mm or less, an order of magnitude smaller than the smaller mesh sizes, and are capable of reaching the alveolated gas exchange portions of the lungs, where they can be absorbed. Hence, respirable PM is always reported in microns. Larger PM, even if inhaled, will be deposited in the upper respiratory tract (URT) where they are removed. PM that is deposited in the nasopharynx behind nasal hairs is carried to the throat. PM that deposits in the tracheobronchial system are carried upward to the epiglottis and then to the throat. From the throat, large PM is ingested. Mostly, we will be interested in volatile liquids, so let's turn our attention from PM to the case of an evaporating pool of liquid.

NO VENTILATION

When a liquid stands in a body of air, such as an open container left in an unventilated room over a weekend, molecules of the liquid will enter a vapor phase over the pool of liquid, and while some of these molecules will randomly go back into the liquid phase, some molecules will diffuse into the air over the pool, called the *headspace*. This scenario assumes that evaporation is taking place at a much greater rate than absorption into a third species, adsorption onto a third species, or any other transformation that may be occurring at the same time to the material. We can safely assume that some evaporation is taking place because some liquid vaporizes whenever a gas comes into contact with it. This situation is calculated as follows in Equation 8.8:

$$C_{sat} = \frac{p_{vap}}{p_{atm}} \times 10^6 \qquad (8.8)$$

where: C_{sat} = saturation concentration, ppm (volume)
         p_{vap} = vapor pressure, mmHg
         P_{atm} = atmospheric pressure, 760 mmHg corrected for altitude.

For pure substances, vapor pressure can be estimated by the Hass and Newton method, if not available in literature,

$$\Delta t = \frac{(273.1+t)(2.8808 - \log p)}{\phi + 0.15(2.8808 - \log p)} \tag{8.9}$$

where: Dt   = temperature added to boiling point,°C
         t   = observed boiling point,°C
         log p = observed vapor pressure, mmHg
         f   = entropy of vaporization at 760 mmHg.

The entropy of vaporization can be determined by

$$\phi = \frac{\Delta H_{vap}}{2.303 RT_b} \tag{8.10}$$

where: DH_{vap} = heat of vaporization, cal/gm
         R   = gas law constant
         T_b = boiling point, K.

The heat of vaporization can be taken from literature. The gas law constant is calculated using the Ideal Gas Law

$$R = \frac{PV}{nT} \tag{8.11}$$

where units are consistent (see Coleman Major in references).
        Another way to calculate vapor pressure is the application of Raoult's Law, which determines the partial pressure of a substance in solution.

$$p_1 = px_1a \tag{8.12}$$

where: p_1 = partial pressure of component 1 in solution, mmHg
         p   = partial pressure of solution, mmHg
         x_1 = mole fraction of component 1 in solution,
         a   = activity coefficient of component 1.

The mole fraction of x_1 is

$$x_1 = \frac{\dfrac{W_1}{M_1}}{\dfrac{W_1}{M_1} + \dfrac{W_2}{M_2} + \ldots \dfrac{W_i}{M_i}} \tag{8.13}$$

for n components, where: W_i = weight fraction of component i, and
M_i = molecular weight of component i, g/g-mole.

For example, a mixture contains 10% by weight ethyl acetate, 40% w/w xylene, and 50% w/w toluene. If the partial pressure of the mixture is 50 mmHg at 25°C, what is the partial pressure of the toluene component?

$$x_1 = \frac{\dfrac{W_1}{M_1}}{\dfrac{W_1}{M_1} + \dfrac{W_2}{M_2} + \ldots \dfrac{W_i}{M_i}} = \frac{\dfrac{0.50}{92.1}}{\dfrac{0.50}{92.1} + \dfrac{0.40}{106} + \dfrac{0.10}{88.1}} = \frac{0.0054}{0.0103} = 0.5224$$

and

$$p_1 = px_1 a = 50(0.5224) = 26.1 \, mmHg$$

assuming a = 1.

The equilibrium partial pressure of the vaporized liquid is the vapor pressure of the liquid at its bulk temperature. This model is severely limited for several reasons. Generally, it grossly overestimates the COC concentration. First, in reality, equilibrium takes a long time to reach and before equilibrium can be reached, some air movement invariably occurs to mix the vapors collecting over the pool. An air-handling unit comes on and blows the vapor away from the pool. Someone walks by dragging an eddy of air behind him or her, effectively blowing the vapor away from over the pool. For another thing, realistic exposure scenarios are few with this model. Someone could walk into a room at the end of a long weekend, during which time the air-handling unit was turned off, before the heat or air-conditioning is turned back on, slowly enough not to mix the vapor with fresh air. But, not very likely. Also, the scenario of peering into the open bung of a drum and taking a sniff is not very likely. Not altogether unlikely, though, as in the mid-eighties I witnessed a person do just that before I could stop him. We were fortunate that day in that the contents of the drum smelled worse than their rather benign health effects. In fairness to all, we had not anticipated anything other than scrap rubber in the drums. Yet, you never know. Nevertheless, this is not an everyday occurrence, so let's look at the next logical modification to our model to make it more realistic.

Let's say for argument's sake that a person is going to fill a partially full drum of waste toluene through an open bung. Inside the drum is several gallons, let's say twenty, of toluene and the headspace contains vapors of toluene. As the employee

TABLE 8.10
Universal Gas Constant, R

8.314 J/g-mole K
1.987 cal/g-mole K
83.14 cm^3 bar/g-mole K
82.5 cm^3 atm/g-mole K
0.7302 atm ft.^{3}/lb.-mole R
10.73 psia ft.^{3}/lb.-mole R
1,545 ft.-lb./lb.-mole R
1.986 Btu/lb.-mole R

begins to fill the drum with additional waste toluene, vapor is displaced in the headspace by the incoming liquid, resulting in toluene emissions from the open bung. Without the benefit of a vapor recovery system, the toluene vapor is free to escape. Let us assume that the splash filling is a rapid operation (90 second duration) and that evaporation during filling is negligible. Let us also assume that the ideal gas law is valid for this situation.

The volumetric rate of air displacement from the headspace is the product of the container volume and the fill rate. The mole fraction of toluene in the headspace is the product of the saturation mole fraction (p_{Avap}/p_{atm}) and the degree of saturation (f). Then the volumetric rate of toluene release is the product of the volumetric rate of headspace air displacement and the volume fraction of toluene in the headspace, or $Vrfp_{Avap}/p_{atm}$. From the ideal gas law, the density of toluene vapor is equal to the product of its molecular weight (MW) and atmospheric pressure divided by the universal gas constant and the absolute temperature. So that the mass rate of toluene release, G_A, is

$$G_A = \frac{Vrp_{Avap}f(MW_A)}{RT}$$

where: V   = volume of the container headspace to be filled,
             4.7 ft.^{3} (35 gallons/7.48 gal/ft.^{3}),
        r   = rate of filling, 0.0111 sec^{-1} (fill in 90 sec),
        p_{Avap} = vapor pressure, 20 mmHg (from Caravanos),
        f   = 1.25 (0.50 to 1.00 for submerged loading; 1.00 to
             1.45 for splash loading per Matthiessen),
        MW_A = 92.1 lb./lb.-mole (molecular weight of toluene),
        R   = universal gas constant, 10.73 psia ft.^{3}/lb.-mole R,
        T   = vapor temperature, 537R assumed.

In this case, G_A = 0.0006 lb./sec. Table 8.10 lists valid values of R in various units. Do not confuse the universal gas constant R with R degrees Rankine.

CLOSED ROOM

The exposure from a source within a closed room is found by completing a material balance in the room. This is called the Equilibrium Box Model by some writers (EPA, Jayjock). If you have a closed room where an evaporating liquid is present, you can consider the room as a box and complete a mass balance around it in order to determine a concentration at any given time, which represents the potential exposure, in the absence of hard monitoring data. With this method, if you have a very large room, you can draw an imaginary box around the contaminant source and the complete the mass balance on a smaller box than the room itself.

The concentration of the contaminant at any time, t_i, in the box, will be equal to the sum of the mass of contaminant that entered the box — consider the source outside the box — during the time interval from some original time, t_0, to t_i, minus the mass of the contaminant that leaves the box during the same interval of time, then (the sum is) divided by the volume of the box, or

$$C_i = \frac{m_{AI} - m_{AO}}{V} \qquad (8.14)$$

where: C_i = concentration of contaminant A at any time t_i,
m_{AI} = mass of A entering the box,
m_{AO} = mass of A leaving the box,
V = volume of the box.

With such a wonderful model, why don't we use this instead of spending money on monitoring and analysis? No equation perfectly models what is happening in our environment, especially the box model. Therefore, models can be used for screening and making preliminary decisions, and can be used in the absence of data, but where lives and health conditions depend on accurate knowledge of exposures, such as in a workplace, then monitoring and analysis is indicated. However, for the purposes of engineers and scientists, who are conducting chemical exposure assessments to make general, or perhaps preliminary, planning decisions, models are useful tools. Also, they can be helpful outdoors, where only a tiny fraction of the necessary monitoring is economically feasible.

More specifically, the chief fault of the closed room model is that it does not take into account imperfect mixing of air inside the room. The model does not tell us that a concentrated pocket of vapors may be located at some point {x,y,z} from the source in the room, as determined in Cartesian coordinates. This could have serious implications for the health and safety of those who enter the room, especially those who happen to pass through the point {x,y,z}. Keep in mind however, that TWA monitoring will not tell you this information either. The only way you would know about the peak concentration at point {x,y,z} is to monitor the room with a direct reading, real time instrument and walk around the room with the probe held at several different heights above the floor until you just happen to pass through the right point in space and time. Does that picture explain why employees sometimes complain of symptoms of overexposure when monitoring results show a concentration lower than the PEL or other exposure standard?

MIXED ROOM

In order to account for eddies and currents in a closed space, which may cause concentrated pockets of the contaminant in the room, we can modify the previous model with a mixing factor (k). The degree of mixing is described by the value of the mixing factor. For instance, a value of 1.0 represents and ideally mixed room, without any pockets of high concentration of the contaminant. Matthiessen reviewed EPA research on characterization of ventilation in buildings and reported that mixing factor values from 0.1–0.333 are typical. Crowl and Louvar refer to the range 0.1–0.5 for the factor. Figure 8.1 shows how the material balance is taken, and the mathematical expression, as developed in Crowl and Louvar, is

$$V \frac{dC}{dt} = G - kQC \qquad (8.15)$$

where: G = vapor generation rate, mass/time,
         Q = ventilation rate, volume/time.

At steady-state, concentration does not change with time:

$$V \frac{dC}{dt} = 0$$

$$0 = G - kQ_v C$$

$$kQ_v C = G$$

$$C = \frac{G}{kQ_v} \qquad (8.16)$$

It has been shown (Matthiessen; Crowl and Louvar) that

$$C_{ppm} = \frac{GRT}{kQ_v PM} \times 10^6 \qquad (8.17)$$

where: R = ideal gas constant,
         P = atmospheric pressure.

In industry, general ventilation rates vary from 500 ft.^{3}/min to 10,000 ft.^{3}/min. Clement, under contract to EPA, found that a typical value is 3,000 ft.^{3}/min. However, trade data for conservative assumptions and use 500 ft.^{3}/min as the worst case, if the ventilation rate is unknown. In the outdoors, ventilation rate was determined as

---

TABLE 8.11
Applications of Closed Room Model

Determine the exposure of a person standing near
- a pool of volatile liquid,
- an opening to a storage tank, or
- an open container of volatile liquid.

26,400 v by Clement, where structures such as buildings are minimal and where v is the wind speed in miles per hour (mph). Clement found an average wind speed of 9 moh, but you can also obtain regional data from the National Weather Service.

Unfortunately, general dilution ventilation does not mix perfectly or instantaneously with contaminated room air. That is why we use the mixing factor k discussed above. Table 8.11 lists the applications of this model.

Keep in mind that the perfect and instantaneous mixing assumed by the model is not a reality. Another limitation is that only one source is considered. This is handled by making the box include only one source. To use this model outside these restrictions, you must go back to the most general calculus expression from which it was derived in order to modify the equation to fit other specific cases:

$$V\frac{dC_v}{dt} = QC_i + S - \left[\left(\frac{Q}{V} + e\right)VC_v\right] \tag{8.18}$$

where: C_v = contaminant concentration in work space, volume COC/volume air,
  Q = volumetric ventilation rate, volume/time,
  C_i = contaminant concentration in infiltration air, v/v,
  S = source generation rate, volume/time,
  e = extinction rate, time^{-1}, and
  V = room rate, volume.

Typically, this expression is simplified by assuming that the extinction rate is negligible. Extinction of the COC results from adsorption, absorption, or chemical transformation processes. This is observed in deposition of the COC on walls and equipment, in the condensation of hot vapors, and in photodegradation. Another simplifying factor that is often assumed is contaminant-free state of infiltration. In most situations, the concentration of the COC at t = 0 is negligible, though, in some workplace situations this may not be the case. Yet another simplifying assumption is that the generation and ventilation rates are constant over time. Again, this may not be true. We have already discussed the mixing assumption. Also, the real situation may not approach equilibrium concentration, as assumed. In a simple working form, Equation 8.17 is reduced to Equation 8.15. The simplest form is

$$C_v = \frac{S}{Q} \tag{8.19}$$

Use of a known volumetric flow rate, Q, is preferred to guessing, obviously.

ACGIH lists mixing rates, k, in its Ventilation Handbook. The best ventilation would have a k value from 0.67 to 1. From 0.5 to 0.67, the mixing is good. Fair ventilation is 0.2 to 0.5, and poor is anything less than 0.2. Generally, the worst case is taken as $k = 0.1$.

TWO CHAMBER MODEL

What do you do in the case where the worker is exposed to a process that ceases operation, but then shuts down and the worker remains in the room, being exposed to a concentration that decays with time? This also applies to the homeowner and his family who are present when a pesticide applicator sprays the residence and then they live in the decaying concentration until the next spray application. Also, the situation fits house painters who apply coatings that emit solvents in high concentration during the application, then a decaying concentration is present until the solvent concentration becomes undetectable at some future time, t_∞. You can use the mixed room model twice with two sets of conditions or you can use a two chamber model as follows. Two periods are established; the first period runs from the time the COC is first released, t_0, to the time it ceases being released, t_1. The second period is the duration from t_1 to the time the worker leaves the room or when the COC has decayed to an undetectable concentration, t_2.

We will assume that the concentration of our COC, the COC concentration in fresh air, and the rate of extinction are all zero at t_0. The source strength is assumed to be constant over the period t_0 to t_1. Ventilation rate, or air changes per hour, is also assumed to be constant and the mixing is considered ideal, $k = 1$. This leads to the following equation:

$$C_{\Delta t} = \frac{S \cdot t_1}{aV} - \frac{S}{a^2 V}\left[1 - e^{-a(t_1 - t_0)}\right]\left[1 - e^{-a(t_2 - t_1)}\right] \tag{8.20}$$

where: $C_{\Delta t}$ = average COC concentration over time interval, ppm,
$\quad\quad\quad$ S   = source strength, ft.^{3}/hr,
$\quad\quad\quad$ a   = number of air changes/hour,
$\quad\quad\quad$ V   = room volume, ft.^{3},
$\quad\quad\quad$ t_0 = time of release, hr,
$\quad\quad\quad$ t_1 = time release ceases, hr, and
$\quad\quad\quad$ t_2 = time exposure ends, hr.

This sometimes is called a multi-chamber or complex mass balance model.

SINKS

The relationship of losses, or sinks, due to reasons other than ventilation is not well known. Currently, a few universities are researching these sinks, but much work remains to be documented. An intuitive model has not successfully been developed, probably because these behaviors are dependent upon complicated relationships.

In general, Jayjock, et al., gives us this equation:

$$K_{s\,sink}C_{Aroom} = K_{adsorb}C_{Aroom}A + K_{desorbf}MA \qquad (8.21)$$

where: K_{sink} = a factor representing loss of concentration due to sinks,
C_{aroom} = the concentration of A in the room,
K_{adsorb} = a factor representing loss of concentration due to adsorption,
A = surface area of sink,
K_{desorb} = a factor representing the increase of concentration due to desorption,
M = mass per unit area.

VAPOR GENERATION

The next task is to determine the vapor generation rate, G. some of the information that is required to model vapor generation is molecular weight, boiling point, structure analogies, octanol-water partition coefficient, Henry's Law constant, thermodynamic activity coefficient, solubility, particle size, and temperature. The so-called openness of a release scenario amounts to the relative surface area involved. In a confined area (do not confuse this with the term confined space) a release is not as open as in a grassy field that surrounds the release point. Caravanos calculates G as the emission rate (ER) in pints per minute,

$$G = ER = \frac{QM}{403\,sp.gr} \qquad (8.22)$$

where: Q = ventilation rate, cfm,
sp. gr. = liquid specific gravity, dimensionless,
M = molecular weight.

Shade and Jayjock, on the other hand, calculate G in lb./hr:

$$G = \frac{13.3792\,MPA}{T}\left(\frac{D_{ab}V_z}{\Delta Z}\right)^{0.5} \qquad (8.23)$$

where M is in lb./lb.-mole, and
P = vapor pressure of liquid, in Hg,
A = pool surface area, ft.^{2},
T = pool temperature,°K,
D_{ab} = diffusion coefficient, ft.^{2}/sec,
V_z = air velocity over pool, ft./sec, and
DZ = length of pool in direction of air flow, ft.

In the absence of information to calculate G, use the Henry's Law constant, H, to estimate the concentration of a vapor in air. This is particularly useful when the No Ventilation model is utilized.

HENRY'S LAW

The mass of a gas dissolved in a given mass of liquid is directly proportional to the pressure of the gas above the solution.

In this case, Henry's Law is expressed mathematically, as

$$C_g = H \times C_l \qquad (8.24)$$

where: H = Henry's Law constant
  C_g = equilibrium concentration of COC in the head space,
  C_l = bulk concentration of COC in the liquid.

Henry's Law applies at low concentrations, but for nearly pure liquids, the partial pressure of the COC is expressed as Raoult's Law:

$$P_{COC} = K x_{COC} \qquad (8.25)$$

RAOULT'S LAW

The partial pressure of a component over a solution is the product of the vapor pressure of that component and the mole fraction of the component.

Chemical solutions that depart significantly from Raoult's Law still obey the law, if the mole fraction of the component of interest, the COC, is small. Therefore, Raoult's Law is a good approximation for the solvent when the solute fraction is small. When the mole fraction of all solutes is very small the solvent obeys Raoult's Law, while the solute obeys Henry's Law. Solutes that obey both laws are called *ideal solutes*. When the solvent obeys Raoult's Law, the ideal solute obeys Henry's Law.

How fast does a liquid evaporate? The rate depends on its saturation vapor pressure. The liquid molecules must gain a minimum amount of kinetic energy in order to overcome intermolecular forces holding them in the liquid state. With enough energy the molecules break through the surface of the liquid, escaping into the atmosphere above it, diffusing through the molecules of the gases that comprise air.

Evaporation *is a different generation process than boiling, where the vapor escapes the liquid surface with sufficient force to displace air and other gases, rather than diffusing through them.*

In stagnant air, the rate of evaporation is proportional to the difference between the saturation vapor pressure and the partial pressure of the vapor in air:

$$G \propto P^{sat} - p_a \qquad (8.26)$$

If air flows over a pool of liquid, with a coordinate system x,y,z such that {0,0,0} is a point just over the pool, with air flowing in the z-direction and parallel with the surface of the liquid, the following equation applies along a volume element, if the surface temperature of the pool is constant, and the heat of vaporization is provided by the surroundings, resulting in a constant evaporation rate.

$$\frac{\delta N_{A,z}}{\delta z} + \frac{\delta N_{A,x}}{\delta x} = 0 \qquad (8.27)$$

where: $N_{A,z}$ = molar flux of COC A in the z-direction, mol/time area,
$N_{A,x}$ = molar flux of COC A in the x-direction, mol/time area,

The flux in the z-direction is assumed to have no concentration changes with respect to time. That is, the flux in the z-direction is assumed to be at equilibrium conditions. At equilibrium, the surface temperature of the evaporating pool will be constant. This condition requires that the pool be sufficiently large and that the heat of evaporation be provided by the surroundings. Also, assume that the diffusion of A in the z-direction is small compared to air velocity and that the concentration, $c_A\{x,0,z\}$, is constant with respect to time. Also, we assume that diffusion at the pool edges in the y-direction are negligible. If air velocity is constant in x- and y-directions and changing only in the z-direction,

$$N_{A,z} = C_A V_z \qquad (8.28)$$

Convective transport in the x-direction is negligible and the area above the pool where the steepest concentration gradient is located has the least mixing, so that c_A is very small.

$$N_{A,x} = -D_{A,B} \frac{\delta C_A}{\delta x} \qquad (8.29)$$

$D_{A,B}$ is the diffusion coefficient of COC A in air (B). Using the liquid pool temperature as the temperature of the air flowing over the pool,

$$V_z \frac{\delta C_A}{\delta z} = D_{A,B} \frac{\delta^2 C_A}{\delta x^2} \qquad (8.30)$$

The boundary conditions are

$$
\begin{array}{lll}
c_A = 0 & \text{at} & z = 0 \\
c_A = C_{A,0} & \text{at} & x = 0 \\
c_A = 0 & \text{at} & x = \infty
\end{array}
$$

C_{A0} is the concentration resulting from the vapor pressure of the liquid at the liquid surface temperature, assuming any temperature gradient between the surface and the air can be neglected. Solving,

$$
\frac{C_A}{C_{A0}} = \text{erfc}\left(\frac{x}{\sqrt{4D_{AB}\,z/V_z}}\right) \tag{8.31}
$$

At any point on the liquid surface, the evaporation rate is

$$
N_{A,x}\big|_{x=0} = -D_{A,B}\frac{\delta c_A}{\delta x} \tag{8.32}
$$

Summing all points over the surface gives the total evaporation:

$$
E = 1/A \iint N_{A,x}\,dzdy = 1/A \iint -D_{AB}\frac{\delta c_A}{\delta x}\,dzdy \tag{8.33}
$$

Differentiation of Equation 8.31, gives a solution to Equation 8.34:

$$
E = N_A \text{total} = \frac{1}{\Delta y \Delta z}\iint \frac{2D_{AB}c_{A,0}}{\sqrt{4\pi D_{Ab}\,z/v_z}}\,dzdy = 2c_{A,0}\sqrt{\frac{D_{AB}v_z}{\pi\Delta z}} \tag{8.34}
$$

Total evaporation rate, E, is in terms of moles per time per area. If E is required in terms of weight, the molecular weight has to be factored into the equation:

$$
E_{mass} = 2\,MWc_{A,0}\sqrt{\frac{D_{AB}v_z}{\pi\Delta z}} \tag{8.35}
$$

but,

$$
c_{A,0} = \frac{P^{vap}}{RT} \tag{8.36}
$$

so, where R = 39.381 in. Hg ft.^{3}/lb. moles Kelvin:

$$
E_{mass} = \frac{13.3792\,MWP^{vap}}{T}\sqrt{\frac{D_{AB}v_z}{\Delta z}} \tag{8.37}
$$

where: MW = molecular weight, lb./lb. moles,
         P^{vap} = vapor pressure, inches Hg,
         T   = temperature, K
         D_{AB} = diffusion coefficient of vapor in air, ft.^{2}/sec,
         v_z = air velocity, ft./min, and
         Δz = pool dimension in direction of air flow, ft.

When a large volume of gas contains molecules of two species, if the temperature, pressure, and velocity gradients are sufficiently small, by kinetic theory we can determine the diffusivity coefficient as

$$D_{AB} = \frac{2}{3}\left(\frac{K^3}{\pi^3}\right)^{1/2}\left(\frac{1}{2m_A}+\frac{1}{2m_B}\right)^{1/2}\frac{T^{3/2}}{p\left(\dfrac{d_A+d_B}{2}\right)^2} \tag{8.38}$$

but, by the Chapman–Enskog theory, (see Bird, Stewart, and Lightfoot):

$$cD_{AB} = 2.2646\times10^{-5}\frac{\sqrt{T\left(\dfrac{1}{M_A}+\dfrac{1}{M_B}\right)}}{\sigma_{AB}^2\Omega_{D,AB}} \tag{8.39}$$

By the Ideal Gas Law, c = p/RT, so

$$D_{AB} = 0.0018583\frac{\sqrt{T^3\left(\dfrac{1}{M_A}+\dfrac{1}{M_B}\right)}}{p\sigma_{AB}^2\Omega_{D,AB}} \tag{8.40}$$

where: T   = temperature, K,
         M_A = molecular weight of A,
         M_B = molecular weight of B,
         p   = pressure, atm.,
         σ_{AB} = Leonard–Jones distance, angstroms,
         $\Omega_{D,AB}$ = a dimensionless function of temperature.

Using the method of Bird, Stewart, and Lightfoot to calculate σ_{AB} and $\Omega_{D,AB}$:

$$\sigma_{AB} = e^{1.02}M^{0.091} \tag{8.41}$$

and

$$\Omega_{D,AB} = e^{0.39}\left(\frac{\sqrt{97e^{2.38}M^{0.76}}}{T}\right)^{0.397} \tag{8.42}$$

Substitutions yields

$$D_{AB} = \frac{4.09 \times 10^{-5} T^{1.9} \sqrt{1/29 + 1/M_A} M_A^{-0.33}}{p} \tag{8.43}$$

where,   Diffusivity is in cm²/sec, and

    p = pressure, atm.,

    T = temperature, K,

    M_A = molecular weight of evaporating liquid.

In terms of the evaporation rate, in Equation 8.34:

$$E_{mass} = \frac{2.79 \times 10^{-3} M^{0.835} \left(P^{vap}\right)^4 \sqrt{1/M_A + 1/29}}{T^{0.05}} \sqrt{\frac{v_z}{\Delta z p}} \tag{8.44}$$

Here,   E is expressed in lb./hr ft.², and

    M = molecular weight,

    p = total pressure, atm.,

    Δz = dimension of pool in direction of air flow, ft,

    T = pool surface temperature, K,

    P^{vap} = vapor pressure of liquid, in. Hg, and

    v_z = velocity of air in z-direction, ft./min.

An alternative empirical expression by Pace Laboratories,

$$E_{mass} = 0.000237 M P^{vap} V_z^{0.625} \tag{8.45}$$

The EPA uses an expression by Mackay,

$$E_{mass} = \frac{0.03197 V_z^{0.78} M^{2/3} P^{vap}}{T} \tag{8.46}$$

where: M = molecular weight,

        P^{vap} = vapor pressure, in. Hg, and

        V_z = air velocity, ft./min.

However, Matthiessen (also Crowl and Louvar) follows a parallel path of logic and refers to this solved equation:

$$G = \frac{M K A \left(P^{sat} - p_a\right)}{R T_L} \tag{8.47}$$

where: K = mass transfer coefficient for area A, length/time,

Often, $P^{sat} >>> p_a$ and, therefore,

$$G = \frac{MKAP^{sat}}{RT_L} \qquad (8.48)$$

Substituting eqn. 8.45 into 8.20 gives

$$C_{ppm} = \frac{KAP^{sat}T}{T_L kQ_v} \times 10^6 \qquad (8.49)$$

However, we can expect that $T_L = T$, so

$$C_{ppm} = \frac{KAP^{sat}}{kQ_v} \times 10^6 \qquad (8.50)$$

Jayjock uses this equation to model a gas source at any point $\{r, \phi\}$ in a hemisphere. This model can be derived from Equation 8.28 by setting initial conditions and using spherical rather than Cartesian coordinates.

$$C = \frac{G}{2\pi Dr}\left[1 - erf\left(\frac{r}{\sqrt{4tD}}\right)\right] \qquad (8.51)$$

where: C = airborne COC concentration, mg/m³,
         G = COC emission rate at steady-state, mg/min,
         D = eddy diffusion coefficient, m²/min,
         r   = distance from source to point of interest, m, and
         t   = elapsed time of release, min.

In the workplace, the distance r is taken from the source to the worker's breathing zone. In developing his model, Jayjock refers to a Carlslaw and Jaeger text (*Conduction of Heat in Solids*, 1959), which is one of the gospels of conduction, to devise an analog to the Fourier law of conduction $q = -k \, \delta x/\delta t$, but any good text on conductive heat transfer, such as Arpaci or Churchill, will give you the physics and mathematics necessary to derive a similar equation to fit your specific circumstances.

Mass transfer has to do with the movement of one material over an interface with another material. With respect to a pool of evaporating liquid, the interphase is the phase boundary, or surface of the pool where it is in contact with air. The movement of vapor molecules and air molecules into and out of the pool of liquid is due to a combination of diffusion and bulk flow. The latter occurs without respect to concentration gradient, but the diffusional contribution is proportional to the difference of concentration of vapor molecules in the liquid and air. (Air molecules also diffuse into the liquid, but that is not of interest to us.) This proportionality can be expressed as a mathematical equality:

$$K = aD^{2/3} \tag{8.52}$$

where D is the diffusional coefficient for the gas-phase and a is a proportionality constant. Very little experimental data exists for mass transfer and diffusional coefficients. Fortunately, data exists for a few species, such as water vapor, and the mass transfer coefficients is proportional to the ratio of diffusional coefficients of any two species.

$$\frac{K}{K_0} = \left(\frac{D}{D_0}\right)^{2/3} \tag{8.53}$$

In physical chemistry the relationship between gas-phase diffusion coefficients is well known as Graham's Law.

GRAHAM'S LAW

Under the same conditions of temperature and pressure, the rates of diffusion of gases are inversely proportional to the square roots of the molecular weights of the gases.

Or,

$$\frac{D}{D_0} = \sqrt{\frac{M_0}{M}} \tag{8.54}$$

Therefore,

$$K = K_0\left(\frac{M_0}{M}\right)^{1/3} \tag{8.55}$$

and D_0 of water is 0.83 cm/sec.

BOILING POOL

If the pool of liquid is boiling, as may be the case with a low boiling substance, the heat transfer rate to and from the surroundings affect boiling rate. Heat is conducted from the ground or underlying surface by conduction. Both conduction and convection mechanisms are responsible for the heat transfer between the surrounding air and the pool. The sun and nearby heat sources, such as a fire, transfer heat to the pool by radiation.

Vaporization and condensation at constant temperature and pressure, according to Himmelblau, are equilibrium processes and the equilibrium pressure is the vapor pressure. A liquid will exert vapor pressure in a room, an evacuated space, or into a cylinder enclosed by a piston. Vapor pressure of any liquid increases with temperature. When a gas comes into contact with a liquid, some of the liquid will evaporate. After some period of time, an equilibrium is reached and no more liquid will evaporate. For pools of liquid at ambient temperatures, this is a long period of time. A well known chemical engineering relationship is that the mole fraction of the vapor, y_A, is a function of pressure:

$$y_A = \frac{P_{Avap}}{P_{atm}} \tag{8.56}$$

which relates to the concentration in ppm:

$$ppm_A = y_A \times 10^6 = \frac{P_{Avap}}{P_{atm}} \times 10^6 \tag{8.57}$$

However, this model will overestimate, because it assumes equilibrium is reached, while, in fact, equilibrium is rarely reached and the head space is not saturated with vapor.

This relationship can be used to estimate the vapors emitted from filling a drum, for instance. If G_A is the mass rate of COC release from the drum, and the rate of headspace displacement is Vr, and the degree of saturation is a fraction, f, the volumetric rate of release is

$$V_A = \frac{Vrfp_{Avap}}{P_{atm}} \tag{8.58}$$

The density of the vapor, based in the Ideal Gas law:

$$\rho_{vap} = \frac{M_A P_{atm}}{RT} \tag{8.59}$$

and

$$G_A = \frac{M_A Vrfp_{Avap}}{RT} \tag{8.60}$$

Matthiessen found that for submerged loading of a drum, where the end of the fill pipe or hose is located beneath the liquid surface, f varies from 0.5 to 1.0. In splash loading, the end of the fill pipe of hose is above the liquid surface, and $1.00 < f < 1.45$, according to Matthiessen.

Using the *Chemical Engineer's Handbook* (cf. P. 3–274) we can predict vapor pressure of any liquid by the Antoine equation:

$$\ln p^{sat} = A - B/(T + C) \tag{8.61}$$

or the Reidel equation:

$$\ln p^{sat} = A - B/T + C\ln T + DT^6 \tag{8.62}$$

To solve the Antoine and Reidel equations, you must refer to literature noted in Perry's (cf. P. 3–274). Another source of information is Reid, *et al.* If only two data points are known, Perry's recommends the Clapeyron equation:

$$\ln p^{sat} = A - B/T \tag{8.63}$$

where: $A = \ln p_1^{sat} + B/T_1$

$$B = \ln\left(\frac{p_2}{p_1}\right)\bigg/\left(\frac{1}{T_1} - \frac{1}{T_2}\right)$$

The limitation of these equations is that they are not accurate over large temperature intervals because they assume that the heat of vaporization, ΔH_{vap}, is constant over the temperature range of interest. However, this assumption does not hold true if the temperature range is more than a few tens of degrees. Therefore, Perry recommends a reduced form of the Reidel equation, if the reduced temperature and pressure are known for the liquid and at least one other data point is available.

$$\ln p_r^{sat} = A^\circ - B^\circ/T_r + C^\circ \ln T_r + D^\circ T_r^6 \tag{8.64}$$

where: p_r^{sat} = reduced vapor pressure, p^{sat}/P_c,
          T_r = reduced temperature, T/T_c,
          $A^\circ = -35Q$,
          $B^\circ = -36Q$,
          $C^\circ = 42Q + \alpha_c$,
          $D^\circ = -Q$, and
          $Q = 0.0838(3.758 - \alpha_c)$.

α_c is determined by

$$a_c = \frac{0.315\psi_1 - \ln p_{lr}^{sat}}{0.0838\psi_1 - \ln T_{lr}} \tag{8.65}$$

where

$$\psi_1 = -35 + 36/T_{lr} + 42 \ln T_{lr} - T_{lr}^6 \tag{8.66}$$

Jayjock, et al., refer to the Wagner equation, which is simpler.

$$\ln p_{Avap} = \frac{a\left(1-T_r\right) + b\left(1-T_r\right)^{1.5} + c\left(1-T_r\right)^3 + d\left(1-T_r\right)^6}{T_r}$$

$$p_{Avap} = \frac{p}{p_{Ac}}$$

$$T_r = \frac{T}{T_{Ac}} \tag{8.67}$$

If the heat of vaporization and a boiling point are known, the Clausius–Clapeyron may be used to estimate vapor pressure:

$$\ln\left(\frac{P_{25}^{vap}}{P_1}\right) = \frac{\Delta H_{vap}}{R}\left(\frac{1}{T_1} - \frac{1}{298}\right) \tag{8.68}$$

where: R $= 8.314$ J/g-mole K, and
ΔH_{vap} = heat of vaporization, J/g-mole.

Most often, in the workplace, contaminants are released into rooms that act as well-mixed containers. Assuming perfect, instantaneous mixing, with low generation rates, and exhaust air at the same concentration as the room air, we can say that the volumetric flow rate in equals the volumetric flow rate out equals Q. Performing a mass balance for the COC, we say that the rate of accumulation equals the rate of the COC into the room minus its rate out.

The rate of accumulation can be expressed

$$V\frac{dc_{Aroom}}{dt} = Qc_{Ain} + G_A - Qc_{Aroom} - K_{sink}c_{Aroom} \tag{8.69}$$

where: V   = room volume, ft.^{3},
c_{Aroom} = mass concentration of A in room air, lb./ft.^{3},
t   = time, sec,
Q   = volumetric air flow rate, ft.^{3}/sec,
c_{Ain} = mass concentration of A incoming with air, lb./ft.^{3},
G_A = mass generation rate, lb./sec, and
K_{sink} = sink coefficient, ft.^{3}/sec.

Unfortunately, sinks have not been well characterized as discussed above.

WELL MIXED ROOM WITH BACKPRESSURE

When the evaporating liquid has a sufficiently large surface area, backpressure exerts and influence on the model. In rooms smaller than twenty cubic feet, any pool with a surface area larger than 1/6 of the surface area of the room is subject to backpressure influence. In larger rooms, those pools with a surface area greater than a few hundred square feet are susceptible to this error inducing phenomenon.

Jayjock, et al., propose this general equation for illustration purposes:

$$G_A = \frac{MW_A KAP^{vap}}{T_L R}$$
(8.70)

where: K = mass transfer coefficient, and
         T_L = bulk liquid temperature.

This equation is not unfamiliar to us and Matthiessen has provided information that EPA estimates the mass transfer coefficient as

$$K = 0.83 \left(\frac{18}{MW_A} \right)^{1/3}$$
(8.71)

where the coefficient is estimated based on the K for water.

Now, we can modify our generation rate for the effect of backpressure. Jayjock, et al., provides:

$$G_A = \frac{MW_A KA \left(P^{vap} - P^{back} \right)}{T_L R}$$
(8.72)

where p^{back} is the partial pressure of the evaporating species:

$$P^{back} = \frac{8.2 \times 10^{-8} CT_L}{MW}$$
(8.73)

At $t = 0$, the initial conditions are:

$$
\begin{aligned}
C &= 0 \\
P^{back} &= 0 \\
G_A &= G_{max}
\end{aligned}
$$

At any time, t:

$$C = \frac{G}{Qm} + \left(C_0 - \frac{G}{Qm} \right) e^{\frac{-QM_t}{V} t}$$
(8.74)

where Qm/V is the mathematical expression of ventilation as air changes per hour. The liquid mass remaining in a pool at any time, t, is

$$L(t) = L_0 \cdot \exp(-kt) \tag{8.75}$$

and the vapor mass emission rate is

$$G(t) = k \cdot L_0 \cdot \exp(-kt) \tag{8.76}$$

For example, if L_0 = 40,000 lb. and k = 0.25 min^{-1}, how much liquid remains and what is the ventilation rate after five and 10 minutes.

Initially, L(0) = 40,000 lb.; G(0) = 10,000 lb./min
After five minutes, L(5) = 18,895 lb.; G(5) = 4,724 lb./min.
After 10 minutes, L(10) = 8,925 lb.; G(10) = 2,231 lb./min.
Therefore, in a well-mixed room with an exponentially decreasing emission rate

$$VdC = kL_0 \exp(-kt)dt - Qc(t)dt \tag{8.77}$$

We have to examine infinitesimal time intervals to avoid large errors. Integrating, we get

$$C(t) = \frac{kL_0}{(kV) - Q}\left[\exp\left(-\frac{Q}{V}t\right) - \exp(-kt)\right] + C_0 \exp\left(-\frac{Q}{V}t\right) \tag{8.78}$$

The maximum concentration is reached at time

$$t_{max} = \frac{V \ln\left(\dfrac{kV}{Q}\right)}{(kV) - Q} \tag{8.79}$$

Take, for example, the case of a five gallon pail of ethyl acetate, a common paint solvent, which has spilled on the floor. The room has six air changes per hour and has a volume of 1,000 ft.^{3}, therefore Q = 100 ft.^{3}/min. The most that could have spilled is five gallons, which is 18,925 mL. The OSHA PEL is 1,400 mg/m^3 and the LEL is 2.0 percent. What danger exists?

The vapor pressure of ethyl acetate is 74 mmHg, so the saturation concentration is (74/760)×10^6 or 97,368 ppm or 356,367 mg/m^3. The saturated room contains (356 g/m^3)×(28.3 m^3) or 10,075 g or 22.2 lb. of ethyl acetate. The volume spilled is (10,075 g)/(0.90 g/ml) or 11,200 mL. This is reasonable as the maximum amount spilled was (5 gal)×(3,785 mL/gal) or 18,925 mL. The time required to evaporate the maximum amount is

$$t_{max} = \frac{V \cdot \ln\left(\frac{k \cdot V}{Q}\right)}{(k \cdot V) - Q} = \frac{28.3 \ln \frac{8.49}{2.83}}{8.49 - 2.83} = 5.5 \, \text{min}$$

and the expected concentration at that time is

$$C(t) = \frac{k \cdot L_0}{(k \cdot V) - Q}\left[\exp\left(-\frac{Q}{V}t\right) - \exp(-kt)\right]$$

$$= \frac{0.3\left(10.1 \times 10^6\right)}{(0.3 \cdot 28.3) - 2.83}\left[\exp\left(-\frac{2.83}{28.3}5.5\right) - \exp(-0.3 \cdot 5.5)\right]$$

$$= 206,000 \, \text{mg/m}^3 = 56,300 \, \text{ppm}$$

The OSHA PEL, the IDLH (10,000 ppm), and the LEL are exceeded.

TWO BOX MODEL

The well-mixed room model underestimates the concentration of a COC close to its source, because the exposure level is higher near the source and lower at extreme distances in the room. The next model breaks the room into two boxes, the near-field and the far-field. The air within each field is assumed to be well-mixed, but air flow between the fields is assumed to be limited as if the far-field is an isolated second room. The inter-field air flow is assumed to be β m³/min. The movement of air into and out of the far-field is assumed to be Q m³/min. The mass rate, G, is assumed to be constant and no sink terms are considered. The mass balances are

$$V_N \cdot dC_N = \left[G \cdot dt + \beta \cdot C_{Fdt}\right] - \beta \cdot C_N \cdot dt \tag{8.80}$$

for the near-field, and

$$V_F \cdot dC_F = \beta \cdot C_N \cdot dt - \left[\beta \cdot C_F \cdot dt + Q \cdot C_F \cdot dt\right] \tag{8.81}$$

for the far-field, where:
V_N = near-field volume, m³
V_F = far-field volume, m³
C_N = near-field concentration, mg/m³
C_F = far-field concentration, mg/m³
β = air flow rate between near- and far-fields, m³/min
Q = room air supply rate, m³/min
dt = infinitesimal time interval, min

With a constant emission rate, the dynamic concentration (not at steady-state) of the near- and far-fields are

$$C_N(t) = \frac{G}{\left(\dfrac{\beta}{\beta+Q}\right)} + G\left(\frac{\beta Q + \lambda_2 V_N(\beta+Q)}{\beta Q V_N(\lambda_1 - \lambda_2)}\right)\exp(\lambda_1 t)$$

$$- G\left(\frac{\beta Q + \lambda_1 V_N(\beta+Q)}{\beta Q V_N(\lambda_1 - \lambda_2)}\right)\exp(\lambda_2 - t) \qquad (8.82)$$

and

$$C_F(t) = \frac{G}{Q} + \left(\frac{\lambda_1 V_N + \beta}{\beta}\right)\left(\frac{\beta Q + \lambda_2 V_N(\beta+Q)}{\beta Q V_N(\lambda_1 - \lambda_2)}\right)\exp(\lambda_1 t)$$

$$- G\left(\frac{\lambda_2 V_N + \beta}{\beta}\right)\left(\frac{\beta Q + \lambda_1 V_N(\beta+Q)}{\beta Q V_N(\lambda_1 - \lambda_2)}\right)\exp(\lambda_2 t) \qquad (8.83)$$

where

$$\lambda_1 = 0.5\left[-\left(\frac{\beta V_F + V_N(\beta+Q)}{V_N V_F}\right)\right] + \sqrt{\left(\frac{\beta V_F + V_N(\beta+Q)}{V_N V_F}\right)^2 - 4\left(\frac{\beta Q}{V_N V_F}\right)}$$

and

$$\lambda_1 = 0.5\left[-\left(\frac{\beta V_F + V_N(\beta+Q)}{V_N V_F}\right)\right] - \sqrt{\left(\frac{\beta V_F + V_N(\beta+Q)}{V_N V_F}\right)^2 - 4\left(\frac{\beta Q}{V_N V_F}\right)}$$

These equations account for the change in mass in the near- and far-fields over the very small time interval dt. Mass enters the near field from the source, G, and from the far-field, $-\beta$. Mass leaves the near-field to the far-field, β. The only mass entering the far-field is from the near-field, β, but mass leaves the far-field by being carried back to the near-field, $-\beta$, or by means of the exhaust airflow, Q.

The equilibrium concentrations for the near- and far-fields, where the emission rate is constant, are:

$$C_{N,eq} = \frac{G}{Q} + \frac{G}{\beta} = \frac{G}{\left(\dfrac{\beta}{\beta+Q}\right)Q} \qquad (8.84)$$

for the near-field, and

$$C_{F,eq} = \frac{G}{Q} \qquad (8.85)$$

for the fair-field. When $\beta < Q$, $C_{N,eq} > C_{F,equation}$ But, when $\beta >> Q$, $C_{N,eq} \cong C_{F,equation}$ In eqn. 8.83, the proportion term in the denominator is called the *local purging flow rate* or the *effective near-field ventilation rate*. Notice that this term is always less the Q, or

$$\left(\frac{\beta}{\beta+Q}\right)Q < Q.$$

In the literature, many equations contain a safety factor, K, and in developing this model, Nicas defines a mixing factor, m, as

$$m = \frac{1}{K},$$

where $0 < m \le 1$.

Further, mQ is the effective room supply air rate. It is intuitive that

$$\left(\frac{\beta}{\beta+Q}\right) = m.$$

In a well-mixed room, $m \rightarrow 1$, the near- and far-field models produce no discernible difference in results. For a given set of circumstances, the model produces a set of decay curves and these can be used to the time of maximum concentration or the concentration at any time, t. Two general equations, similar to 8.81 and 8.82 describe the decay of concentration starting with an initial concentration in the near-field, $C_{N,0}$, and an initial concentration in the far-field, $C_{F,0}$.

$$C_N(t)\frac{\beta(C_{F,0}-C_{N,0})-\lambda_2 V_N C_{N,0}}{V_N(\lambda_1-\lambda_2)}\exp(\lambda_1 t)$$

$$+\frac{\beta(C_{N,0}-C_{F,0})+\lambda_1 V_N C_{N,0}}{V_N(\lambda_1-\lambda_2)}\exp(\lambda_2 t) \qquad (8.86)$$

and

$$C_F(t)=\frac{(\lambda_1 V_N+\beta)\left[\beta(C_{F,0}-C_{N,0})-\lambda_2 V_N C_{N,0}\right]}{\beta V_N(\lambda_1-\lambda_2)}\exp(\lambda_1 t)$$

$$+\frac{(\lambda_2 V_N+\beta)\left[\beta(C_{N,0}-C_{F,0})+\lambda_1 V_N C_{N,0}\right]}{\beta V_N(\lambda_1-\lambda_2)}\exp(\lambda_2 t) \qquad (8.87)$$

where: $C_{N,0}$ = near-field concentration at time $t = 0$,
         $C_{F,0}$ = far-field concentration at time $t = 0$,
         λ_1 = see eqn. 8.82, and
         λ_2 = see eqn. 8.82.

Based on decay from a uniform room concentration the decay slope is $-mQ/V$, the effective ventilation rate or room air flow is mQ or $V|\text{slope}|$, and the exposure estimate, C_{eq}, at steady state is G/mQ, or $G/V|\text{slope}|$.

The EPA uses a complex model based on the simple mass balance model:

$$C = \frac{1.7 \times 10^5 T_a G}{MQk} \tag{8.88}$$

where: C = COC concentration, ppm,
         T_a = ambient air temperature, K,
         G = vapor generation rate, g/sec,
         M = molecular weight, g/g-mole,
         Q = ventilation rate, ft.³/min, and,
         k = mixing factor.

No, the units for Q are not a misprint. The EPA equation mixes unit systems. From this equation, however, the EPA derives a more complex model as shown here:

$$C(t) = \frac{St_1}{aV} - \frac{S}{a^2V}\left[1 - e^{-a(t_1-t_0)}\right]\left[e^{-a(t_2-t_1)}\right] \tag{8.89}$$

where: C = concentration, ppm,
         S = source strength, cm³/hr,
         a = number of air changes per hour, hr⁻¹, and
         V = room volume, m³.

The solution of this model is:

$$C = \left(\frac{St_1}{Vm} + C_0\right)\left(1 - e^{-\left(\frac{m}{r_t}\right)t}\right) + C_b e^{-\left(\frac{m}{r_t}\right)t} \tag{8.90}$$

where: C = concentration, g/m³,
         S = source generation rate, g/hr,
         r_t = residence time of air in room, hr,
         m = mixing factor,
         V = room volume, m³,
         C_0 = outdoor concentration, g/m³,
         C_b = initial (beginning) concentration in room, g/m³, and
         t = time, hr.

or, in other terms,

$$C = \left[\frac{2.1 \times 10^6}{Q} + (1000d) \right] G \tag{8.91}$$

where: C = concentration, mg/m^3,
         Q = ventilation rate, $ft.^3/min$,
         d = distribution factor, sec/m^3, and
         G = vapor generation rate, g/sec.

Where ventilation is from the worker's side, d ranges from 0.1 to 0.3 according to EPA research. The range is from 0.6 to 3 when the ventilation is from the rear of the worker.

If, instead of being in a room, a worker is filling drums or bags, the expression used by EPA to estimate exposure is:

$$G = \frac{fMVrP^{vap}}{3600RT_L} \tag{8.92}$$

Haven't we seen this before? Here,

       G   = vapor generation rate, g/sec,
       f   = saturation factor,
       M   = molecular weight, g/g-mole,
       V   = volume of container, cm^3,
       r   = fill rate, hr^{-1},
       P^{vap} = vapor pressure of pure substance, atm,
       R   = 82.05 atm cm^3/g-mole K, and
       T_L = bulk liquid temperature, K.

For submerged loading, where the end of the fill pipe is below the surface level of the liquid in the drum, the saturation factor, f, varies from 0.5 to 1. In splash loading the end of the fill pipe is over the bulk liquid and therefore the added material splashes onto the surface. The saturation factor in the latter case ranges from 1.0 to 1.45.

OPENING IN PIPE

For fluids moving through a pipe, we can apply the Bernoulli equation, at least the mechanical energy balance portion of it. Consider a volume of fluid flowing through a pipe of constant cross-sectional diameter. With friction loss and, therefore, an irreversible process, the Bernoulli equation becomes

$$\frac{Z_1 g}{g_c} + \frac{V_1^2}{2g_c} - (p_2 - p_1)v - F + W_e = \frac{Z_2 g}{g_c} + \frac{V_2^2}{2g_c} \tag{8.93}$$

Liquids are incompressible fluids, therefore eqn. 8.68 simplifies

$$\frac{Z_1 g}{g_c} + \frac{V_1^2}{2g_c} - \frac{(P_2 - P_1)}{\rho} - F + W_e = \frac{Z_2 g}{g_c} + \frac{V_2^2}{2g_c} \qquad (8.94)$$

In these equations, v is specific volume, reciprocal of density.

If a pipe develops a small hole, as the fluid within escapes through the hole, its pressure is converted to kinetic energy. The velocity of the escaping fluid will be affected by the conversion of some of its kinetic energy into friction losses between the moving fluid and the surface of the hole and between the fluid particles as they contact each other. Crowl and Louvar refer to this overall situation of fluid exiting a hole as a limited aperture release.

Within the pipe, the pressure, measured in gauge P_g, is constant until the bitter end, and the external pressure is atmospheric, so $\Delta P = P_g$. No shaft work is done. Velocity of the fluid within the pipe is considered negligible. Also, for the purposes of our model, we assume no change in the elevation of the fluid within the pipe. Friction is handled by a constant discharge coefficient, such that

$$-\frac{\Delta P}{\rho} - F = C_D^2 \left(-\frac{\Delta P}{\rho} \right) \qquad (8.95)$$

Substituting these modifications into the energy balance gives

$$\bar{u} = C_D \sqrt{a} \sqrt{\frac{2g_c P_g}{\rho}} \qquad (8.96)$$

The α is constant, and Crowl and Louvar define

$$C_o = C_D \sqrt{a} \qquad (8.97)$$

Therefore, the result is a formula widely used for escape velocity, which is:

$$\bar{u} = C_o \sqrt{\frac{2g_c P_g}{\rho}} \qquad (8.98)$$

The amount of fluid leaving the hole is

$$G = \rho \bar{u} A = A C_o \sqrt{2\rho g_c P_g} \qquad (8.99)$$

TABLE 8.12
Suggested Values of C_o

Use $C_o =$	when this condition exists:
0.61	sharp-edged orifice
0.81	Reynolds number > 30,000 short pipe section attached to a vessel
1.00	(L/D ≥ 3) well-rounded nozzle unknown or uncertain

Multiple the latter answer by the amount of time (duration of spill) to determine the total amount spilled. Values of C_o suggested in literature (Crowl and Louvar) are given in Table 8.12.

OPEN VALVE

Solving for the velocity and amount of material exiting an open valve is standard engineering practice. Valves, and other fittings, in a run of piping offer resistance to flow, thus contributing to friction losses. In decreasing order of resistance, the relative resistance of valves are: globe and angle valves offer high resistance; gate, ball, plug, and butterfly valves offer lower resistance. You probably won't see a butterfly valve on the end of a pipe transferring fluids. However, a situation involving a butterfly valve is not outside the realm of possibilities. For instance, what if a line were broken for maintenance and the temporary end of it were a butterfly?

When solving for flow through a valve we assume negligible kinetic energy changes, no pressure drop, and no shaft work, so the energy balance becomes

$$\frac{g}{g_c} \Delta z + F = 0 \tag{8.100}$$

where

$$F = \frac{2fL\overline{u}^2}{g_c d} \tag{8.101}$$

where L is the equivalent length of pipe. For piping systems, the pipe length is adjusted to allow for additional friction losses due to fittings such as elbows, tees, valves, couplings, unions, and other hardware. Fittings and valves create disturbances in normal flow, called turbulence, which contribute to friction. A short run of pipe with many fittings and valves may very well have more friction loss than a long run of straight pipe with few, if any, fittings and valves.

$$L = L_{straight} + \Sigma L_{equivalent} \tag{8.102}$$

TABLE 8.13
Reiteration of Flow Calculations

Pick a value for friction factor, f.
Determine average velocity, u.
Determine Reynolds number, Re.
Calculate f by Colebrook method.
Repeat until f converges.

The term f is the Fanning friction factor, which is a function of the Reynolds number, R_e, a dimensionless indicator of the degree of turbulence, and the roughness of pipe, ϵ. For laminar (smooth, nonturbulent) flow:

$$f = \frac{16}{Re} \tag{8.103}$$

For turbulent flow, use the Colebrook equation:

$$\frac{1}{\sqrt{f}} = -4\log\left(\frac{1}{3.7}\frac{\epsilon}{d} + \frac{1.255}{Re\sqrt{f}}\right) \tag{8.104}$$

The factor, f, is independent of Re for fully developed turbulent flow:

$$\frac{1}{\sqrt{f}} = -4\log\left(3.7\frac{d}{\epsilon}\right) \tag{8.105}$$

For smooth pipes, where e = 0:

$$\frac{1}{\sqrt{f}} = 4\log\frac{Re}{1.255} \tag{8.106}$$

Finally, when Re <100,000 in smooth pipe, the Blausius approximation is used:

$$f = 0.079\,Re^{-1/4} \tag{8.107}$$

Flow rate and total mass out of a valve require reiterative calculations, based on the procedure in Table 8.13.

OPENING IN TANK

Corrosion, damage from vehicular collision, or cracks developing from stresses on welds by pipe runs can open holes in bulk liquid tanks. When a head h_L exists above the opening, the rate of liquid escape can be calculated. Assuming the gauge pressure in the tank is P_g, external pressure is atmospheric, shaft work W_s is zero, and fluid velocity in the tank is zero, Bernoulli's equation yields

$$-\frac{\Delta P}{\rho} - \frac{g}{g_c}\Delta z - F = C_1^2\left(-\frac{\Delta P}{\rho} - \frac{g}{g_c}\Delta z\right) \tag{8.108}$$

where, C_1 is a dimensionless discharge coefficient. We can find the average instantaneous leak velocity by

$$\bar{u} = C_1\sqrt{a}\sqrt{2\left(\frac{g_c P_g}{\rho} + gh_L\right)} \tag{8.109}$$

Crowl and Louvar define a new discharge coefficient C_0 as

$$C_0 = C_1\sqrt{a} \tag{8.110}$$

Plugging this back into eqn. 8.84

$$\bar{u} = C_0\sqrt{2\left(\frac{g_c P_g}{\rho} + gh_L\right)}$$

As the leaking tank drains, the liquid height, the escape velocity, and mass flow rate decrease. The instantaneous mass flow is

$$Q_m = \rho\bar{u}A = \rho A C_0\sqrt{2\left(\frac{g_c P_g}{\rho} + gh_L\right)} \tag{8.111}$$

where A is the area of the leak opening.

INERTED TANK

Suppose the tank has a blanket of inerting gas on it, such as when the liquid has an explosive vapor potential. For a tank of cross-sectional area A_t the mass of liquid above the leak is

$$m = \rho A_t h_L \tag{8.112}$$

The rate of change of mass is

$$\frac{dm}{dt} = -Q_m \tag{8.113}$$

From this we obtain

$$\frac{dh_L}{dt} = \frac{C_0 A}{A_t} \sqrt{2\left(\frac{g_c P_g}{\rho} + gh_L\right)}$$

(8.114)

Rearranging and integrating from an initial h_L^0

$$\int_{h_L^0}^{h_L} \frac{dh_L}{\sqrt{\frac{2g_c P_g}{\rho} + 2gh_L}} dt = \frac{C_0 A}{A_t} \int_0^t dt$$

(8.115)

which integrates to

$$\frac{1}{g}\sqrt{\frac{2g_c P_g}{\rho} + 2gh_L} - \frac{1}{g}\sqrt{\frac{2g_c P_g}{\rho} + 2gh_L^0} = -\frac{C_0 A}{A_t} t$$

(8.116)

The liquid height at any time t is

$$h_L = h_L^0 + \frac{C_0 A}{A_t}\left(\sqrt{\frac{2g_c P_g}{\rho} + 2gh_L^0}\right)t + \frac{g}{2}\left(\frac{C_0 A}{A_t} t\right)^2$$

(8.117)

This liquid height can be used to find mass discharge at t

$$Q_m = \rho C_0 A \sqrt{2\left(\frac{g_c P_g}{\rho} + gh_L^0\right)} - \frac{\rho g C_0^2 A^2}{A_t} t$$

(8.118)

where the first right-hand term is the initial mass discharge rate and the second is the decay term. The vessel will stop discharging when the fluid level reaches the opening at some time t_e, when Q_m is zero:

$$t_e = \frac{1}{C_0 g}\left(\frac{A_t}{A}\right)\sqrt{2\left(\frac{g_c P_g}{\rho} + gh_L^0\right)}$$

(8.119)

but at atmospheric pressure, this simplifies to

$$t_e = \frac{1}{C_0 g}\left(\frac{A_t}{A}\right)\sqrt{2gh_L^0}$$

(8.120)

Kumar, Vatcha, and Schmelzle assume that $C_0 = 0.7$ and using metric units get

---

TABLE 8.14
Specific Heat Ratios for Selected Gases

$\gamma = 1.333$	$\gamma = 1.4$
Carbon Dioxide	Air
Combustion products, hot	Carbon Monoxide
Superheated Steam	Hydrogen
	Nitrogen
	Oxygen

$$Q_m = 0.7A\rho\sqrt{19.6h_L^0 + 2P_g/\rho} \qquad (8.121)$$

for eqation 8.93. Of course this ignores the decay term and assumes constant mass flow from the leak opening.

PRESSURIZED TANK

Where the tank is pressurized and contains a fluid in the gaseous state, Kumar, Vatcha, and Schmelzle have shown that

$$Q_m = A\sqrt{\gamma P_t \rho \left(\frac{2}{\gamma+1}\right)^{\frac{\gamma+1}{\gamma-1}}} \qquad (8.122)$$

where: P_t = tank pressure
         γ = ratio of specific heats.

Table 8.14 gives values for γ for several gases.

POURING OR DUMPING

A pouring or dumping model assumes a source rate of R lb./day, if continual pouring occurs, or R' lb., if a batch dumped all at once or over one brief time period after which the exposure begins.

RUNAWAY REACTIONS

Many chemical reactions create gases and vapors, and, therefore, increase the pressure of their containment vessels. In runaway reactions, a chemical reaction is beyond the control of its operators , and too much pressure is created to continue containing the reaction. Factors that cause runaway reactions are: loss of cooling, loss of mixing power, and catalyst overload. The Boyles and Charles Laws apply for governing reactors. As the vessel relieves itself, two-phase flow occurs. The venting process is tempered because a volatile liquid is vaporizing, or flashing, thus removing energy from the exothermic reaction that is driving the runaway. Treat such a reactor as an adiabatic system. The reactor fluid increases temperature due to overpressure and sensible heat accumulates. Yet, energy is removed as liquid vaporizes and discharges. Choked, two-phase flow through a hole is

$$Q_m = \frac{\Delta H_v A}{v_{fg}} \sqrt{\frac{g_c}{C_p T_s}}$$ (8.123)

where: ΔH_v = heat of vaporization,
v_{fg} = change of specific volume of flashing liquid,
C_p = heat capacity of the fluid, and
T_s = absolute saturation temperature.

Mass flux is

$$G_T = \frac{Q_m}{A}$$ (8.124)

Sizing a vent for a reactor has been studied by several researchers. Crowl and Louvar derive this equation for the vent area A:

$$A = \frac{m_0 q}{G_T \left[\sqrt{\frac{V}{m_0} \frac{\Delta H_v}{v_{fg}}} + \sqrt{C_v \Delta T} \right]^2} = \frac{m_0 q}{G_T \left[\sqrt{\frac{V}{m_0} T_s \frac{dP}{dT}} + \sqrt{C_v \Delta T} \right]^2}$$ (8.125)

where: m_0 = mass in reactor,
q = exothermic heat released per unit mass,
V = volume of vessel, and
C_v = liquid heat capacity at constant volume.

If the vessel venting is homogenous, that is single phase, and the variation of G_T is insignificant, and q is constant, and physical characteristics (C_v, ΔH_v, and v_{fg}) are constant, the system is referred to as a tempered reactor system, then

$$q = \frac{1}{2} C_v \left[\left(\frac{dT}{dt} \right)_s + \left(\frac{dT}{dt} \right)_m \right]$$ (8.126)

where S refers to set pressure and m to maximum turnaround pressure.

FIRE

Fire is the rapid oxidation of a material during which heat and light are emitted. Rapid oxidation is fire or combustion whereas slow oxidation (of metals usually) without heat and light is corrosion. Combustion is an exothermic, self-sustaining reaction between oxygen and a fuel. Anything that burns is a fuel. Solid fuels may glow or smolder but gaseous fuels emit visible flames.

Combustion is initiated when a fuel in the presence of oxygen reaches a minimum temperature called ignition temperature. Piloted ignition occurs when some external source such as a flame, spark, ember or heat causes the ignition. Autoignition occurs when there is no external source. Other names for autoignition are spontaneous combustion and spontaneous ignition. The autoignition temperature of any given material is higher than its ignition temperature.

Fire, therefore, requires four things to be sustained: fuel, oxygen, heat, and a chain reaction. Without any one of these components, the collective of which is called the Fire Pyramid, the fire will go out. Fire, whether raging or smoldering, is chemically the rapid, exothermic oxidation of an ignited fuel. The actual combustion reaction occurs only in the vapor phase, liquids and solids do not actually burn. Your eyes tell you differently, but the liquids are volatilized and the vapor burns. Solids either melt and are subsequently volatilized prior to combustion or decompose directly to vapor just prior to combustion in a reaction called pyrolysis. The vapors are produced by deep seated heat in the solid object and no chain reaction occurs. That's why you think the solid is burning, you are observing significant heat damage, but the flame, if you observe carefully, dances just slightly above the solid, where the vapor is being released.

The flame is premixed if air was mixed with the fuel on purpose, such as in a stove burner, or in an internal combustion engine. The fires of concern for exposure assessment have diffusion flames where escaping vapors mix with air by diffusion.

Liquid fuels, such as gasoline and diesel fuel, are ubiquitous in the United States. In industrial plants, many solvents can be found which are also liquid fuels: acetone, ethers, acetates, ketones, xylene, toluene, to name a few. Gaseous fuels that are common include acetylene, propane, carbon monoxide, and hydrogen. Solid fuels include such materials as wood, sawdust, fibers, plastics, fabric, paper, and metal particles.

Oxygen is supplied directly by the surrounding air, but also from compressed air, and bottled oxygen. Gaseous fluorine and chlorine also act as oxidizers in a fire. Acids such as hydrogen peroxide, nitric acid, and perchloric acid are oxidizers also. Solid oxidizers include metal peroxides and ammonium nitrate.

Temperatures required to sustain combustion depend on the fuel. Ordinary hydrocarbons burn around 300°F. When you think of heat, as part of the fire pyramid, think of ignition sources, which as a source of heat can add energy to the fuel and oxygen combination and make the combustion process initiate.

The chain reaction is not too important with solid fuels, but in the burning of vapors and gases it is critical. This step involves the production of hydroxyl (OH^-) radicals, which only last about one millisecond.

Fire will occur only if all the elements of the Fire Triangle (fuel, oxygen, heat), or chain reaction for liquids and vapors, is present at the same time. If fuel is not present, fire will not occur. If the amount of fuel present is insufficient, the fire will start and go out, assuming the heat produced does not ignite some other fuel nearby in the presence of oxygen. Fire will not occur if oxygen is not present. Even if full

TABLE 8.15
Sources of Flash Points in Literature

Chemical Process Safety: Fundamentals with Applications
[Crowl & Louvar, Englewood Cliffs, NJ: Prentice–Hall, 1990.]
The Condensed Chemical Dictionary
[Hawley, ed., New York: Van Nostrand Reinhold, 1981.]
Engineering Design for Control of Workplace Hazards
[Wadden & Scheff, New York: McGraw–Hill Book Company, 1987.]
The Merck Index: An Encyclopedia of Chemicals, Drugs, and Biologicals
[Windholtz, ed., Rahway, NJ: Merck & Co., 1983.]

TABLE 8.16
Sources of Flammability Limits in Literature

CRC Handbook of Chemistry and Physics
[Weast, ed., Boca Raton, FL: CRC Press, latest edition.]

combustion does not occur, enough oxygen may be present to smolder, or keep embers glowing for awhile. If the oxygen that is consumed is not replenished during combustion, the oxygen concentration can drop to the point where combustion ceases. The ignition source must have sufficient energy to initiate the fire or fire does not occur. Once fully started, the fire supplies its own heat, and the ignition source is no longer relevant.

The *autoignition temperature (AIT)* is the fixed temperature for each unique mixture of compounds above which its vapors are capable of extracting sufficient energy from the surroundings to self-ignite. The *flash point (FP)* is the lowest temperature at which a liquid gives off enough vapor to form an ignitable mixture with air over the bulk liquid. Vapor at the flash point will ignite and burn momentarily until the vapor is consumed. Table 8.15 and 8.16 list some sources for experimentally determined flash points and flammability limits respectively. Note that in other English speaking countries, the terms *inflammable* and *inflammability* mean flammable and flammability respectively. In our American brand of English, these in-terms have fallen into disuse, and some confusion exists by those who think inflammable means not flammable, when, in fact, it means the opposite.

To sustain combustion of vapors over a liquid, the bulk temperature of the liquid must be raised to the point where the vapor generation by evaporation is equal to the vapor consumption by combustion. This higher temperature is called the *fire point.* Definitions of hazardous by EPA, OSHA, and DOT are based on the flash point, but, it is the fire point that interests the modeling engineer. Vapor–air mixtures only ignite and burn if both the temperature and the concentration is right. In fact, for a given fuel vapor–air mixture, only a specific range of compositions will burn. Anything beneath that range is called *fuel lean* or *air rich* and any composition beyond the range is called *fuel rich* or *air lean.* The range itself is called the

TABLE 8.17
Stages of Fire

Incipient Stage
Significant combustion particles, but no smoke, flame, or radiant heat. Invisible smoke particles obey gas
  laws; rise to ceiling where they can be detected by ionization detectors.
Smoldering Stage
Combustion particles increase in size; become visible smoke, detectable by human eye/photoelectric
  detectors. No flame or significant heat has developed yet.
Flame Stage
Point of ignition occurs; flames start. Visible smoke decreases, but heat increases, giving off infrared
  energy, detectable by infrared detectors.
Heat Stage
Large amounts of heat, flame, smoke, and toxic gases are generated. Thermal detectors are used to detect
  heat.

flammability limits. The point at which the mixture becomes fuel lean is called the
lower flammability limit *(LFL)*. The upper flammability limit *(UFL)* is the upper
end of the range where the mixture is too rich in fuel.

Fire has four stages, shown in Table 8.17.

An explosion is a very rapid fire, generally lasting for the duration of milliseconds. A fire can result from an explosion or an explosion may result from a fire.
They are merely two species of the same physical and chemical phenomena.

The Le Chatelier equation is used to calculate the LFL or UFL of a flammable
mixture under limited conditions:

$$LFL_{mix} = \frac{1}{\sum_{i=1}^{n} \frac{y_i}{LFL_i}} \tag{8.127}$$

and

$$UFL_{mix} = \frac{1}{\sum_{i=1}^{n} \frac{y_i}{UFL_i}} \tag{8.128}$$

where y_i is the mole fraction of the ith component.

Flammability does depend on temperature as pointed out by reviewers Crowl
and Louvar (p. 164, 1990):

$$LFL_T = LFL_{25}\left[1 - 0.75(T - 25)/\Delta H_c\right] \tag{8.129}$$

and

$$UFL_T = UFL_{25}\left[1 + 0.75(T - 25)/\Delta H_c\right] \tag{8.130}$$

where DH_c is the net heat of combustion (kcal/mole), and the temperature T is given in Celsius.

The relationship of flammability to pressure was also reviewed by Crowl and Louvar, and only the UFL has any appreciable dependence on pressure:

$$UFL_p = UFL + 20.6\left(\log^{P+1}\right) \tag{8.131}$$

For fuel at stoichiometric concentration, C_{st}, in % volume:

$$C_{st} = \frac{LFL}{0.55} = \frac{UFL}{3.50} \tag{8.132}$$

Explosions can be mechanical or the result of rapid chemical reaction. Damage by explosions is done by the pressure or shock wave. The terms *lower explosive limit (LEL)* and *upper explosive limit (UEL)* are identical to LFL and UFL respectively and do not differ in meaning or numerical value. Mechanical explosions are caused by the rupture of a vessel due to pressure build up of a nonreactive gas. Hazard literature refers to this as the sudden release of pressure. An exploding compressed air bottle would be an example of a sudden release of pressure or mechanical explosion. However, be careful, because not every sudden release of pressure may be considered a mechanical explosion.

The difference between a mechanical explosion and a chemical explosion is that the former does not involve fire. Also, mechanical explosions are always confined, whereas chemical explosions may be confined or unconfined. A confined explosion occurs within a vessel or building, while unconfined explosions occur in the open. Unconfined explosions are not common because the vapor–air mixture is usually quickly diluted by air and the LFL is not reached, however, unconfined explosions tend to be very destructive since large quantities of fuel vapor must be involved to achieve an LFL in open air. A subcategory of chemical explosions is the dust explosion consisting of fine solid particles dispersed in air as an aerosol. Any solid material, if ground fine enough to provide sufficient surface area to absorb heat rapidly, and the heat source is sufficient to provide the ignition energy required, will explode. Steel dust will explode, for example, if finely divided and sufficient ignition energy is provided.

Explosions are also subdivided according to degree of damage potential. A *deflagration* occurs when the resulting shock wave moves at a velocity less than the speed of sound in the unreacted medium. A *detonation* is an explosion that produces a shock wave that moves faster than the speed of sound in the unreacted medium. The shock wave itself is a pressure wave that moves through the air. In open air, the shock wave is followed by a strong gust of wind caused by air replacing the vacuum immediately behind the pressure front. The combined shock wave and following wind is the blast wave. The pressure increase in the shock wave happens so fast that the process is adiabatic, that is, no heat is transferred to or from the surroundings. The pressure exerted on an object by the impact of a blast wave is

TABLE 8.18
Effects of Overpressure

If Overpressure Is:	Expect:
0.5–1.0 psi	glass to shatter
1.0 psi	people to be knocked down
2.0–3.0 psi	8–12 inch block to shatter
	8–12 inch concrete wall to shatter
5.0 psi	utility poles to snap
	some eardrums to rupture
7.0 psi	loaded rail car to overturn
11.0 psi	some lung damage
15.0 psi	50% of eardrums to rupture

called overpressure and provides a measure of the strength of the blast. Damage by overpressure is as follows in Table 8.18.

BLEVE

A special kind of explosion, the nightmare of every fireman and emergency responder, is the BLEVE (pronounced BLEH–vee). The acronym BLEVE stands for Boiling Liquid Expanding Vapor Explosion, which is descriptive of the chain of events leading to the explosion. An external flame or radiant heat from an external fire impinges on the surface of the tank, heating the liquid contents and raising its vapor pressure. As the walls heat up, the structural integrity of the tank is weakened. Eventually, the liquid reaches its boiling point and the vapor pressure is atmospheric or greater depending on whether some vapor escape rapidly enough. When the last of the liquid is boiled off, the vapor is superheated and, with the right amount of oxygen, an explosive mixture is created. The tank then becomes a huge bomb. The BLEVE itself is the explosive vaporization of the remaining vessel contents, often followed by the combustion of the vaporized cloud that preceded it. If you have never seen a BLEVE , several good videos are available from training video vendors and fire departments. First, the tank explodes, then, depending on whether the liquid is combustible, the vapor cloud explodes. Double whammy of catastrophic proportions. As said earlier, BLEVEs are every emergency responder's worst nightmare.

The volume of vapor formed when a liquid evaporates is found by relating the densities of vapor and liquid:

$$V_v = \frac{8.33 \times \text{sp.gr.}}{0.075 \times \text{vapordensity}} = \frac{111 \times \text{sp.gr.}}{\text{vapordensity}} \tag{8.133}$$

where V_v is in ft.^{3}/gallon of liquid.

QUESTION (ANSWERS BEGIN ON PAGE 401)

1. What concentration of benzene is expected from a container of a solvent mixture that contains 30 mole percent toluene and 40 mole percent benzene, if the vapor pressure of toluene is 29.9 mmHg and of benzene is 95.9 mmHg respectively and the toluene concentration measures 150 ppm?

REFERENCES

American Conference of Governmental Industrial Hygienists. *Industrial Ventilation: A Manual of Recommended Practice.* 21st Ed. 1988.

Arpaci, Vedat S. *Conduction Heat Transfer.* Reading, MA: Addison–Wesley Publishing Co., 1966.

Ayers, Kenneth W. *et al. Environmental Science and Technology Handbook.* Rockville, MD: Government Institutes, Inc. 1994.

Bauer, Roger L. *Safety and Health for Engineers.* New York: Van Nostrand Reinhold, 1990.

Bird, R. Byron, Warren E. Stewart, and Edwin N. Lightfoot. *Transport Phenomena.* New York: John Wiley & Sons. 1960.

Burmaster, David E. and Jeanne W. Appling. "Introduction to Human Health Risk Assessment, with an Emphasis on Contaminant Properties." In: *Environment Reporter.* April 7, 1995, pp. 2431–40.

Caravanos, Jack. *Quantitative Industrial Hygiene: A Formula Workbook.* Cincinnati: American Conference of Governmental Industrial Hygienists. 1991.

Churchhill, Ruel V. *Fourier Series and Boundary Value Problems.* 2d ed. New York: McGraw–Hill Book Company, 1969. [The original was published in 1941.]

Clement Associates. *Mathematical Models for Estimating Workplace Concentration Levels: A Literature Review.* Washington, D.C.: Office of Toxic Substances, U.S. Environmental Protection Agency, Contract 68-01-6065, 1981.

Crane Technical Paper Number 410. *Flow of Fluids through Valves, Fittings, and Pipe.* New York: Crane Co. 1986.

Crowl, Daniel A. and Joseph F. Louvar. *Chemical Process Safety: Fundamentals with Applications.* Englewood Cliffs, NJ: Prentice–Hall, Inc. 1990.

Daneshyar, H. *One-Dimensional Compressible Flow.* New York: Pergamon Press, 1976.

Desmarais, Anne Marie C. and Paul J. Exner. "The Importance of The Endangerment Assessment in Superfund Feasibility Studies." *Hazardous Materials Control Monograph Series: Health Assessment.* Silver Spring, MD: Hazardous Materials Control Research Institute.

EPA. 40 CFR 300 National Oil and Hazardous Substances Pollution Contingency Plan. *47 FR 31203.* July 16, 1982.

Hass, H.B. and R.F. Newton. "Correction of Boiling Points to Standard Pressure." In *CRC Handbook of Chemistry and Physics.* 56th Ed. Cleveland: CRC Press, 1974. p. D–176.

Jayjock, Michael. *Introduction to Health Risk Assessment.* A Presentation given as Professional Development Course No. 306 at the 1996 American Industrial Hygiene Conference & Exposition, May 18, 1996, Washington, D.C.

Jayjock, Michael and Neil C. Hawkins. "A Proposal for Improving the Role of Exposure Modeling in Risk Assessment." *American Industrial Hygiene Association Journal.* December 1993, pp. 733–741.

Jayjock, Michael, Chris Keil, Mark Nicas, and Patricia Reinke. *A Tool Box of Mathematical Models for Occupational Exposure Assessment.* Part of a presentation given as Professional Development Course No. 402 at the 1996 American Industrial Hygiene Conference & Exposition, May 19, 1996, Washington, D.C.

Karam, Joseph G. and Martha J. Otto. "Ocean Disposal Risk Assessment Model." *Hazardous Materials Control Monograph Series: Health Assessment.* Silver Spring, MD: Hazardous Material Control Research Institute.

Kay, Robert L. and Chester L. Tate, Jr. "Public Health Significance of Hazardous Waste Sites." *Hazardous Materials Control Monograph Series: Health Assessment.* Silver Spring, MD: Hazardous Materials Control Research Institute, pp. 65–71.

Kumar, Ajay, N.S. Vatcha, and John Schmelzle. "Estimate Emissions from Atmospheric Releases of Hazardous Substances." *Environmental Engineering World.* November–December 1996, pp. 20–3.

Major, Coleman. "Values of the Gas Constant, R." In *CRC Handbook of Chemistry and Physics.* 56th Ed. Cleveland: CRC Press, 1974. p. F–232.

Matthiessen, R. Craig. "Estimating Chemical Exposure Levels in the Workplace." *Chemical Engineering Progress.* April 1986, pp. 30–4.

McCabe, Warren L. and Julian C. Smith. *Units Operations of Chemical Engineering.* 2d Ed. New York: McGraw–Hill Book Company, 1967.

McElroy, Frank E. ed.-in-chief. *Accident Prevention Manual for Industrial Operations: Administration and Programs (Vol. I).* 8th. ed. Chicago: National Safety Council, 1981.

National Safety Council. *Accident Prevention Manual for Industrial Operations: Engineering and Technology (Vol. II).* 9th. ed. Chicago: National Safety Council, 1988.

Perry, Robert H., Don W. Green, and James O. Maloney. *Perry's Chemical Engineers' Handbook.* 6th ed. New York: McGraw–Hill Book Company, 1984.

Preuss, Peter W., Alan M. Ehrlich and Kevin G. Garrahan. "The U.S. Environmental Protection Agency's Guidelines for Risk Assessment." *Hazardous Materials Control Monograph Series: Risk Assessment Volume I.* Silver Spring, MD: Hazardous Materials Control Research Institute, pp. 6–13.

Quagliano, James V. *Chemistry.* Englewood Cliffs, NJ: Prentice– Hall, Inc. 1958.

Reid, Robert C., John M. Prausnitz, and Bruce E. Poling. *The Properties of Gases & Liquids.* 4th ed. New York: McGraw–Hill Book Company, 1987.

Reinke, Patricia, Michael Jayjock, Chris Keil, and Mark Nicas. *Workplace Exposure Modeling: A Toolbox of Mathematical Models for Occupational Exposure Assessment.* A Presentation given as Professional Development Course No. 402 at the 1996 American Industrial Hygiene Conference & Exposition, May 19, 1996, Washington, D.C.

Rodricks, Joseph V. "Comparative Risk Assessment: A Tool for Remedial Action Planning." *Hazardous Materials Control Monograph Series: Risk Assessment Volume I.* Silver Spring, MD: Hazardous Materials Control Research Institute, pp. 45–8.

Schewe, George J., Joseph Carvitti and Joseph Velten. "Human Exposure Estimates Using U.S. EPA Guideline Models: An Integrated Approach." *Hazardous Materials Control Monograph Series: Health Assessment.* Silver Spring, MD: Hazardous Materials Control Research Institute, 9–13.

Smith, J.M. and H.C. Van Ness. *Introduction to Chemical Engineering Thermodynamics.* 3rd ed. New York: McGraw–Hill Book Company, 1975.

Valsaraj, Kalliat T. *Elements of Environmental Engineering: Thermodynamics and Kinetics.* Boca Raton, FL: CRC Press/Lewis Publishers, 1995.

Wadden, Richard A. and Peter A. Scheff. *Engineering Design for the Control of Workplace Hazards.* New York: McGraw–Hill Book Co., 1987.

9 Fundamentals of Pathway Transport Assessment

In order to determine whether a gas, liquid, or solid in the workplace may result in an injury, an apparently reasonable pathway linking the chemical substance to the injury site must be assumed. However, in the environment, we generally have to show that a pathway exists, although in some cases one is merely assumed, especially in a worst case scenario. A pathway, then, is any route the COC may take from its environmental source to a receptor. Table 9.1 shows the five main parts of an exposure pathway.

Earlier we discussed the source and release of COCs, and later we'll discuss exposure points and routes of entry. Now, we will examine transport or exposure media. The medium in which a contaminant is present determines the human exposure by providing the pathway through which a dose is received (daily human dose — DHD).

Contaminants that end up in water will be ingested directly by drinking, or indirectly by eating fish or other aquatic flora/fauna. Exposure by inhalation may occur, if the contaminated water is boiled, and to a lesser degree in the shower. Contaminants that evaporate into the atmosphere will be inhaled. Contaminants in soil will be directly absorbed through the skin, by children who play in dirt, and by farmers, gardeners, and others who work in it. Soil contamination may be taken up by plants which are eaten by livestock and end up in milk and meat. People eat the meat, the plants too, and drink milk.

Demonstration that a feasible pathway exists is prerequisite to demonstrating injury but evidence of a pathway alone is not conclusive. For instance, in the environment, if a prey species such as a mouse has a certain level of poison we can conclude that some predator species such as hawks has a plausible pathway linking it to the poison but we cannot thereby conclude that the predator species has been injured. The potential exists however. Or if, in a workplace, one operator of a process presents signs and symptoms of overexposure to a particular chemical, we cannot say that all operators of that process have been similarly injured though we may definitely conclude that the potential exists.

Pathway analysis includes evaluating the series of events by which a substance has been transported from a source such that it comes into direct contact with a human or some natural resource. A complete pathway has no permanent barriers preventing or minimizing exposure. Without a complete pathway, no exposure exists. Determination of pathway may also include the evaluation of the sequence of events by which a substance has been transported from the source and has caused an indirect impact on a human or natural resource. For instance, oil may be transported from a spill site by ocean currents, wind and wave action and cause reduced populations of some fish species which leads to starvation of birds which prey on those fish.

TABLE 9.1
Parts of an Exposure Pathway

A source of the COC (chemical of concern),
A release mechanism,
A transport and/or exposure medium,
AN exposure point with receptors present or potentially present, making exposure possible, and
A route of entry.

An important consideration about the contaminant species is whether it is persistent and stable in the environment, at its source, along some pathway, and at some remote receptor site. Persistence and stability are the topic of Chapter 3 relative to human exposure and Chapter 4 relative to the environment. Any number of physical, chemical and biological processes may act to assist or hinder the transfer of a contaminant. The degradation and distribution of a contaminant in the environment and the quantity which is predicted to arrive at the remote receptor is what this chapter is about.

The historical assumption, made by public leaders and industry regarding the protection of public health with respect to chemical exposure, was that natural processes degrade a chemical into a relatively harmless state or reaction product at a rate which eliminates the original hazard. This is the case to some degree. For some chemicals however, natural processes only reduce, not eliminate the hazard. For some, the rate of transformation is so slow that the natural processes do little good for humans and the environment. For yet other chemicals, natural processes actually exacerbate the original hazards.

Therefore, highly persistent species are a particular concern because they have the opportunity to bioaccumulate in the food chain. Persistence is an even greater concern where transport is feasible. In these situations, even low levels of a toxic chemical can bioaccumulate. Persistence is evaluated based on the half-life of the contaminant, and on the sorption of the contaminant to soils and sediments.

The half-life, $t_{1/2}$, of a substance is calculated as follows:

$$t_{1/2} = \frac{1}{1/h + 1/b + 1/r + 1/p + 1/v} \tag{9.1}$$

where: h = hydrolysis half-life,
         b = biodegradation half-life,
         r = radioactive half-life,
         p = photolysis half-life, and
         v = volatilization half-life.

The component half-lives are estimated from available data in handbooks and other resources.

Photolysis is one mechanism by which organic chemical transport is affected. Sunlight transforms some organic chemicals in or on media, producing other products, which may be more or less toxic than the original species. Microbes play a major role in transport, as the instruments of *biodegradation*. The product of a biodegradation process is called a *biodegradate*. *Hydrolysis* produces *hydrolytic products*. Sorption to soils and sediments is based on the logarithm of the octanol–water partition coefficient.

An important characteristic of chemicals is their solubility in water, which is everywhere in nature. More soluble chemicals proportionally more mobile in the environment than lesser soluble chemicals. However, the lesser soluble chemicals are more likely to be accumulated in the environment, to bioaccumulate, to volatilize. Highly soluble chemicals are more likely to undergo biodegradation as the microbes can typically deal more effectively with weak aqueous solutions than concentrated chemicals that may kill them. Adsorption is more likely to bind chemicals with low solubility in water.

Exposure to chemicals depends on one of two classes of pathways: transport and exposure pathways. Transport pathways involve the displacement of the chemical, henceforth called contaminant, from some source of contamination to a receptor site. Such pathways include movement through water column, water surface, sediments, beaches, floodplain, groundwater, soil, air, bioaccumulation and food-chain uptake. Exposure pathways are those avenues by which humans directly absorb or react to contamination through contact, inhalation or ingestion. Transport pathways are typically found in environmental settings and exposure pathways in the workplace but instances where the reverse is true are possible.

Procedures for exposure pathway analysis includes interviews with exposed individuals, collection of site data and, typically, measurement of the contaminant in the environment. Procedures for transport pathway analysis include field observations, bench studies, modeling, and literature searches, alone or in combination. The relationship of transport pathways to exposure pathways is shown in Table 9.2. These pathways may be subcategorized as direct or indirect. When the COC is received in the same medium as it was released at its source, a direct pathway exists. For example, if a stack is emitting PM-10 and residential neighbors are inhaling PM-10 contaminated air blowing across the factory property onto theirs, the pathway is direct. If the receptor takes in the COC in a medium different from the release, the pathway is indirect. For instance, if a mill releases dioxin into a river, fish ingest contaminated water, a fisherman catches contaminated fish, his wife prepares the fish as a meal, and the fisherman, his wife and children ingest dioxin at dinner — the pathway was indirect.

TRANSPORT FUNDAMENTALS

Chemicals partition between environmental media based on their chemical and physical properties. The mechanisms for transport of molecules of contaminants are two: diffusion and advection. Diffusion is the transportation of molecules from regions of higher to lower concentration. Advection is the transport of molecules from regions of greater to lesser pressure or temperature. According to Silka (in Ayers, et al.), advection accounts for the bulk of transport over environmental pathways, mixing by advection being perhaps a million times faster than by diffusion.

TABLE 9.2
Transport v. Exposure Pathways

Transport Pathways	Work	Drink Water	Swim/Bath	Land Recreate	Indoor Air	Outdoor Air	Food	Fire and/or Explosion
GW	I,C	S	C,S,I		I			H
SW	I,C	S	C,S,I			I	S	H
Soil/ Sediment	I,C	S	C,S,I	C,S,I		I,S	S	
Dust	I,C,S	S		S,I	I,S	I,S	S	
Rain	S	S				I,S	S	
Leaky Drums	I,C,S	S	C,S	C,S,I		I	C	H
Other	I,C,S		C,S	C,S,I			S	H

Note: GW = groundwater; SW = surface water.

Exposure pathways: I = inhalation, C = skin/eye contact; S = ingestion, H = safety hazard.

Diffusion is the mechanism that disperses molecules of solvent in the air from an open container in a closed, unventilated storeroom. When a worker opens the storeroom after a long weekend the concentration of the solvent inside the storeroom is high enough to give her an acute dose sufficient to cause her to pass out.

In a slightly different scenario a ventilation system is left operational over a long weekend and a much lower concentration is immediately detected all over the plant when it is opened on Monday morning. The same amount of solvent is evaporated but advection currents created by the ventilation system spread the same number of molecules farther, faster and the overall concentration of solvent in the air is lower than the buildup near the open container by diffusion.

DIFFUSION MECHANISM

Whenever one fluid is in another, concentration gradients occur as a result of directional movements of cross currents and diffusion of mass through space. Diffusion is the tendency of a fluid (gas, vapor or liquid) to disperse into and mix with another fluid. As stated above this process is driven by a concentration gradient and the spread of molecules by diffusion is random and called mass flux, described as mass per unit time and unit area. The resultant concentration gradient is described in terms of mass per unit volume unit length. A proportionality constant, called *diffusivity*, is the ratio of the flux and concentration gradient, described in terms of area per unit time. This describes the increasing surface area of a growing hemisphere of a mass of pollutant diffusing into a fluid.

Say, for example, that a source G emits X molecules of a pollutant at time t. One hundred of the molecules from G are assumed to be on the surface of a hemisphere containing all molecules emitted at time t. This sphere grows as the molecules disperse and the density of molecules decreases. A corresponding decrease in concentration of X occurs as we move away from the source. Diffusivity is the rate of growth of the surface area of the hemisphere.

Molecular diffusion is driven strictly by the concentration gradient. It is slow and does not typically occur in workplaces. The molecular diffusivity of trichloro-ethylene is 0.0003 m^2/min, for instance.

Turbulent diffusion is another matter. The turbulent motion of any fluid is called eddies, or eddy movement, or eddy motion. Eddies transport mass as they move in a fluid. Intuitively, large eddies cause rapid mass transport, while small eddies cause slower mass transport. Mass transport by eddies is called diffusivity, D. The values of this property for any species of concern is dependent on the fluid through which it disperses and cannot be estimated, but must be determined experimentally. Liter-ature values, according to Keil, range from 0.05 to 11.5 m^2/min, with typical values ranging from 0.3 to 3 m^2/min. Keil calls the study of diffusivity *one of the last great frontiers of Newtonian mechanics.* Most of the work being done on the subject is limited to restricted fluid flow.

When the transport of molecules is governed by diffusion under a concentration gradient in a stagnant fluid (such as air or water) it is called *Fickian* motion from *Fick's second law* which says that diffusion occurs at a specific rate depending upon the diffusion coefficient of the species being diffused and the volume into which it is diffusing and this law is expressed in steady-state terms in Cartesian coordinates as:

$$\frac{C_{(x,y,z)}}{t} = \frac{D_x^2 C}{x^2} + \frac{D_y^2 C}{y^2} + \frac{D_z^2 C}{z^2} \tag{9.2}$$

where: C = concentration of the chemical
D = diffusion coefficient, subscript indicates direction
x = distance traveled in x-direction,
y = distance traveled in y-direction, and
z = distance traveled in z-direction.

Let us define our system such that its outer boundary in one dimension has a concentration of zero, Eqn. 9.2 becomes

$$\frac{C_{z,t}}{C_{0,0}} = \text{erfc}\left[\frac{z}{4\,Dt^{0.5}}\right] \tag{9.3}$$

Here, $C_{z,t}$ is the concentration at distance z and time t and $C_{0,0}$ is the initial concen-tration and erfc is the complimentary error function.

Neil, referring to Carslaw and Jaeger, presents a slightly different version of this equation in radial coordinates:

$$C = \frac{G}{2\pi Dr}\left[1 = \text{erfc}\left(\frac{r}{\sqrt{4Dt}}\right)\right] \tag{9.4}$$

where: C   = concentration at time t,
        G   = generation rate of COC,
        D   = eddy diffusivity,
        r   = radial distance from source, and
        erfc = error function.

The error function is available in tables, can be calculated from probability tables, or can be estimated, so Keil gives us Equation 9.4 estimated as

$$C = \frac{G}{2\pi Dr}\left[1 - \sqrt{\left(1 - e^{-\frac{r^2}{\pi tD}}\right)}\right] \qquad (9.5)$$

which, at steady state becomes,

$$C = \frac{G}{2\pi Dr} \qquad (9.6)$$

If, for instance, a trichloroethylene degreaser loses one liter of material per shift when operational what is the expected concentration of at a workstation one meter away at 1, 10, 30, and 60 minutes after the degreaser is uncovered? You determine that $D = 0.5$ m²/min from a handbook.

$$C = \frac{3}{2\pi(0.5)(1)}\left[1 - \text{erfc}\left(\frac{r}{\sqrt{4(0.5)t}}\right)\right] = 0.955\left[1 - \text{erfc}\left(\frac{r}{\sqrt{2t}}\right)\right]$$

time, min	erfc	C, mg/m³
1	0.683	0.515
10	0.224	0.907
30	0.145	0.937
60	0.103	0.946

So, in this case, the diffusion occurs rapidly in the first ten minutes, then levels off
    Along any transport pathway, the concentration of a contaminant relative to everything else is controlled by any one or more of four processes: partitioning, dispersion, adsorption and degradation. Table 9.3 summarizes the diffusion system.
    Separation or partitioning between co-existing phases (air–water, water–soil, air–soil, chemical–air, chemical–water, chemical–soil) is controlled by the equilibrium concentration of the species in each phase.

$$K_{1-2} = \frac{C_1}{C_2} \qquad (9.7)$$

TABLE 9.3
Diffusion System

Loading	Inflow rates of toxics.
Speciation	Acid-base equilibrium (pH)
	Sorption
Transport	Precipitation/dissolution
	Advection
	Volatilization
	Sedimentation
Transformation	Biodegradation
	Photolysis
	Hydrolysis
	Radioactivity
	Oxidation/reduction
Bioaccumulation	Bioconcentration
	Biomagnification

TABLE 9.4
One Gram-Mole Dry Air at 77°F, 1 atm

Volume = 24.5 liters = 48.4 ft.^{3}
Mass = 28.97 grams = 0.0638 lb.

where: K_{1-2} = partition coefficient between phases 1 and 2
C_1 = equilibrium concentration in phase 1
C_2 = equilibrium concentration in phase 2

K_{1-2} becomes the soil-water partition coefficient for organic carbon for the COC (K_{oc}) in soil. Modeling of contaminant plumes from air sources such as stacks or spills or leaks of chemicals is based on partitioning and dispersion.

Consider a pool of any chemical species, even water, on the ground. At any given temperature the amount of chemical in the air over the chemical pool is a function of its partial pressure. Generally, the amount of chemical is expressed in the concentration term mole fraction (m_a):

$$m_a = \frac{P_a}{P_T} \tag{9.8}$$

where: P_a = vapor pressure of the chemical at temperature and
P_T = total pressure of the head space, 1 atm in the open.

Table 9.4 gives information about one mole of dry air at 77°F and 1 atm.

A contaminant "plume" spreads into uncontaminated air or water by means of heterogeneities in the movement of the air or water. This mixing with uncontaminated media (air or water) is called dispersion. Changes in temperature and pressure in the atmosphere creates turbulence or wind which disperses a contaminant plume. Ocean currents created by temperature differences can disperse contaminants over a wide range of ocean. Dispersion occurs in three dimensions and a convention is followed when describing dispersion in Cartesian coordinates.

x-direction: parallel with flow (down stream/up stream)
y-direction: perpendicular to flow but in the same plane (across stream)
z-direction: perpendicular to flow and perpendicular to the plane of flow (vertical)

When there is no flow or no turbulence (laminar flow) dispersion ceases. However, even in no flow situations heterogeneities other than flow can create a driving force for dispersion.

A mass balance around a differential element above a liquid pool gives

$$G = \frac{13.379 M P^{vap} A}{T} \left(\frac{D_{ab} v_z}{\Delta z} \right)^{0.5}$$

(9.9)

where: G = generation rate, lb./hr,
M = molecular weight, lb./lb.mole,
P^{vap} = vapor pressure, in. Hg,
A = area, ft.^{2},
D_{ab} = diffusion coefficient, ft.^{2}/sec, a through b,
v_z = air velocity, ft./min,
T = temperature, K, and
Δz = pool length along flow direction z.

The diffusion coefficient is for the movement of species a, some COC, through species b, air (or water or soil).

Once evaporation of a pool reaches equilibrium, the surface temperature is constant and the heat of vaporization is provided by surrounding materials so that the evaporation rate is constant. Let the direction that is normal to the surface of the pool be the x direction and the direction of air flow across the pool be the z direction. According to Bird, Stewart, and Lightfoot,

$$\frac{\delta N_{A,z}}{\delta_z} + \frac{\delta N_{A,x}}{\delta x} = 0$$

(9.10)

where: $N_{A,z}$ = molar flux of A in z direction, moles/time area, and
$N_{a,x}$ = molar flux of A in z direction, moles/time area.

In the z direction, we assume no changes in concentration, $C_A(x,z)$, occur as the result time, and neglect the contribution from diffusion of A in the z direction assuming it is small compared to air velocity. Diffusion at the edges of the pool in the direction normal to the air flow (y direction) is assumed to be negligible too. The air velocity is assumed to be constant with respect to the x direction and to change only in the z direction. Eqn. 9.7 becomes

$$N_{A,z} = C_A v_z \qquad (9.11)$$

With no net flow in the x direction, we can neglect $N_{A,x}$. So,

$$N_{A,z} = -D_{a,b} \frac{\delta C_A}{\delta x} \qquad (9.12)$$

At the bulk liquid temperature,

$$v_z \frac{\delta C_A}{\delta z} = D_{a,b} \frac{\delta^2 C_A}{\delta x^2} \qquad (9.13)$$

where the boundary conditions are

$$\begin{aligned} C_A &= 0 &&\text{at } z = 0 \\ C_A &= C_{A,0} && x = 0 \\ C_A &= 0 && x = \infty \end{aligned}$$

The solution to Equation 9.10 is

$$\frac{C_A}{C_{A,0}} = 1 - \frac{2}{\pi} \int_0^{\frac{x}{\sqrt{4D_{AB}z/v_z}}} e^{-\eta^2} d\eta = \text{erfc}\left(\frac{x}{\sqrt{4D_{AB}z/v_z}} \right) \qquad (9.14)$$

The complementary error function (erfc) is shorthand for the integral in Equation 9.11. The vapor concentration is a function of x and z. At any point of the liquid surface, the evaporation rate is

$$N_{A,x}\big|_{x=0} = -D_{AB} \frac{\delta C_A}{\delta x} \qquad (9.15)$$

but the total evaporation is

$$R_{evap} = 1/\left(A_{pool}\right) \int \int N_{A,x} dz dx = 1/\left(A_{pool}\right) \int \int -D_{AB} \frac{\delta C_A}{\delta x} dz dx \qquad (9.16)$$

EPA solved the partial derivative in Equation 9.13 by differentiating Equation 9.11:

$$N_A \text{ total} = \frac{1}{\Delta y \Delta z} \int \int \frac{2D_{AB}C_{A,0}}{\sqrt{4\pi D_{AB} z/v_z}} dz dy = 2C_{A,0}\sqrt{\frac{D_{AB}v_z}{\pi \Delta z}} \qquad (9.17)$$

yielding moles/time area. In terms of mass,

$$R_{evap} = 2MWC_{A,0}\sqrt{\frac{D_{AB}v_z}{\pi \Delta z}} \qquad (9.18)$$

but

$$C_{A,0} = \frac{P^{vap}}{RT} \qquad (9.19)$$

so that, in lb./hr ft.^{2}

$$R_{evap} = \frac{13.379 MW P^{vap}}{T}\sqrt{\frac{D_{AB}v_z}{\Delta z}} \qquad (9.20)$$

where: MW   = molecular weight,
         P^{vap} = vapor pressure, in. Hg,
         T   = temperature, K,
         D_{AB} = diffusion coefficient of vapor in air, ft.^{2}/sec,
         v_z = air velocity, ft./min, and
         Δz = pool length along z-direction, ft.

Bird, Stewart and Lightfoot give the mass diffusion coefficient in terms of the Chapman–Enskog theory, at low densities,

$$D_{AB} = 0.0018583 \frac{\sqrt{T^3\left(\frac{1}{M_A} + \frac{1}{M_B}\right)}}{p\sigma_{AB}^2\Omega_{D,AB}} \qquad (9.21)$$

where: D_{AB} is given in cm^2/sec and
         T   = temperature, K,
         M_A = molecular weight of species A,
         M_B = molecular weight of air, 29,
         p   = pressure, atm,
         σ_{AB} = Lennard–Jones distance, angstroms, and
         Ω_{AB} = function of Lennard–Jones potential.

Data for calculating the Lennard–Jones potential may be found in Bird, Stewart and Lightfoot. The Lennard–Jones distance and potential are estimated by:

$$\sigma_{AB} = e^{1.02}(MW)^{0.091}$$ (9.22)

and

$$\Omega = e^{0.39}\left(\frac{\sqrt{97e^{2.38}(MW)^{0.76}}}{T}\right)^{0.397}$$ (9.23)

Substituting these into Equation 9.18 yields

$$D_{AB} = \frac{4.09 \times 10^{-5}T^{1.9}\sqrt{1/29+1/MW_A}(MW_A)^{-0.33}}{P}$$ (9.24)

where D_{AB} is given in cm^2/sec. From this, we find for flowing air

$$R_{evap} = \frac{2.79 \times 10^{-3}(MW)^{0.835}(P^{vap})^4\sqrt{1/MW_A+1/29}}{T^{0.05}}\sqrt{\frac{v_z}{\Delta zp}}$$ (9.25)

which EPA uses to predict the evaporation rate of liquid in flowing air.

In a regression analysis of data observed at Pace Laboratories, EPA uses this expression for all chemicals except the low vapor pressure alcohols: n-hexanol; n-heptanol; and i-octanol:

$$R_{evap} = 0.000237MWP^{vap}v_z^{0.625}$$ (9.26)

where: R_{evap} is determined in lb./hr ft.^{2} and
    MW = molecular weight
    P^{vap} = vapor pressure at bulk liquid temperature, in. Hg,
    v_z = air velocity, ft./min.

Finally, EPA also uses another expression for evaporation rate, also in terms oflb./hr ft.^{2}, that was developed by Mackay:

$$R_{evap} = \frac{0.03197v_z^{0.78}MW^{2/3}P^{vap}}{T}$$ (9.27)

An analysis the prediction of the equation developed from the Pace data, the prediction of the theoretical equation, and the prediction of the Mackay equation show that the Mackay equation consistently underestimated actual evaporation rate, while the theoretical equation and the regression analysis equation fit about equally as well, both overpredicting sometimes and underpredicting others and about the same number of times.

Silka (see Ayers) has found a relationship between dispersion and contamination: z-dispersion = 0.1 y-dispersion = 0.1 x-dispersion = 0.1 distance of the contamination problem.

In adsorption, a molecule or ion affixes to the surface of another substance. Usually this process is reversible, given the right conditions. Adsorption is very important in studying the transport of contaminants through soil and sediments. Both metals and organics adsorb to a greater or lesser degree onto organic matter and clay particles in the soil and sediment. Soils are a sink (that is, have a storage capacity) for cationic metals and organic compounds which have low solubility in water and high sorption coefficients.

Several processes break down or degrade molecules of contaminants into less toxic molecules or ions. Some degradation processes are hydrolysis, photolysis and biodegradation.

Topography/Hydrography

Make a thorough description of the land and water features of the surface. How will the contaminant plume travel over the surface? Are there natural barriers? Natural short circuits? Identify the overland flow and storm runoff patterns. This is particularly important where surface water is used as the potable water supply. Popular water recreational and commercial fishing areas are also very sensitive to contamination. Another way in which surface water becomes a pathway for contaminants is through consumption of wildlife and domestic animals which use it as a drinking hole and when farmers use it to irrigate crops.

Meteorology/Climate

To understand the potential for exposure to airborne contamination we must understand the meteorology of the site and the regional climate. Prevailing wind direction and average speed are important but monthly and/or seasonal wind roses prepared by the National Weather Service are also valuable. Rainfall and other precipitation data are vital for storm runoff analysis and gives some clue about natural scrubbing of contamination from the air.

To compute net precipitation determine monthly precipitation and evapotranspiration for the site location using local measured monthly averages. When this data is not available locally, use monthly averages from the nearest National Oceanographic and Atmospheric Administration (NOAA) weather station that has a similar geographic setting or collect data from the two nearest stations on either side (north–south or east–west or northeast–southwest) of the site and take the average of their averages.

When measured monthly evapotranspiration is not available, the monthly potential evapotranspiration (E_i inches/month) is calculated as follows:

$$E_i = 0.6F_i\left(\frac{10T_i}{I}\right)^a \tag{9.28}$$

where: F_i = monthly latitude adjustment, dimensionless (Table 9.5)

T_i = monthly mean temperature, °C, and

$$I = \sum_{i=1}^{12}\left(T_i/5\right)^{1.514} \tag{9.29}$$

and

$$a = 6.75\times10^{-7}I^3 - 7.71\times10^{-5}I^2 + 1.79\times10^2 I + 0.49239 \tag{9.30}$$

Select the latitude adjustment for each month from Table 9.5 and extrapolate if the latitude of the site is greater than 50° North or 20° South and interpolate if the latitude is between those limits but not an even value found in the table.

Next, calculate monthly net precipitation by subtracting monthly evapotranspiration or the estimated potential from monthly precipitation. If the evapotranspiration or estimated potential exceeds the monthly precipitation for any given month, use the value of zero. Calculate the annual net precipitation by summing the monthly net values.

AIR TRANSPORTATION

The air pathway has been studied for many years and we have several models to draw from.

SINGLE-SOURCE ADVECTION–DIFFUSION MODEL

In the situation of cross-drafts, eddy diffusion is important. The estimation of indoor advection-diffusion is similar to Gaussian plume models in air pollution modeling outdoors. The air speed, u, and x, the downwind component of r, are necessary additional variables. In the steady-state case, Keil has developed:

$$C_r = \frac{G}{2\pi Dr}e^{-\frac{u}{2D}(r-x)} \tag{9.31}$$

Look familiar?

Dispersion models such as the Gaussian describe how COCs are transported from a source to a receptor. Particularly in the outdoors this is an important link. Indoors, as mentioned above, the link is assumed. In worst cases the link can be assumed outdoors, but to make such an assumption one will be forced to defend the indefensible since dispersion implies dilution.

In air (or water), the dispersion of a chemical from a source follows a plume with the maximum concentration at the release point and the minimum detectable

TABLE 9.5
Monthly Latitude Adjustment

Latitude	January	February	March	April
≥50 N	0.74	0.78	1.02	1.15
45 N	0.80	0.81	1.02	1.13
40 N	0.84	0.83	1.03	1.11
35 N	0.87	0.85	1.03	1.09
30 N	0.90	0.87	1.03	1.08
20 N	0.95	0.90	1.03	1.05
10 N	1.00	0.91	1.03	1.03
0	1.04	0.94	1.04	1.01
10 S	1.08	0.97	1.05	0.99
20 S	1.14	0.99	1.05	0.97

Latitude	May	June	July	August
≥50 N	1.33	1.36	1.37	1.25
45 N	1.28	1.29	1.31	1.21
40 N	1.24	1.25	1.27	1.18
35 N	1.21	1.21	1.23	1.16
30 N	1.18	1.17	1.20	1.14
20 N	1.13	1.11	1.14	1.11
10 N	1.08	1.06	1.08	1.07
0	1.04	1.01	1.04	1.04
10 S	1.00	0.96	1.00	1.02
20 S	0.96	0.91	0.95	0.99

Latitude	September	October	November	December
≥50 N	1.06	0.92	0.76	0.70
45 N	1.04	0.94	0.79	0.75
40 N	1.04	0.96	0.83	0.81
35 N	1.03	0.97	0.89	0.85
30 N	1.03	0.98	0.89	0.88
20 N	1.02	1.00	0.93	0.94
10 N	1.02	1.02	0.98	0.99
0	1.01	1.04	1.01	1.04
10 S	1.00	1.06	1.05	1.09
20 S	1.00	1.08	1.09	1.15

concentrations along the farthest edges of the plume. Determining the boundaries of a plume is called mapping it. A material that is being continuously released will form a continuous plume from the source to some point downwind where it is no longer detectable. A puff formed by an instantaneous release of material, on the other hand, will move downwind as a complete unit until it dissipates to the point of being undetectable by mixing with fresh (unpolluted) air.

TABLE 9.6
Conditions of Atmospheric Stability

Stable	Neutral	Unstable
Light winds	Some gusting	Very gusty
Little turbulence	Changes in wind direction	Rapid velocity changes
		Rapid direction changes

PLUME VARIABLES

The variables that affect the plume are wind speed, atmospheric stability, ground conditions, height of release, momentum of initial release, and buoyancy of initial release.

Wind Speed. Plumes increase in length and decrease in width as wind speed increases. Higher wind speeds carry the COC downwind faster, but also dilute it faster by mixing it with a larger quantity of fresh air.

Atmospheric Stability. The amount of vertical mixing of air is directly related to atmospheric stability. Most of the time, especially during the daylight hours, atmospheric temperature and pressure decrease with increasing altitude, providing two gradients for vertical mixing. The rate of decrease of temperature with altitude is called the *lapse rate*. However, vertical motion in the atmosphere is unaffected by the temperature gradient when the *adiabatic lapse rate* exists. Vertical air motion is actually suppressed when the temperature decrease is slower than the adiabatic lapse rate or if temperature increases with altitude. The atmosphere is called *stable* when the lapse rate is less than adiabatic. An *inversion* exists when temperature increases with altitude. The inversion atmosphere is very stable, with the higher temperature layer acting as a cap on the lower temperature layer of air. This condition gives rise to the accumulation of smoke, photochemical/ozone haze, fog, and other opaque aerosols that we have come to call smog. Originally, smog was the acidic and deadly combination of SMoke and fOG in London. When the actual lapse rate is equivalent to the adiabatic lapse rate, the atmosphere is said to be *neutral*. An *unstable* atmosphere occurs when the decrease in temperature with altitude exceeds the adiabatic lapse rate. Table 9.6 summarizes our common perceptions of these conditions. When solar heating is strong, an unstable produces moderate winds with sudden gusts that change direction and velocity often. Under cloudy skies, or when winds are strong, a neutral atmosphere is produced. While overall wind speed may be greater, less gusting with respect to changes of speed and direction occurs. Typically the evening allows the earth's surface to cool faster than the atmosphere and a stable atmosphere is produced.

Ground Conditions. Topographic features are capable of affecting the air turbulence and therefore the mixing. The mixing is enhanced by trees, shrubbery, and structures. Open, grassy areas decrease mixing. Calm water in harbors and bays produce the least mixing effect. Changes in elevation of the land affect mixing also. Release clouds that are denser than air flow around hills, while lighter releases will climb hills. Dense clouds will seek the lowest topographical point downwind from the release.

Height of Release. Ground level concentration is directly affected by the height of release. As release height increases, the downwind distance, where the plume touches the ground, increases. Therefore, ground level concentration is inversely proportional to release height because the increased distance allows for greater dispersion, which reduces concentration at ground level.

Momentum. Obviously, the more momentum a gas cloud has coming out of a stack or other release situation, the further it will travel before air turbulence begins to affect it.

Buoyancy. Buoyancy and momentum work hand in hand to overcome any initial influence by air turbulence. The buoyancy of the cloud is the property that causes it to rise in air and is directly related to the cloud/air density gradient and indirectly to the cloud/air temperature gradient. Once the initial momentum and buoyancy dissipate, turbulence becomes the dominant factor for mixing.

PLUME MODELS

As mentioned above, plume models describe steady state concentrations of COCs released from continuous sources.

Gaussian. The downwind transport of a chemical in air has traditionally been modeled using a Gaussian dispersion equation. The COC cloud is assumed to be neutrally buoyant as compared with the atmosphere. Consider a statistical variation of concentration about a mean value at any point [x,y,z] downwind from a continuous release point. The general Gaussian distribution of concentration is:

$$\overline{X}(x,y,z) = \frac{M}{2\pi\sigma_y\sigma_z u}\left[\exp\left(-\frac{y^2}{2\sigma_y^2}\right)\right]\cdot\left[\exp\left(\frac{-(z-h)^2}{2\sigma_z^2}\right)+\exp\left(\frac{-(z+h)^2}{2\sigma_z^2}\right)\right] \quad (9.32)$$

where: X = average concentration at point (x,y,z),

M = mass rate of continuous release,

h = height of release,

σ_y = standard deviation of distribution of crosswind direction,

σ_z = standard deviation of distribution in vertical direction, and

u = mean wind velocity.

The greatest concentrations occur along the centerline of the plume, but release rate, distance to (x,y,z), wind velocity, and atmospheric stability are the primary influences on concentration.

Steady State, No Wind. Crowl and Louvar have worked out several situations of Gaussian releases. Where the release rate has reached steady state and the wind velocity is zero, they solve Equation 9.32 to get

$$\overline{X}(x,y,z) = \frac{M}{4\pi K * \sqrt{x^2+y^2+z^2}} \quad (9.33)$$

where: $K^* =$ eddy diffusivity constant independent of direction.

Non-steady State, No Wind. The preceding situation does not happen very often in the real world, so the first enhancement we want to make to our model is as follows (see Crowl and Louvar for solution steps):

$$\overline{X}(x,y,z,t) = \frac{M}{4\pi K^* \sqrt{x^2 + y^2 + z^2}} \text{erfc}\left(\frac{\sqrt{x^2 + y^2 + z^2}}{2\sqrt{K^* t}}\right) \tag{9.34}$$

As t→∞, this equation reduces to the steady state solution.

Steady State, Wind. More common than the preceding cases, is the one where release conditions have steadied, but wind velocity is greater than zero. If the wind blows in the x (horizontal downwind) direction only,

$$\overline{X}(x) = \frac{M}{4\pi K^* x} \tag{9.35}$$

Not too horrendous a model.

Steady State, Wind, K = f(x,y,z). Even more common is the case where the eddy diffusivity varies with direction, but is constant for any given direction. Solving along the centerline, where y = z = 0:

$$\overline{X}(x) = \frac{M}{4\pi \sqrt{K_y K_z}} \tag{9.36}$$

Steady State, Source on Ground Level. With the release source on the ground, an impervious barrier to the plume, the solution of Equation 9.36 is doubled:

$$\overline{X}(x,y,z) = \frac{M}{2\pi x \sqrt{K_x K_y}} \exp\left[-\frac{u}{4x}\left(\frac{y^2}{K_y} + \frac{z^2}{K_z}\right)\right] \tag{9.37}$$

Steady State, Source at Height h. The ground is still an impervious barrier, but now it is h distance from the centerline of the plume, and:

$$\overline{X}(x,y,z) = \frac{M}{4\pi x \sqrt{K_x K_y}} \exp\left(-\frac{uy^2}{4K_y x}\right) \tag{9.38}$$

$$\times \left\{\exp\left[-\frac{u}{4K_z x}(z-h)^2\right] + \exp\left[-\frac{u}{4K_z x}(z+h)^2\right]\right\}$$

As h→0, Equation 9.38 becomes Equation 9.37.

Puff, No Wind. For those who wish to model a one-time accidental release, the puff can be modeled. Keep in mind, though, that puffs can occur on an intermittent, but regular, basis, not merely on an accidental basis. Here, we have a fixed release, M^*, and assume that eddy diffusivity is independent of direction. At any time t, the status of the puff is:

$$\overline{X}(x,y,z,t) = \frac{M^*}{8(\pi K^* t)^{3/2}} \exp\left[-\frac{\left(x^2 + y^2 + z^2\right)}{4K^* t}\right] \tag{9.39}$$

Note, that as t→∞, Equation 9.39 approaches steady state continuous release with no wind.

Puff, No Wind, $K = f(x,y,z)$. The solution of Crowl and Louvar is:

$$\overline{X}(x,y,z,t) = \frac{M^*}{8(\pi t)^{3/2}\sqrt{K_x K_y K_z}} \exp\left[-\frac{1}{4t}\left(\frac{x^2}{K_x} + \frac{y^2}{K_y} + \frac{z^2}{K_z}\right)\right] \tag{9.40}$$

Puff, Wind. The simplest form considers x direction only.

$$\overline{X}(x,y,z,t) = \frac{M^*}{8(\pi t)^{3/2}\sqrt{K_x K_y K_z}} \exp\left\{-\frac{1}{4t}\left[\frac{(x-ut)^2}{K_x} + \frac{y^2}{K_y} + \frac{z^2}{K_z}\right]\right\} \tag{9.41}$$

Puff, No Wind, Source on Ground. Crowl and Louvar find:

$$\overline{X}(x,y,z,t) = \frac{M^*}{4(\pi t)^{3/2}\sqrt{K_x K_y K_z}} \exp\left[-\frac{1}{4t}\left(\frac{x^2}{K_x} + \frac{y^2}{K_y} + \frac{z^2}{K_z}\right)\right] \tag{9.42}$$

Fully Variable Diffusivity. Typically, the eddy diffusivity, rather than being constant, varies according to position, time, wind speed, and weather conditions making it difficult to determine experimentally. Theoreticians such as Sutton have devised the *dispersion coefficient*, the standard deviations of concentrations in the downwind, x, crosswind, y, or vertical, z, directions. Dispersion coefficients, σ, are functions of atmospheric conditions and distance downwind. Sutton's definition of dispersion coefficient is:

$$\sigma_x^2 = \frac{1}{2}\overline{X}^2(ut)^{2-n} \tag{9.43}$$

In order to determine σ, the atmospheric stability classes of Table 9.7 are used. The dispersion coefficient σ_x is calculated directly. Then one of the equations below are

TABLE 9.7
Atmospheric Stability Classes for Dispersion Coefficient

	Daytime			Nighttime	
Wind Speed, ft./sec	Strong Radiation	Medium Radiation	Slight Radiation	Cloudy	Calm & Clear
<5	A	A–B	B		
5–10	A–B	B	C	E	F
10–15	B	B–C	C	D	E
15–20	C	C–D	D	D	D
>20	C	D	D	D	D

A,B: unstable; C,D: neutral; E,F: stable

TABLE 9.8
Calculating σ_z

Stability Class/Downwind	Range x, meters
A1	100–300
A2	300–3000
B1	100–500
B2	500–100000
C	100–100000
D1	100–500
D2	500–100000
E1	100–500
E2	500–100000
F1	100–500
F2	500–10000

used to calculate σ_y and σ_z based on atmospheric stability class (A–F) and distance in the x-direction. For σ_y, one of the sister equations of Equation 45 is used:

$$\sigma_y(A) = 0.493x^{0.88}$$
$$\sigma_y(B) = 0.337x^{0.88}$$
$$\sigma_y(C) = 0.195x^{0.90}$$
$$\sigma_y(D) = 0.128x^{0.90}$$
$$\sigma_y(E) = 0.091x^{0.91}$$
$$\sigma_y(F) = 0.067x^{0.90} \tag{9.44}$$

Then, for various downwind x-direction ranges and stability classes, see Table 9.8, σ_z is calculated using the appropriate Equation 9.45 sister equation.

TABLE 9.9
Dispersion Coefficients for Puff Releases

	x = 100m		x = 4000m	
Stability Condition	σ_y (m)	σ_z (m)	σ_y (m)	σ_z (m)
Unstable	10	15	300	220
Neutral	4	3.8	120	50
Stable	1.3	0.75	35	7

The Pasquill–Gifford model modifies the Gaussian model by taking these dispersion coefficients into account. Crowley and Louvar do an excellent job of presenting the derivations of these dispersion coefficients. Here we shall merely satisfy ourselves with making practical use of them.

$$\sigma_z(A1) = 0.087x^{1.10}$$
$$\log_{10}\sigma_z(A2) = -1.67 + 0.902\log_{10}x + 0.181\left(\log_{10}x\right)^2$$
$$\sigma_z(B1) = 0.135x^{0.95}$$
$$\log_{10}\sigma_z(B2) = -1.25 + 1.09\log_{10}x + 0.00181\left(\log_{10}x\right)^2$$
$$\sigma_z(C) = 0.112x^{0.91}$$
$$\sigma_z(D1) = 0.093x^{0.85}$$
$$\log_{10}\sigma_z(D2) = -1.22 + 1.08\log_{10}x - 0.061\left(\log_{10}x\right)^2$$
$$\sigma_z(E1) = 0.082x^{0.082}$$
$$\log_{10}\sigma_z(E2) = -1.19 + 1.04\log_{10}x - 0.070\left(\log_{10}x\right)^2$$
$$\sigma_z(F1) = 0.057x^{0.80}$$
$$\log_{10}\sigma_z(F2) = -1.91 + 1.37\log_{10}x - 0.119\left(\log_{10}x\right)^2 \tag{9.45}$$

Finally, dispersion coefficients for puff releases are summarized in Table 9.9. Louvar do an excellent job of presenting the derivations of these dispersion coefficients. Here we shall merely satisfy ourselves with making practical use of them.

Pasquill–Gifford Model 1. With a continuous plume, from a ground level source emitting at steady state, with constant wind speed in x-direction, we have this solution:

$$\overline{X}(x,y,z) = \frac{M}{\pi\sigma_y\sigma_z u}\exp\left[-\frac{1}{2}\left(\frac{y^2}{\sigma_y^2} + \frac{z^2}{\sigma_z^2}\right)\right] \tag{9.46}$$

Pasquill–Gifford Model 2. Raising the source to a height H above ground level, gives this equation:

$$\overline{X}(x,y,z) = \frac{M}{2\pi\sigma_y\sigma_z u} \exp\left[-\frac{1}{2}\left(\frac{y}{\sigma_y}\right)^2\right]$$

$$\times \left\{\exp\left[-\frac{1}{2}\left(\frac{z-H}{\sigma_z}\right)^2\right] + \exp\left[-\frac{1}{2}\left(\frac{z+H}{\sigma_z}\right)^2\right]\right\} \tag{9.47}$$

Pasquill–Gifford Model 3. In the case of an instantaneous release, a puff, from ground level with wind constant in direction and velocity:

$$\overline{X}(x,y,z,t) = \frac{M*}{\sqrt{2}\pi^{3/2}\sigma_x\sigma_y\sigma_z}\exp\left\{-\frac{1}{2}\left[\left(\frac{x-ut}{\sigma_x}\right)^2 + \frac{y^2}{\sigma_y^2} + \frac{z^2}{\sigma_z^2}\right]\right\} \tag{9.48}$$

Pasquill–Gifford Model 4. A puff from height H:

$$\overline{X}(x,y,z,t) = \frac{M*}{\sqrt{2}\pi^{3/2}\sigma_x\sigma_y\sigma_z}\exp\left[-\frac{1}{2}\left(\frac{y}{\sigma_y}\right)^2\right]$$

$$\times\left\{\exp\left[-\frac{1}{2}\left(\frac{z-H}{\sigma_z}\right)^2\right] + \exp\left[-\frac{1}{2}\left(\frac{z+H}{\sigma_z}\right)^2\right]\right\} \tag{9.49}$$

where the coordinate system on the ground moves with the puff.

$$\overline{X}(x,y,z,t) = \frac{M*}{(2\pi)^{3/2}\sigma_x\sigma_y\sigma_z}\exp\left[-\frac{1}{2}\left(\frac{y}{\sigma_y}\right)^2\right]$$

$$\times\left\{\exp\left[-\frac{1}{2}\left(\frac{z-H}{\sigma_z}\right)^2\right] + \exp\left[-\frac{1}{2}\left(\frac{z+H}{\sigma_z}\right)^2\right]\right\} \tag{9.50}$$

$$+ \exp\left[-\frac{1}{2}\left(\frac{x-ut}{\sigma_x}\right)^2\right]$$

Hot gases being ejected from a tall stack cool as they mix with air and become diluted. At some point the temperature gradient is not discernible and the gas–air mixture reaches neutral buoyancy, at which time the stream movement is consistent with air flow. The Holland formula is used to estimate the rise of the gas stream from the stack to the point of neutral buoyancy due to buoyancy and the momentum of the release itself.

$$\Delta H = \frac{\overline{u}_s d}{\overline{u}} \left[1.5 + 2.68 \times 10^{-3} Pd \left(\frac{T_s - T_a}{T_s} \right) \right] \tag{9.51}$$

where: ΔH = correction to release height,
        u_s = stack gas velocity, m/s,
        u = wind speed, m/s,
        P = atmospheric pressure, mb,
        T_s = stack gas temperature, K, and
        T_a = air temperature, K.

Non-Flat Obstructed Terrain. When a plume is dispersing in terrain where hills, valleys, escarpments, and mountains are present, it is distorted. Distinct regions are produced by such terrain where the plume may stagnate, loft skyward, descend earthward, meander in channels, or disperse more rapidly due to shear enhanced mixing. Thermal currents force airflows over mountains and valleys. Rugged terrain produces thermal circulation, drainage flow, mountain waves, and downslope wind storms.

Another complicating factor is encountered in the city where buildings create more complex flow and dispersion of air than terrain does. Factors to consider are building geometry, spacing between buildings, local terrain surface, local meteorology, and release dynamics. Buildings produce backwash, divert plumes along alleys and street canyons, create channeling between neighboring structures, and produce local vortices.

The effects of non-flat obstructions may be neglected when the terrain gradient is less than 1 in 10 in neutral and unstable conditions and less than 1 in 100 in stable conditions. It does not take much of a gradient for the effects on dispersion to become noticeable. The presence of an upwind ridge may be neglected in neutral or unstable conditions if the release height is one and one-half times the height of the ridge or higher or if the upwind distance to the ridge is at least twenty times its height in neutral conditions or more than ten times its height in very unstable conditions. An isolated hill upwind of the source in neutral or unstable conditions may be ignored when the release height is greater than one and one-half the hill elevation or if the distance to the hill is greater than seven times its height. The presence of an obstacle upwind of the source in stable conditions may be ignored when the release height is greater than the height of the obstacle or the upwind distance is greater than forty times the obstacle height in slightly stable conditions or greater than one hundred times the obstacle height in very stable conditions. When a hill or ridge or any obstacle is located downwind of the source, its presence may be neglected under any atmospheric condition if either the release height is greater than the obstacle elevation plus the vertical plume standard deviation, $\sigma_z(x)$, or $\sigma_z(x)$ is greater than the obstacle height.

Non-buoyant Dispersion. Dense gases are not buoyant and do not obey the equations discussed above. In these cases, a release Richardson number, Ri_0, must be calculated. This dimensionless number relates the density-driven dispersing forces in the released plume, or puff, to the dispersing forces of the atmosphere. Large Ri_0 values mean the dense gas effect is important, whereas a small values means the effect may be safely neglected.

Dense gas dispersion is susceptible to the surface roughness parameter, which increases dilution of the plume or puff. In the near field, dense gas plumes are density-driven and partially shielded from atmospheric meander. Jet momentum is critical when the plume is flammable and less critical for toxic gases. Typically, the release most interesting to the exposure analyst is the dense gas plume because it slumps back to earth from the release point.

The maximum concentration of the plume, in %v/v, averaged over the first few minutes of touchdown after release from a vent is:

$$C_{max} = 3.44C_s \left(\frac{D}{2H + h_s} \right)^{1.95} \left(\frac{v_s}{u} \right) \times 10^{-6} \qquad (9.52)$$

where: C_S = stack gas concentration of COC, %v/v,
   D = stack exit diameter, mm,
   H = plume rise, m,
   h_S = stack height, m,
   v_S = stack gas exit velocity, m/s, and
   u = average wind speed, m/s.

The plume rise is calculated as

$$H = 1.32D \left(\frac{v_s}{u} \text{sp.gr.} \right)^{0.33} Fr^{0.67} \times 10^{-3} \qquad (9.53)$$

where the specific gravity is determined as

$$\text{sp.gr.} = \frac{MW(T_A)}{29T_S} \qquad (9.54)$$

and

$$Fr = \frac{31.62v_S}{\left[gD(\text{sp.gr.} - 1)/\text{sp.gr.} \right]^{0.5}} \qquad (9.55)$$

The specific gravity is for the stack gas in its entirety, not just for the COC component. If the vent has reached sonic velocity, these equations may not be used. Also, the results are doubtful if the specific gravity of the plume is 1.15 or less.

WATER COLUMN TRANSPORT MODEL

In water, anthropogenic pollutants such as heavy metals, hydrocarbons, chlorinated hydrocarbons, radionuclides and others are adsorbed onto suspended particulate matter and may therefore eventually settle to the bottom of the lake, river, or ocean. Adsorption, desorption and settling dynamics are the mechanisms of interest in water columns.

Some species dissolve in water, however, and so the total picture includes a distribution of pollutants between adsorbed and dissolved phases. The simplest way to treat this problem is to assume that linear partitioning exists and that the two phases are equally distributed. That is, one-half of the pollutant is adsorbed and one-half is dissolved. Let the fraction of the pollutant in the dissolved form be designated by f which is a function of the chemical species adsorbed-dissolved partitioning coefficient, K_p, which is dimensionless and the concentration of suspended solids in water, C_{ss}, also dimensionless. The dissolved fraction can be expressed:

$$f = 1/\left(K_p C_{ss} + 1\right) \tag{9.56}$$

In addition to particulate adsorption and settling, diffusion processes contribute to a long-term flux of pollutants from the water column to the floor of the water body.

Pollutants that adhere to particulate matter with non-zero settling velocities continue to settle, just as the particle did before, but modified by random diffusion. Dissolved-phase pollutants and pollutants that adhere to non-settling particulate matter are neutrally buoyant. Movement of neutrally buoyant particles depends on horizontal advection by the current movements with random diffusion in the horizontal and vertical planes. Typical values of these parameters in the ocean are given in Table 9.10.

The simplest modeling of chemicals in a water column assumes no chemical decay rate. That is, the transport model depends only on the product of the adsorbed-dissolved partitioning coefficient and the fraction of suspended solids in the water:

$$C_{COC} = K_p C_{ss} \tag{9.57}$$

In oceanic modeling Karam and Otto have examined two specific media for water column transport: the continental shelf (water column $\leq$ 200 m) and deep ocean (water column > 200 m).

A water column transport model is based on

$$C_{COC} = RC_a e^{-365\gamma} \log\left(K_p C_{ss}\right) \tag{9.58}$$

Surface Water Transport. Surface water is classified into four categories: rivers, lakes, oceans and coastal tidal pools. Threats due to surface water migration are contamination of drinking water, the human food chain and environmental damage.

TABLE 9.10
Ocean Transport Parameters

Vertical dispersion coefficients	100 cm²/sec
Upper water column	10 cm²/sec
Horizontal dispersion coefficient	400 cm²/sec
Particle settling velocity	0.002 m/sec
10%	0.001 m/sec
20%	0.0001 m/sec
50%	0.00001 m/sec
Suspended solids fraction	0.3×10^{-6} mg/mg

TABLE 9.11
Surface Water Migration Pathway

Categories of surface water
   Rivers
   Lakes
   Oceans
   Coastal water tide pools
Migration components
   Overland/flood migration
   Groundwater to surface migration
Surface water threats
   Drinking water
   Human food chain
   Environmental damage

After the National Contingency Plan.

These three threats must be assessed for two migration components: overland/flood migration and groundwater to surface migration. Table 9.11 summarizes this information.

Rivers are year-round flowing waters with one terminus at some point of origin or source (not to be confused with origin or source of contamination) and the other terminus being the ocean. A river may flow into a coastal tidal basin instead of directly into the ocean and wetlands which are in contact with the river are included as part of the definition. Anthropogenic ditches are included if they have a constant flow into the river. Intermittently flowing waters and ditches are included in the definition of river only in areas with less than twenty inches of mean annual precipitation.

Lakes include both natural and anthropogenic bodies of water including impoundments that lie along rivers, but not the Great Lakes, which are defined as ocean-like. Isolated, but perennial, lakes, ponds, and wetlands are included, as are

TABLE 9.12
Data Requirements for Evaluation
of Surface Water Pathway

Distance to nearest surface water from source
Rainfall intensity
Net precipitation
Potential for surface (soil) erosion
Surface (soil) permeability
Potential for flooding
Containment effectiveness

static water channels and oxbows located alongside rivers. Small rivers without dikes, which merge with adjacent wetlands, are included in the lake category. Also, wetlands adjacent to other bodies of water defined as lakes are included as lakes.

Oceans are the areas seaward from the baseline of the Territorial Sea, the Great Lakes, and wetlands contiguous to the Great Lakes. Coastal tidal waters include embayments, harbors, sounds, estuaries, back bays, lagoons, and wetlands.

The overland/flood migration pathway involves the movement of a contaminant from a source downgradient to a point of entry to surface water. The in-water segment of this pathway continues in the direction of flow for a river until the target distance is reached. For lakes, oceans, coastal tide waters, and the Great Lakes, the target distance arc is the limit used. When the in-water segment includes both rivers and other water bodies, the target distance arc is applied as a limit for the combined pathways. This is done for each watershed encountered. The target distance extends fifteen miles downriver from the point of entry or in a fifteen mile arc from the point of entry to another body of water. If flow within the pathway is reversed by tides, extend the target distance upstream only as far as the tidal run can reach, which will be less than fifteen miles.

In order to evaluate surface water as a pathway of concern, we must gather information such as given in Table 9.12.

Precipitation contacting contaminated soil dissolves some fraction of the contaminant into the water. Likewise, when precipitation contacts a pool of liquid or solid chemical some fraction is dissolved. When the only process governing transport is the uptake of the contaminant by water then equilibrium solubilities will overpredict observed concentrations in bulk surface water. However, entrained particles of the contaminant, or a microscopic suspension of colloidal nature, can increase the potential concentration of the contaminant in water by tenfold (that means $10 \times$ solubility). Finally, some of the contaminant can adsorb to particulate matter suspended in the water and further increase its potential maximum concentration.

As the suspended sediments in surface water settle out, the metal ions and organic contaminants adsorb onto the settling particles. Thus, the contamination problem is transferred to another media: bottom sediment. Many metal ions will precipitate out of solution, as insoluble metal oxides, or as oxyhydroxides. Heavy metals will also coprecipitate with iron readily. Again the problem is transferred to the bottom

TABLE 9.13
Half-life Factors

Chemical	Factors	Half-life
Benzene	$P_a = 0.125$ atm	5.0 hrs
	Column = 1 meter	
Naphthalene	$P_a = 0.0001$ atm	100 hrs
	Column = 1 meter	
Anthracene	35° north latitude	1.6 hrs
	Summer UV-radiation	
Anthracene	35° north latitude	4.8 hrs
	Winter UV-radiation	

sediment. Organics with higher vapor pressures (volatile organic chemicals or VOCs) will evaporate to some extent from the water. The more turbulence and mixing with air, the more evaporation takes place. Solar energy is capable of breaking the molecular bonds of many organic chemicals to form other chemicals or shorter carbon chains. The relationship of these processes to the half-lives of common organics is shown in Table 9.13.

Groundwater to surface water migration is another concern we must examine. Contamination of surface water can occur when the saturated zone contaminant plume intersects a stream or lake. Humans can be exposed when they ingest surface water or consume fish from it. Contamination of the surface water may also lead to other aquatic organisms bioaccumulating the contaminant.

Let us assume that contaminated groundwater completely intersects a body of surface water. This is an upland watershed and there is complete mixing within the stream. Transport within the stream is assumed to follow a first-order decay with biological and chemical degradation and volatilization. Surface water transport is

$$C_s = \frac{\dot{m}_g}{Q_s} \tag{9.59}$$

Sediment Model. Pollutant particles eventually settle within the water column and ends up on the floor of the ocean, river, or lake. Two alternative fates are transportation outside the domain (near complete dilution) or decay to background levels. When the water column is less than three hundred feet deep, settled particles may be reintroduced to the column through resuspension. Currents and wave action make the water column turbulent. Storm generated turbulence is governed by wave height and period frequency distribution. Sediments are the detrital, inorganic or organic particles that ultimately settle to the bottom of a body of water. Sediments have four volumetric components. Interstitial water makes up fifty percent (the actual volume is subject to wide variation) or more of the sediment. An inorganic component, consisting of geological detritus from terrestrial sources, is the next larger component. A low volume, but important, component is organic matter that control the sorption rate and bioavailability of contaminants. The anthropogenic component consists of contaminants from human sources and topsoil eroded due to human activity.

Vertical pollutant transport within sediments is governed by:

$$\frac{\delta C}{\delta t} = D\frac{\delta^2 C}{\delta z^2} - kC \qquad (9.60)$$

where: C = pollutant concentration
         D = dispersion coefficient including bioturbation effects
         k = pollutant decay rate
         t = time
         z = depth (positive downward distance)

Karam and Otto examined two media in the ocean bottom: continental shelf and deep ocean sediments.

Metals are partitioned between oxides and acid volatile sulfides (AVS), which control their availability in anoxic sediments. When the metal exceeds the binding capacity of the sediment AVS pool, the remainder is bioavailable:

$$C_{Bioavailable} = \frac{C_M}{C_{AVS}} - 1 \qquad (9.61)$$

where: C_M = molar concentration of transition metal, and
         C_{AVS} = molar concentration of AVS.

Flux, the rate of transfer from the water column to sediment and vice versa, is defined as follows if the chief mechanism of transport is advection:

$$F_a = \varnothing CU \qquad (9.62)$$

where: F_a = advective flux, lb./ft.^{2}-hr,
         $\varnothing$ = porosity of sediment,
         C = concentration of contaminant, lb./ft.^{3}, and
         U = velocity, ft./hr.

The sediment porosity is its relative pore volume, which is equal to the areal porosity.

Molecular diffusion flux is described as

$$F_d = \varnothing\theta(dC/dz) \qquad (9.63)$$

where: θ = diffusion coefficient, and
         dC/dz = vertical concentration gradient.

A depositional flux is defined as

$$D_f = -\emptyset SC \tag{9.64}$$

where S is the deposition rate of material onto the sediment surface.
Total flux is simply

$$F_{tot} = \emptyset\left[\theta(dC/dz) + UC - SC\right] \tag{9.65}$$

COCs are bioaccumulated when their concentration in some organism is measurable relative to the concentration in the sediment. The bioaccumulation factor, BAF, is

$$BAF = C_{tss}/C_S \tag{9.66}$$

where: C_{tss} = concentration of COC in tissue, steady state, µg,g,
C_S = concentration of COC in sediment, µg/g.

$$C_{tss}/L = AF(C_S/TOC) \tag{9.67}$$

where: L = lipid content, g/g,
AF = accumulation factor, g carbon/g lipid, and
TOC = total organic carbon in sediment, g/g.

More precisely, bioaccumulation is the net of uptake and elimination kinetics. A first-order model has been devised to predict bioaccumulation in fish:

$$dC_t/dt = k_1 C_w - k_2 C_t \tag{9.68}$$

where: C_t = tissue residue, µg/g tissue,
t = time, hr,
k_1 = uptake rate constant, mL water/g tissue-hr,
k_2 = elimination rate constant, hr^{-1}, and
C_w = COC concentration in water, µg/g water.

The elimination rate includes, not only depuration, but also the metabolic degradation of the compound. Also, many scientists prefer the sediment uptake rate coefficient or sediment uptake clearance, k_s, to the k_1 term. This makes Equation 9.68 become

$$dC/dt = k_s C_S - k_2 C_t \tag{9.69}$$

where C_S is the concentration of the COC in the sediment. When the COC concentration is constant the concentration in tissue residues at time t is

$$C_t(t) = C_S k_s / k_2 \left(1 - e^{-k_2 t}\right) \tag{9.70}$$

The maximum accumulation in tissue is

$$C_{t\,max} = C_S k_s / k_2 \tag{9.71}$$

and the corresponding bioaccumulation factor is

$$BAF = C_{t\,max} / C_S = k_s / k_2 \tag{9.72}$$

Apparent steady state occurs at ninety-five percent of tissue equilibrium residue, and

$$S = 3.0 / k_2 \tag{9.73}$$

where S is the time to steady state tissue residue.

For growing organisms, the tissue residue is

$$C_t(t) = C_S k_s / \left(k_2 + k_3\right) \left[1 - e^{-(k_2 + k_3)t}\right] \tag{9.74}$$

where k_3 is the growth rate constant. Now, maximum tissue residue is

$$C_{t\,max} = C_S k_s / \left(k_2 + k_3\right) \tag{9.75}$$

Groundwater. Hydrological and hydrogeological data are needed to understand how the contamination will migrate from the release point to groundwater. The physical, chemical, biological and geological processes taking place in contaminated groundwater are complex. The chief concern is ultimate contamination of drinking water supplies. The relationship between surface water and groundwater define the routes of migration and connections to other environmental media.

The topography of the site in question gives clues about groundwater behavior. *Headslopes*, the inside slopes of bends or sharp turns of valleys, concentrate surface runoff. That's obvious, if you think about it. Surface runoff is dispersed by the slopes at the ends of ridges, called *noseslopes*. For any hill, or slope, the highest point is the *summit*, the *shoulder* is where the slope breaks downward, the *backslope* is the main slope of the hill between the shoulder and the *footslope* where silt and debris accumulate. The *toeslope* is a relatively minor slope in the valley proper where silt and debris has collected. The greatest level of infiltration occurs at the footslopes and toeslopes, where runoff slows and starts to pond, in some cases. The next significant level of infiltration occurs on the noseslopes and *interfluves*, or ridges, where the slopes are gentle, but runoff is higher than at the footslope or toeslope. Infiltration is least on the backslopes and shoulders, where slopes, and therefore runoff, is greatest. Again, this is intuitive and we find no surprises here. Surface runoff, therefore, is greatest on steep slopes and least on flat surfaces. Regardless of the topographic feature, surface runoff is maximized by saturated soil.

Concave sideslopes (slope rounded downward) concentrate more water in soil that *convex* sideslopes (rounded upward) do. More groundwater is typically found in alluvial fill, the silty area in the valleys between hills, than on the interfluves. Subtle changes in topography can typically be identified by comparing relative greenness of vegetation. Where groundwater is more plentiful, vegetation will be greener than at a dryer spot.

Water flows through soil in accordance with the Darcy equation

$$Q = K \frac{\delta h}{\delta x} \tag{9.76}$$

where: Q = water flow, m^3/sec,
         $\delta h/\delta x$ = hydraulic potential.

However, groundwater does not obey Darcy's equation when conduit flow occurs through rock fractures, or when the typical flow in Karst formations occur. Unexpected things happen when surface water drains into these formations. The only thing you can do is to conduct groundwater flow experiments, and that may be prohibitively expensive, depending upon the scope and budget of your assessment project.

The hydraulic potential is a gradient that consists of a matric, gravitational, and pressure component. The *matric* potential (P_m) is the component that consists of the attraction of water to subsurface soils. Underground water will adsorb to the surface of solids and is also affected by capillary action in the pores between soil particles. Both phenomena contribute to the matric potential. The forces causing this potential and effect on water are called *matric suction*, which is inversely proportional to soil particle size and pore diameter. The *gravitational* potential (P_g) is calculated

$$P_g = GhP_T \tag{9.77}$$

where: G = the acceleration of gravity, and
         h = height above reference elevation.

The pressure potential, called the *Osmotic* potential (P_O), is caused by dissolved constituents in groundwater. Solute ions are attracted to water, which reduces the energy potential of the water. The total energy potential of water, giving it impetus to flow, is

$$P_T = P_g + P_m + P_O + \ldots \tag{9.78}$$

leaving an opening for the inclusion of other less significant potentials. The matric and Osmotic potentials are negative while the gravitational potential is positive, so in order to have flow, the following relationship must be true:

$$Pg > P_m + P_O \qquad (9.79)$$

Matric and Osmotic forces dominate in unsaturated zones and gravitational forces dominate in saturated zones.

Subsurface water is generically divided into the Zone of Aeration just beneath the surface level down to some point where it enters the Zone of Saturation. The aeration or *vadose* zone is where hygroscopic water pressure is less than atmospheric pressure. The saturated zone is where groundwater collects, also called the *phreatic water zone*. Here, the pore water pressure is greater than atmospheric pressure. The interface between the top of this latter zone and the vadose zone is the *water table* or groundwater surface. The vadose zone is subdivided into three zones. The *soil–water zone* is the unsaturated zone where water adsorbs to soil particles and roots. The quality and quantity of moisture fluctuates often in the soil–water zone. Transpiration and evaporation are key mechanisms affecting the water balance in this zone. The *intermediate vadose zone* is the next lower part of the overall vadose zone. The intermediate vadose contains residual moisture depending on the matric potential, which holds little water in coarse-grained sands and gravels, but much more water in fine-grained clays. The intermediate vadose contains a lot of air trapped in pore spaces, which slows down the passage of water to the saturated zone until it reaches the *capillary fringe*, the boundary with the lowest sub-zone. The *capillary zone* is the transition between the vadose zone and the saturated zone. Capillary action is caused by the attractive forces of solids suspended in the water and the surface tension of the water. The capillary zone is not deep, about one foot in sandy soils and as deep as three or more feet in loamy soils. Contaminants move more slowly in the vadose zone than in the saturated zone simply because the movement of water is freer in the saturated zone.

Vertical flow of water in the ground is described as

$$Q = -K\left(\frac{\delta y}{\delta x} - 1\right) \qquad (9.80)$$

where flow is expressed as meters water per meter depth.

In non-deformable soil,

$$\frac{\delta q}{\delta t} = \frac{\delta Q}{\delta x} - S \qquad (9.81)$$

where: q = volumetric water content, m^3/m^3,
        t = time, and
        S = a sink.

The Richard equation defines specific water capacity:

$$Cw \frac{\delta y}{\delta t} = \delta/\delta x [K \delta y/\delta x] - \delta K/\delta x - S \tag{9.82}$$

where: $Cw = dq/dy$.

Soil temperature parallels the flow of water in the soil and the heat flow is described as:

$$q_h = -K \delta T/\delta x + c * T * Q \tag{9.83}$$

where: q_h = heat flux density, W/m^2,
        K = thermal conductivity, $W/m°C$,
        T = temperature, $°C$,
        x = depth, m,
        c^* = specific heat capacity of water, $J/kg°C$, and
        Q = water flux, $kg/m^2 sec$.

The heat content of the soil, H, is

$$\delta H/\delta t = -\delta q_h/\delta x + Sh \tag{9.84}$$

Sh is heat sources.

If the top of the saturated zone is at atmospheric pressure, then it is an *unconfined aquifer*. The elevated pressure of a *confined aquifer* results from some permeable geologic layer capping the aquifer, this creating a back pressure when the hydraulic head is higher than the level of the confining layer, which is classically called an *aquitard* or *aquiclude*. *Perched water tables* are aquifers lying on impermeable strata with unsaturated flow underneath. *Regional aquifers* have a point of surface discharge. Hydraulic properties are uniform in *homogeneous aquifers*, but differ in either horizontal or vertical directions in *heterogeneous aquifers*. The hydraulic conductivity is the heterogeneous variable in *anisotropic* aquifers. An *isotropic* aquifer has the same hydraulic conductivity in all directions.

Not all soils but soils in general have three pools of organic material. Added organic matter is material such as manure. Biomass includes the plant material and animals living within the soil. Soil organic matter is rotting vegetation and other sources of organics not included as biomass or added matter. Organic matter decays as first order reaction kinetics with the rate constant dependent on clay content, soil temperature, and the pressure potential of soil water. For a reaction which is first order:

$$R = -kC \tag{9.85}$$

TABLE 9.14
Spatially Variable Hydrologic Parameters

Coefficient of Storage (*storativity*)
Horizontal Hydraulic Conductivity (*transmissivity*)
Porosity (especially *effective porosity*)
Retention
Specific Yield
Vertical Hydraulic Conductivity (*leakage*)

TABLE 9.15
Spatial and Temporally Variable
Hydrologic Parameters

Potential (*hydraulic head*)
Stress (*recharge* and *discharge*)

If the initial concentration of organics is C_0,

$$\frac{C}{C_0} = \exp(-kt) \tag{9.86}$$

A pristine, *native*, aquifer is a system which has established chemical equilibrium between the groundwater it contains and the geological matrix through which it flows. Anthropogenic contamination disturbs the equilibrium. Migration of the contaminant is controlled by adsorption, dispersion, volatilization and degradation as discussed above. The residence time of the contaminant in the aquifer is controlled by adsorption while dispersion affects the shape of the plume. Volatilization and degradation are the processes which actually break down the contaminant.

Groundwater flow can be estimated using analytical, numerical, or physical models but empirical modeling based on laboratory and field measurements of hydrologic parameters is more accurate and the most common approach taken by engineers and scientists who are studying the migration of hazardous substances released to the environment. Measured groundwater parameters are either spatially variable or spatial and temporally variable.

Table 9.14 lists the former and 9.15 lists the latter parameters.

In unconfined aquifers of porous media, specific yield $\approx$ porosity $\approx$ storativity, with a range of 0.05 to 0.25. For very permeable materials, effective porosity and porosity are equal. Table 9.16 lists porosities of common materials. Porosity is the ratio between openings in rock and the total rock volume and is either expressed as a percentage or decimal. Porosity is the amount of water than can be stored in a saturated rock formation. Gravity drainage will not release the entire volume of stored water from a formation, but the amount that drains is the specific yield. The

---

TABLE 9.16
Porosity of Common Materials

Most porous

Clays	0.40–0.70
Silts	0.35–0.50
Basalt, fractured	0.05–0.50
Karst limestone	0.05–0.50
Sands	0.15–0.48
Volcanic tuff	0.30–0.40
Gravels	0.25–0.40
Chalk	0.05–0.40
Sandstones	0.05–0.40
Dolomite limestone	0–0.20
Shales	0–0.10
Slates	0–0.10
Mica-schists	0–0.10
Granite, no fractures	0–0.02
Gneiss, no fractures	0–0.02
Quartzites	0–0.01

Least porous

After Nielsen.

---

amount that does not drain is the specific retention or simply the retention. Porosity, then, equals specific yield plus retention, but, in unconfined aquifers porosity = specific yield.

Storativity is the amount of water that can be released or taken into storage per surface area per change in head. Storativity for confined aquifers ranges from 0.00001 to 0.001 and the range widens for fractured media. For confined porous media, effective porosity and specific yield are easily estimated. In fractured rock, however, especially in the case where transmissivity is low, effective porosity and specific yield are less predictable. Accuracy of field data is ±2% minimum in these cases. Direct measurement of transmissivity is required in fractured rock so a monitoring well is needed which is designed to allow for reliable, representative measurement of hydraulic conductivity, potential, and *chemical concentrations.*

Permeability (P) and hydraulic conductivity (K) interchangeably refer to the ability of water to move through soil or a water table under saturated conditions. Technically, hydrogeologists define intrinsic permeability (k) as a property of the porous media, having little to do with water or its energy potential. Nielsen defines hydraulic conductivity as the quantity of water that will flow through a cross-sectional area of porous material per time with a unit hydraulic gradient and specified temperature. The transmissivity is the hydraulic conductivity of the aquifer, which is the capacity of a volume of the aquifer to yield water, or the rate at which water of a certain kinematic viscosity travels through the width of the aquifer under a hydraulic gradient. Transmissivity is determined by

$$TR = kh = \frac{Q}{jB} \qquad (9.87)$$

where: TR = transmissivity, ft.^{3}/day-ft.,
        k   = coefficient of permeability, ft./day,
        h   = saturated thickness of aquifer, ft.,
        Q   = groundwater flow, ft.^{3}/s,
        j   = hydraulic gradient, and
        B   = width of aquifer, ft.

Well discharge from a confined aquifer is calculated by:

$$Q = 14.96\pi hk \left(\frac{x_1 - x_2}{\ln \frac{r_2}{r_1}} \right) \qquad (9.88)$$

where: Q = steady-state discharge, gal/s,
        h = depth of aquifer, ft.,
        k = coefficient of permeability, ft./s,
        x = drawdowns at observation wells, ft., and
        r = distance of observation wells from pumped well, ft.

Specific well capacity, S_c, is

$$S_c = \frac{Q}{x} \qquad (9.89)$$

where: S_c = specific capacity, ft.^{3}/day-ft.,
        Q = well discharge, ft.^{3}/s, and
        x = drawdown, ft.

Degradation of chemicals in a groundwater environment comes in the form of biological action and abiotic chemical action. Biological activity normally proceeds at rates that are several orders of magnitude faster than abiotic processes. How pH extreme, oxidation-reduction cells, ionic strength or elevated temperature can increase the rate of abiotic reactions. Transformation may be complex depending on the environmental conditions and the degree of heterogeneity in the aquifer.

One of the more important chemical processes occurring in groundwater is the oxidation-reduction or redox reaction, which are reactions where electrons are transferred. In oxidation, at least one atom or ion has its valence number algebraically increased, meaning electrons are lost. This is coupled by the reduction reaction in which at least one atom or ion has its valence algebraically decreased, meaning electrons are gained. The reactions do not happen in isolation, rather they are

---

TABLE 9.17
Common Groundwater Redox Couples

$O_2 - H_2O$; $SO_4^{2-} - H_2S$; $CO_2 - CH_4$; $NO_3^- - N_2$; $N_2 - NH_4$

TABLE 9.18
Factors Favoring Disequilibrium
of Redox Couples

Flow; Biological Activity;
Redox reactions of light bioactive elements;
Electrochemical reactions of exposed mineral surfaces

---

coupled, hence the name redox. The oxidizing agent is the substance that causes oxidation and gets reduced in the process, whereas the reducing agent is the one that causes reduction and gets oxidized. Soluble phase redox couples found in groundwater are shown in Table 9.17. Redox potential measurements cannot be accurately taken unless the couple is at thermodynamic equilibrium, which may take from ten to one thousand years to reach. Processes in the environment that cause disequilibrium are shown in Table 9.18.

The elements carbon (C), hydrogen (H), oxygen (O), and sulfur (S) are light bioactive elements, which typically react by breaking covalent bonds, a slow process. Mineral based electrochemistry adversely affects the redox potential of a redox couple.

Inorganic oxidation reactions that consume dissolved oxygen in groundwater are: ferrous oxidation, sulfide oxidation, managanese (II) oxidation, and nitrification. These redox couples, given below, are common in self-purifying aquifers.

Ferrous Oxidation

Reduction:	$O_2(g) + 4H^+(w) + 4e^- \_ 2H_2O$
Oxidation:	$4FeCO_3(s) + 8H_2O \_$
	$4FeOOH(s) + 4HCO_3^-(10^{-3}) + 8H^+(w) + 4e^-$
	$O_2 + 4FeCO_3 + 6H_2O \_ 4FeOOH + 4HCO_3^- + 4H^+$

Free energy change:	$DG^0(w) = -21.0$ kcal
Redox potential:	$pE^0(w) = 15.42$

Sulfide Oxidation

Reduction:	$2O_2(g) + 8H^+(w) + 8e^- \_ 8H_2O$
Oxidation:	$HS^- + 4H_2O \_ SO_4^{2-} + 9H^+(w) + 8e^-$
	$2O_2 + HS^- \_ SO_4^{2-} + H^+$

Free energy change:	$\Delta G^0(w) = -23.8$ kcal
Redox potential:	$pE^0(w) = 17.50$

Manganese (II) Oxidation

Reduction:	$O_2(g) + 4H^+(w) + 4e^- \_ 4H_2O$
Oxidation:	$2MnCO_3(s) + 4H_2O \_$
	$2MnO_2(s) + 2HCO_3^- (10^{-3}) + 6H^+ + 4e^-$

$$O_2 + 2MnCO_3 + 2H_2O \_ 2MnO_2 + 2\ HCO_3^- + 2H^+$$

Free energy change:	$\Delta G^0(w) = -7.2$ kcal
Redox potential:	$pE^0(w) = 5.75$

Nitrification

Reduction:	$2O_2(g) + 8H^+(w) + 8e^- \_ 4H_2O$
Oxidation:	$NH_4^+ + 3H_2O \_ NO_3^- + 10H^+(w) + 8e^-$

$$2O_2 + NH_4^+ \_ NO_3^- + H_2O + 2H^+$$

Free energy change:	$DG^0(w) = -10.3$
Redox potential:	$pE^0(w) = 7.59$

Iron sulfide oxidation is pretty complicated, compared to the other couples, and is found where acid mine waters or other acidic sources lower the pH of the aquifer.

Iron Sulfide Oxidation

Reduction:	$2O_2(g) + 8H^+(w) + 8e^- \_ 4H_2O$
Pyrite Oxidation:	$4FeS_2(s) + 4H_2O + 14O_2(g) \_$
	$8H^+(w) + 8SO_4^{2-}(s) + 4Fe^{2+}(s)$
Iron Oxidation:	$4Fe^{2+}(s) + O_2(g) + 4H^+(w) \_ 4Fe^{3+}(s) + 2H_2O$
Iron (III)	$4Fe^{3+}(s) + 12H_2O \_$
Precipitation:	$4Fe(OH)_3(s) + 12H^+(w)$
	$17O_2 + 4FeS_2 + 10H_2O + 8e^- \_$
	$4Fe(OH)_3 + 8SO_4^{2-} + 8H^+$
Electrical Neutralization:	$4H_2O \_ 2O_2 + 8H^+$
Redox Couple:	$15O_2 + 4FeS_2 + 14H_2O \_$
	$4Fe(OH)_3 + 8SO_4^{2-} + 16H^+$

Some groundwater redox reactions consume organic matter. For instance, where sewage has seeped into the ground, or a spill of organic chemicals, including fuels, or where organic sediments have been left in the ground from earlier times, organics can couple with inorganic ions to create a redox potential. The reduction of nitrate is beneficial in that it fixes nitrogen in soil in a form usable by plants for growth. Fermentation is a redox reaction in which an organic substance is reduced by oxidizing another organic substance. Though more complicated than shown below, fermentation produces much less free energy than some other coupled reactions do.

Aerobic Respiration

Reduction:	$O_2(g) + 4H^+(w) + 4e^- \_ 2H_2O$
Oxidation:	$CH_2O + H_2O \_ CO_2(g) + 4H^+(w) + 4e^-$
	$CH_2O + O_2 \_ CO_2 + H_2O$

Free energy change: $DG^0(w) = -29.9$ kcal
Redox Potential: $pE^0(w) = 21.95$

Denitrification

Reduction: $4NO_3^- + 24H^+(w) + 20e^- \_ 2N_2(g) + 12H_2O$

Oxidation: $\underline{5CH_2O + 5H_2O \_ 5CO_2(g) + 20H^+(w) + 20e^-}$

               $5CH_2O + 4NO_3^- + 4H^+ \_ 5CO_2 + 2N_2 + 7H_2O$

Free energy change: $DG^0(w) = -28.4$ kcal

Redox potential: $pE^0(w) = 20.85$

Nitrate Reduction

Reduction: $NO_3^- + 10H^+(w) + 8e^- \_ NH_4^+ + 3H_2O$

Oxidation: $\underline{2CH_2O + 2H_2O \_ 2CO_2(g) + 8H^+(w) + 8e^-}$

               $2CH_2O + NO_3^- + 2H^+ \_ 2CO_2 + NH_4^+ + H_2O$

If the release has not happened, what is its likelihood of release? What containment is provided if the substance escapes its original container or process? What is the net precipitation in the vicinity? Calculate net precipitation as shown above.

Fermentation

Reduction: $CH_2O + 2H^+(w) + 2e^- \_ CH_3OH$

Oxidation: $\underline{CH_2O + H_2O \_ HCOO^- + 3H^+(w) + 2e^-}$

               $2CH_2O + H_2O \_ CH_3OH + HCOO^- + H^+(w)$

Free energy change: $DG^0(w) = -6.4$ kcal

Redox potential: $pE^0(w) = 4.67$

What is the depth to the aquifer? This is the distance from the lowest point of contamination in the ground (not necessarily from ground level) to the top of the aquifer. In other words, it is the distance from the surface to the top of the aquifer minus the distance from the surface to the lowest known point of contamination. Depth to aquifer is only calculated within a two mile radius of the source or center of the plume, if source is unknown. If the plume is wider than two miles, measure the depth to aquifer in several locations and take the least value as the depth. How long will it take for the release to reach the aquifer? See the travel time discussion below. These are some of the questions to be answered about the source in order to assess the groundwater pathway.

Table 9.19, adapted from the National Contingency Plan, lists in order of descending magnitude certain deficiencies about the source which exacerbate the groundwater threat.

Receptor distance (EPA's target distance limit) is about four miles for groundwater. Wells with contaminants from the source are considered within the receptor distance regardless of distance. From a contaminated groundwater plume with no known source, measure the four mile radius from the center of the plume.

If an aquifer interconnection exists within two miles of the source or plume center, combine the aquifers into a single hydrological unit. Extend the distance to any well where contamination is observed. Otherwise, treat each aquifer as a separate unit.

Restrict aquifer boundaries if aquifer discontinuities are established within four miles of the source or center of the plume. A discontinuity is a geologic, topographic

TABLE 9.19
Threat to Groundwater

Greatest Threat

Evidence of migration from source & containment structures.
Underground tank.
No liner for impoundment or lagoon.
Buried containers.
No secondary containment.
No engineered cover for closed source.
No functioning, maintained run-on/run-off system.
No leachate collection above liner of impoundment/lagoon.
Free liquids present with no dike, unsound dike, or dike that is not regularly inspected and maintained.

Lesser Threat

Cover + run-on/run-off + leachate + GWM for closed source.
Liner + leachate + GWM for active impoundment.
Run-on/run-off + vegetative cover for active land treatment.
Impervious containment ($\geq 10\%$ volume) + run-on + GWM +
inspections + good management for container storage.
Free liquids present but contained and well managed in a container storage.

Least Threat

Double liner + leachate + GWM for surface impoundment.
Double liner + leachate + impervious containment ($\geq 10\%$ volume) for container storage.

No Threat

Source inside or under maintained structure so that neither run-off nor leachate is generated; no free
 liquids stored; run-on prevented.
Land treatment in compliance with 40 CFR 264.280.

HS = hazardous substance; GWM = groundwater monitoring.

---

or other structure which entirely transects an aquifer thereby creating a continuous boundary to groundwater flow within the unit. Exclude the portion of the aquifer beyond the discontinuity when assessing the pathway. If contamination is found in a well on the other side of the discontinuity, but within the four mile distance, you are back to considering the entire formation as a unit.

If a Karst aquifer underlies any portion of the sources consider depth to aquifer, travel time, mobility of COC, nearest well and potential contamination very carefully. Assign a thickness of 0 feet to a karst aquifer underlying any portion of the sources of contamination.

Evaluate the travel time through the ground to the aquifer based on a geological assessment of the intervening earth between the lowest point of contamination and the top of the aquifer. Determine hydraulic conductivity of each layer by *in situ* or laboratory testing. Alternatively, you may use Table 9.20 which is adapted from EPA's National Contingency Plan.

EPA assigns a mobility of 1 to contaminants that have shown up in analysis of water taken from an aquifer. Obviously, the contaminant is already arrived and the

TABLE 9.20
Hydraulic Conductivity of Geological Materials

Type of Material	Hydraulic Conductivity (cm/sec)
Clay; low permeability till (compact unfractured till); shale; unfractured metamorphic and igneous rocks	10^{-8}
Silt; loesses; silty clays; sediments that are predominantly silts; moderately permeable till (fine-grained, unconsolidated till, or compact till with some fractures); low permeability limestones and dolomites (no karst); low permeability sandstone; low permeability fractured igneous and metamorphic rocks	10^{-6}
Sands; sandy silts; sediments that are predominantly sand; highly permeable till (coarse-grained, unconsolidated or compact and highly fractured); peat; moderately permeable limestones and dolomites (no karst); moderately permeable sandstone; moderately permeable fractured igneous and metamorphic rocks	10^{-4}
Gravel; clean sand; highly permeable fractured igneous and metamorphic rocks; permeable basalt; karst limestones and dolomites	10^{-2}

calculation of the movement is trivialized. Solubility of the COC in water and its distribution coefficient are used to calculate mobility for other scenarios.

The distribution coefficient, K_d mL/g, is estimated as

$$K_d = K_{oc} f_s \tag{9.90}$$

where f_s is the sorbent content of soil, that is, the fraction of clays plus organic carbon. The value of f_s ranges from 0.03 to 0.77 which establishes the lower and upper limits of the distribution coefficient. The geometric mean of the K_d upper and lower limits is used.

Solubility data is available in several handbooks (for instance, *Perry's Chemical Engineering Handbook* printed by McGraw–Hill or *Handbook of Chemistry and Physics* printed by CRC Press). For a metal or metalloid with no reported solubilities or only qualitative data, determine the overall range of solubilities for compounds of the substance for which water solubility data is available (not necessarily compounds which are present). Calculate the geometric range of the highest and lowest solubility.

Hwang suggests the following three-dimensional transport model under the conditions of a continuous point source discharge after Turner's heat transfer algorithm:

$$C(x, y, z, t) = \frac{C_o Q}{4 \pi \varepsilon R \sqrt{D_y D_z}} e^{\frac{V_x}{2 D_x}} \left[e^{\frac{UR}{2 D_x}} \mathrm{erfc}\left(\frac{Ut + R_d R}{\sqrt{4 R_d D_x t}} \right) \right.$$

$$\left. + e^{\frac{UR}{2 D_x}} \mathrm{erfc}\left(\frac{R_d R - Ut}{\sqrt{4 R_d D_x}} \right) \right] \tag{9.91}$$

where

$$R^2 = x^2 + \frac{D_x}{D_y}y^2 + \frac{D_x}{D_z}z^2 \tag{9.92}$$

and

$$U = V\left(1 + \frac{4D_x R_d k}{V^2}\right)^{1/2} \tag{9.93}$$

where: C = concentration of contaminant at (x,y,z), lb./ft.³
V = seepage velocity, ft./sec
C_o = concentration of contaminant in leachate entering groundwater, lb./ft.³
Q = leachate flow into groundwater, ft.³/sec
R_d = retardation factor
D_x = diffusion coefficient in x-direction, ft.²/sec
D_y = diffusion coefficient in y-direction, ft.²/sec
D_z = diffusion coefficient in z-direction, ft.²/sec
t = time, sec
k = biodegradation constant, 1/sec
erfc = complementary error function.

C_oQ represents pollutant release rate in groundwater.

$$V = \frac{v}{\varepsilon} \tag{9.94}$$

where: v = Darcy's velocity
ε = porosity

$$D_x = \alpha_x V$$
$$D_y = \alpha_y V \tag{9.95}$$
$$D_z = \alpha_z V$$

where α is the dispersivity in the appropriate direction. Steady-state concentrations are calculated:

$$C(x,y,z) = \frac{C_oQ}{2\pi\varepsilon R_d\sqrt{D_yD_z}} e^{\frac{1}{2}\left(\frac{Vx}{D_x} - \frac{UR}{D_x}\right)} \tag{9.96}$$

The vertically averaged concentration is

$$C(x,y,t) = \frac{C_oQ}{\varepsilon AB\,R_d^{1.5}H} \int_0^{t_d} e^{-k(t-t')}Fd't \tag{9.97}$$

where AB represents the dimensions of the source in the direction and cross-direction of groundwater flow.

$$F = \left[\text{erf}\left(\frac{x - V^*(t-t')}{\sqrt{4D_x^*(t-t')}} \right) - \text{erf}\left(\frac{x - v^*(t-t') - A}{\sqrt{4D_x^*(t-t')}} \right) \right]$$
$$\times \left[\text{erf}\left(\frac{y + B/2}{\sqrt{4D_y^*(t-t')}} \right) - \text{erf}\left(\frac{y - B/2}{\sqrt{4D_y^*(t-t')}} \right) \right]$$

(9.98)

In this equation $V^* = V/R_d$; $D_x^* = D_x/R_d$; $D_y^* = D_y/R_d$; and $D_z^* = D_z/R_d$. Also t_d is the length of time the source discharges in seconds and H is depth of the aquifer in feet.

By averaging the concentrations in the aquifer over its depth, this three-dimensional solution is possible:

$$C(x, y, z, t) = \frac{C_o Q}{4\varepsilon R_d AB \sqrt{\pi D_z (t-t')}}$$

(9.99)

or

$$C(x, y, z, t) = \frac{C_o Q}{4\varepsilon R_d AB} \int_0^{t_d} \frac{1}{\sqrt{\pi D_z (t-t')} \text{erf}\left(\frac{H}{\sqrt{4D_z^*(t-t')}} \right)} e^{\frac{-z^2}{4D_z^*(t-t')} - k(t-t')}$$

(9.100)

$$- F' dt$$

Biodegradation processes are aerobic, anaerobic and facultative. The rate of aerobic degradation is controlled by the concentration of oxygen. Oxygen can be introduced to groundwater at the water table interface or oxygen-rich groundwater can pass through zones of absorbed contamination from other areas. Vance offers the following heuristic:

3 lbs oxygen degrades 1 lb. hydrocarbon

The typical *in situ* aerobic decay rate is 35 mg/l day or about one-half ounce per cubic yard of aquifer matrix per day. Aerobic degradation stops organic acids are produced at low oxygen concentration. DO should exceed 2 mg/l for optimum aerobic degradation. At less than 2 mg/l DO the degradation of aromatics slows dramatically.

When the oxygen is consumed the aquifer becomes an anaerobic system. Anaerobic degradation of species in groundwater is limited by the concentration of appropriate electron receptors: nitrate, sulfate or iron. With insufficient electron receptors present the anaerobic degradation ceases with the production of aliphatic and aromatic acids. Under complete anaerobic conditions nitrate reducing bacteria will continue to degrade hydrocarbons. When the DO concentration is 0.1–0.4 mg/l the anaerobic degradation rate is about 1–2% of optimum.

At the edge of the plume where the concentration of the contaminant is less than 1 ppm, first order decay kinetics apply. Within the plume the situation is more complex. Attenuation over time can be measured at the edges of the plume using monitoring wells once per quarter for at least one year. Plot the well data on semilog paper as log concentration vs. time. The slope of the curve is the first order decay constant in %/day.

Decay with distance from the source depends mostly on the heterogeneity of the aquifer. Data from at least three monitoring wells along the axis of the plume are needed to develop a semilog plot of log concentration vs. distance. The slope of this curve is the decay constant divided by groundwater velocity.

where: C_s = near-field stream concentration, lb./ft.^{3},
            m_g = contaminant mass flux entering the stream, lb./day,
            Q_s = stream flow, ft.^{3}/day.

The mass flux can be determined by

$$\dot{m}_g = \dot{m}_L e^{-\frac{R\lambda_v L}{V_s}} e^{-\frac{R'\lambda_s x_s}{V_s'}} \tag{9.101}$$

where: R = retardation factor for vadose zone, dimensionless,
            R′ = retardation factor for saturated zone, dimensionless,
            l_v = overall decay coefficient within vadose zone, ft.^{3}/day,
            l_s = overall decay coefficient within saturated zone, ft.^{3}/day,
            L = distance from lowest point of source to water table, ft,
            x_s = distance traveled within saturated zone, ft.
            V_s = seepage velocity in vadose zone, ft.^{3}/day, and
            V_s' = seepage velocity in saturated zone, ft.^{3}/day.

$$\dot{m}_L = C(0,t)IA \tag{9.102}$$

where: I = infiltration rate from the source, ft.^{3}/day, and
            A = area of the source, ft^2.

The concentration (C_R) in the river anywhere downstream from the groundwater intercept is

$$C_R = C_S e^{-\frac{\lambda_R x_R}{V_R}} \tag{9.103}$$

TABLE 9.21

Data Requirements for

Groundwater as a Pathway

Depth to groundwater from base of source

Permeability of unsaturated zone

Potential to short circuit pathway to water table

Infiltration potential

Containment effectiveness

---

where: l_R = decay coefficient with the stream, dimensionless,

x_R = distance traveled within the stream, ft., and

V_R = velocity of the stream.

Suppose the river is already contaminated?

$$C_{RM} = \frac{(C_{GM} \times Q_{GM}) + (C_{RB} \times Q_{RB}/2)}{Q_{GM} + Q_{RB}/2}$$ (9.104)

where: C_{RM} = polluted concentration of contaminant in the river, lb./ft.^{3},

C_{RB} = background concentration of contaminant in the river, lb./ft.^{3},

C_{GM} = maximum concentration in groundwater, lb./ft.^{3},

Q_{GM} = maximum flow of groundwater, mgd, and

Q_{RB} = full flow of river, mgd.

Soil Transportation. Both the resident population and nearby populations are threatened by contaminated soils. Whether a pure chemical stream contacts soil or contaminated water the chemicals do migrate outward and downward. The data requirements for studying soil or groundwater as a pathway of concern are given in Table 9.21. A description of the geology underneath the source is needed to evaluate the migration and/or stabilization of the contaminant in the porous media. Soil is composed of two kinds of particles: minerals and organics. An unsaturated zone exists from the surface level of the ground down to the underground water table which is the saturated zone. Above the water table the unsaturated zone of soil is called the vadose zone and generically the water there is called vadose water. Figure 9.1 shows the divisions of water under the surface. Vadose water is composed of three types of water: soil water, intermediate vadose zone water and capillary water. For the purposes of our discussion this is all we need to know of underground water types for the moment.

Flow of water in the vadose zone is considered steady and one dimensional for modeling purposes. The governing equation is

$$\frac{d}{dz}\left[KK_{rw}\left(\frac{\delta\psi}{\delta z} - 1\right)\right] = 0$$ (9.105)

where: z   = vertical coordinate pointing downwards, ft.,
        K   = saturated hydraulic conductivity of vadose zone, ft./day,
        K_{rw} = relative permeability of the vadose zone, dimensionless, and
        ψ = pressure head, ft.

The boundary conditions are

$$V(0,t) = 1$$
$$\psi(L,t) = 0$$

where: V = Darcy velocity in vadose zone, ft./day,
        t = duration of pulse source, days, and
        L = distance from contamination to water table, ft.

We know that

$$K_{rw} = \frac{\left(S_w - S_{wr}\right)^n}{\left(1 - S_{wr}\right)^n} \tag{9.106}$$

where: S_w = water phase saturation in vadose zone, dimensionless,
        S_{wr} = residual water phase saturation for the vadose zone, dimensionless,
        n   = an empirical coefficient in the relationship

and

$$\frac{S_w - S_{wr}}{1 - S_{wr}} = \frac{1}{\left[1 + \left(\alpha|\psi - \psi_a\right)^\beta\right]^\gamma} \tag{9.107}$$

for $\psi < \psi_a$ but for $\psi \geq \psi_a$ the expression is equal to unity. In this equation α, β and γ are empirical coefficients. To solve the flow problem we must specify the relationships between the relative permeability and water saturation and between pressure head and water saturation.

The governing equation for vadose zone transport is also treated in one dimension. Longitudinal dispersion, linear adsorption, chemical and biochemical first order decay and advection are included.

$$R\frac{\delta C}{dt} = D\frac{\delta^2 C}{dz^2} - V_s\frac{\delta C}{dz} - \lambda_v C \tag{9.108}$$

where

$$R - 1 + \frac{\rho_b K_d}{\theta} \tag{9.109}$$

where: R = retardation factor for vadose zone, dimensionless,
 C = concentration of contaminant, lb./ft.^{3},
 D = longitudinal hydrodynamic dispersion in vadose zone, ft.^{2}/day
 V_S = seepage velocity into vadose zone, ft./day,
 l_V = overall decay coefficient within vadose zone, ft.^{3}/day.

Source boundary conditions are either constant concentration:

$$C(z,t) = 0$$
$$C(0,t) = C_o$$

(9.110)

or, decaying concentration:

$$P(x) = I - e^{\left[-A(I+Sx)^2\right]}$$

(9.111)

or, or a constant pulse:

$$C(0,t) = C_o\left[I - s(t - T)\right]$$
$$C(\infty,t) = 0$$

(9.112)

where: C_o = initial concentration at the source, lb./ft.^{3}
 T = duration of pulse source, days.

 Contamination enters the saturated zone either by direct leaching through the soil if the source is deep enough or where there is no vadose zone. The transport model for the saturated zone assumes uniform flow in a homogenous aquifer of uniform thickness with three-dimensional dispersion, that is, dispersion in the longitudinal (x), transverse (y) and vertical (z) directions. Let us also assume linear adsorption, first-order decay and dilution due to direct recharge into the contamination plume. The flow domain is semi-infinite in the longitudinal, infinite in the transverse and finite in the vertical direction. Contaminant concentration is assumed to be Gaussian in the lateral direction from the source and uniform over a vertical mixing or penetration depth. The vadose zone and saturated zone are coupled by requiring conservation of mass at the water table.
 The governing equation is

$$D_{xx}\frac{\delta^2 C}{\delta x^2} + D_{yy}\frac{\delta^2 C}{\delta y^2} + D_{zz}\frac{\delta^2 C}{\delta z^2} - V'_S\frac{\delta C}{\delta x} = R\frac{\delta C}{\delta t} + R\lambda_S C + RqC \qquad (9.113)$$

where D_{xx}, D_{yy} and D_{zz} are the hydrodynamic dispersion coefficients in the x, y and z-directions respectively and which are expressed in ft.^{2}/day. Also,

V'_s = seepage velocity in the saturated zone, ft./day,
λ_s = decay coefficient within the saturated zone, ft.³/day,
q   = infiltration into groundwater plume, ft./day,

and

$$\lambda_s = \frac{\lambda_1\theta + \lambda_2\rho_b K_d}{\theta + \rho_b K_d} + \lambda'_b \tag{9.114}$$

where: λ_1 = liquid phase chemical decay, ft.³/day,
λ_2 = solid phase chemical decay, ft.³/day,
θ = porosity of the saturated zone, dimensionless,
ρ_b = bulk density of the soil, lb./ft.³
λ'_b = biological decay coefficient for the saturated zone, ft.³/day, and
K_d = distribution coefficient for the chemical in the liquid and solid phase.

Initial and boundary conditions for these equations are

$$C(x, y, z, 0) = 0 \tag{9.115}$$

and

$$C(0, y, z, t) = C_0 e^{-y^2/2\sigma^2}$$
$$0 \le z \le H \tag{9.116}$$

where: σ = standard deviation of the Gaussian contaminant source, ft., and
H = depth of source penetration within the saturated zone, ft.

Also, these conditions apply at boundaries

$$C(x, \pm\infty, z, t) = 0 \tag{9.117}$$

$$C(\infty, y, z, t) = 0 \tag{9.118}$$

$$\frac{\delta C}{\delta z}(z, y, 0, t) = 0 \tag{9.119}$$

and

$$\frac{\delta C}{\delta z}(x, y, B, t) = 0 \tag{9.120}$$

where B is the thickness of the saturated zone in feet.

TABLE 9.22
VOC in Soil

VOC vapor adsorbed	VOC liquid adsorbed
VOC vapor dissolved	VOC liquid dissolved

TABLE 9.23
Typical Porosities for
Unconsolidated Sediments

Unconsolidated Sediment	ε (%)
Clay	45–55
Silt	35–50
Sand	25–40
Gravel	25–40
Sand/Gravel Mixes	10–35
Glacial Till	10–35

The mean travel time of water in an aquifer is determined by

$$T = \frac{dn}{ki} \qquad (9.121)$$

where: d = distance to well,
         n = effective porosity of the aquifer,
         k = hydraulic conductivity of the aquifer, and
         i = hydraulic gradient of the aquifer.

VOCs entering the soil partitions among liquid and vapor phases. Some VOC dissolves into soil water and some adsorbs onto the surfaces of soil particles so that the following possibilities exist (Table 9.22). Each possibility is a function of volatility, solubility, soil moisture content, type of soil and amount of soil solids. The latter two properties give some clue as to the porosity of the soil. Some typical porosities are given in Table 9.23 for unconsolidated sediments and in Table 9.24 for consolidated rocks. Fractured bedrock or permeable soil and rock can be critical as a pathway from the source to groundwater.

How do VOCs partition between pure liquid and soil gas? If allowed to reach equilibrium we can calculate the mole fraction of vapor in the head space over the liquid assuming constant temperature:

$$m_a = \frac{P_a}{P_T} \qquad (9.122)$$

TABLE 9.24
Typical Porosities for Consolidated Rocks

Consolidated Rock	ε (%)
Sandstone	5–30
Limestone/Dolomite	1–20
Shale	0–10
Fractured Crystalline Rock	0–10
Vesicular Basalt	10–50
Dense, Solid Rock	<1

where: m_a = mole fraction of component a
P_a = vapor pressure of component a
P_T = total air pressure

With that separation out of the way, how do VOCs partition between soil gas and soil moisture? Once again we assume equilibrium is reached and temperature is constant and known so that Henry's Law applies:

$$K_H = \frac{C_G}{C_L} \qquad (9.123)$$

where: K_H = Henry's Law constant for the contaminant at a specified temperature
C_G = concentration of the contaminant in soil gas
C_L = concentration of the contaminant in soil moisture

Another expression for Henry's constant is:

$$K_H = \frac{16.04 P_a M_a}{T C_L} \qquad (9.124)$$

where: M_a = gram molecular weight of contaminant a
T = temperature, Kelvin

REFERENCES

Abdel-Magid, Isam Mohammed, Abdel-Wahid Hago Mohammed, and Donald R. Rowe. *Modeling Methods for Environmental Engineers.* Boca Raton, FL: CRC Press/Lewis Publishers, 1997.

Anderson, H.V. *Chemical Calculations.* New York: McGraw–Hill Book Company, 1955.

Ayers, Kenneth W., *et al. Environmental Science and Technology Handbook.* Rockville, MD: Government Institutes, Inc. 1994.

Bird, R. Byron, Warren E. Stewart, and Edwin N. Lightfoot. *Transport Phenomena*. New York: John Wiley & Sons, 1960.

Bodurtha, Frank T., Jr. "Vent Heights for Emergency Releases of Heavy Gases." *Plant/Operations Progress*, April 1988, pp. 122–6.

Boulding, J. Russell. *Practical Handbook of Soil, Vadose Zone, and Ground-Water Contamination: Assessment, Prevention, and Remediation*. Boca Raton, FL: CRC Press/Lewis Publishers, 1995.

Burton, G. Allen, Jr. Ed. *Sediment Toxicity Assessment*. Chelsea, MI: Lewis Publishers, 1992.

Clement Associates. "Methods for Estimating Workplace Exposure to PMN Substances." A Report to EPA, 1982.

Crowl, Daniel A. and Joseph F. Louvar. *Chemical Process Safety: Fundamentals with Applications*. Englewood Cliffs, NJ: Prentice Hall, 1990.

Crynes, B.L. and H.S. Foglar. Eds. *AIChE Modular Instruction Series E: KINETICS, Volume 1: Rate of Reaction, Sensitivity, and Chemical Equilibrium*. American Institute of Chemical Engineers, 1981.

Daugherty, Jack E. *Industrial Environmental Management: A Practical Handbook*. Rockville, MD: Government Institutes, Inc., 1996.

Desmarais, Anne Marie C. and Paul J. Exner. "The Importance of The Endangerment Assessment in Superfund Feasibility Studies." *Hazardous Material Control Monograph Series: Health Assessment*. Silver Spring, MD: Hazardous Material Control Research Institute, pp. 59–62.

EPA. 40 CFR 300 National Oil and Hazardous Substances Pollution Contingency Plan. *47 FR 31203*. July 16, 1982.

Goyal, Ram K. and Nassir M. Al-Jurashi. "Gas Dispersion Models." *Professional Safety*, May 1990, pp. 23–33.

Guinnup, Dave. "Non-buoyant Puff and Plume Dispersion Modeling Issues." *Plant/Operations Progress*, January 1992, pp. 12–5.

Hubbard, Amy E., Robert J. Hubbard, John A. George and William A. Hagel. "Quantitative Risk Assessment as the Basis for Definition of Extent of Remedial Action at the Leetown Pesticide Superfund Site." *Hazardous Material Control Monograph Series: Risk Assessment I*. Silver Spring, MD: Hazardous Material Control Research Institute, pp. 33–9.

Hummel, Albert A. "Derivation of Equation for Evaporation from Open Surfaces." A Report to EPA, Aug. 18, 1990.

Hwang, Seong T. "Assessing Exposure to Groundwater Contaminants Migrating from Hazardous Waste Facilities." *Hazardous Material Control Monograph Series: Health Assessment*. Silver Spring, MD: Hazardous Material Control Research Institute, pp. 21–5.

Jayjock, Mike, Chris Keil, Mark Nicas, and Patricia Reinke. *A Tool Box of Mathematical Models for Occupational Exposure Assessment*. Professional Development Course 402. 1996 American Industrial Hygiene Conference & Exposition. Washington, D.C., May 19, 1996.

Jorgensen, S.E., B. Halling-Sorensen, and S.N. Nielsen. *Handbook of Environmental and Ecological Modeling*. Boca Raton, FL: CRC Press/Lewis Publishers, 1996.

Karam, Joseph G. and Martha J. Otto. "Ocean Disposal Risk Assessment Model." *Hazardous Material Control Monograph Series: Health Assessment*. Silver Spring, MD: Hazardous Material Control Research Institute, pp. 1–8.

Kay, Robert L., Jr. and Chester L. Tate, Jr. "Public Health Significance of Hazardous Waste." *Hazardous Material Control Monograph Series: Health Assessment*. Silver Spring, MD: Hazardous Material Control Research Institute, pp. 65–71.

Lesso, William G., C. Dale Zinn, and Dwight B. Pfenning. *Hazard Assessment and Risk Analysis for Process Industries*. The University of Texas at Austin, College of Engineering, Mechanical Engineering Department, Sept. 24–27, 1990.

Manahan, Stanley E. *Environmental Chemistry.* 6th. Ed. Boca Raton, FL: CRC Press/Lewis Publishers, 1994.

McKown, Gary L., Ronald Schalla and C. Joseph English. "Effects of Uncertainties of Data Collection on Risk Assessment." *Hazardous Material Control Monograph Series: Risk Assessment Volume I.* Silver Spring, MD: Hazardous Material Control Research Institute, pp. 57–60.

Meroney, Robert N. "Dispersion in Non-Flat Obstructed Terrain and Advanced Modeling Techniques." *Plant/Operations Progress,* January 1992, pp. 6–11.

MRI. "Occupational Exposures from Bagging and Drumming." A Report to EPA, 1986. National Safety Council. *Accident Prevention Manual for Industrial Operations: Engineering and Technology.* 9th. Ed. Chicago: National Safety Council. 1988.

Ney, Ronald E., Jr. *Fate and Transport of Organic Chemicals in the Environment: A Practical Guide.* 2nd Ed. Rockville, MD: Government Institutes, Nielsen, D.M. ed. *Practical Handbook of Groundwater Monitoring.* Chelsea, MI: Lewis Publishers, 1991.

Pace Laboratories. "Evaporation Rates of Volatile Liquids." A Report to EPA, 1989.

PEI Associates. "Effectiveness of Local Exhaust Ventilation for Drum-filling Operations." A Report to EPA, 1988.

—— . "Releases during Cleaning of Equipment." A Report to EPA, 1988.

Plog, Barbara A., George S. Benjamin and Maureen A. Kerwin. *Fundamentals of Industrial Hygiene.* 3rd. Ed. Chicago: National Safety Council. 1988.

Quagliano, James V. *Chemistry.* Englewood Cliffs, NJ: Prentice–Hall, Inc., 1958.

Rodricks, Joseph V. "Comparative Risk Assessment: A Tool for Remedial Action Planning." *Hazardous Materials Control Monograph Series: Risk Assessment Volume I.* Silver Spring, MD: Hazardous Materials Control Research Institute, pp. 45–8.

Salhotra, Atul M., David R. Gaboury, Peter S. Huyakorn and Lee Mulkey. "A Multimedia Exposure Assessment Model for Evaluating Land Disposal of Hazardous Waste." *Hazardous Material Control Monograph Series: Health Assessment.* Silver Spring, MD: Hazardous Material Control Research Institute, pp. 118–124.

Smith Ellen D., Lawrence W. Barnthouse, Glenn W. Suter II, James E. Breck, Troyce D. Jones and Dee Ann Sanders. "Improving The Risk Relevance of Systems for Assessing The Relative Hazard of Contaminated Sites." *Hazardous Materials Control Monograph Series: Risk Assessment Volume I.* Silver Spring, MD: Hazardous Materials Control Research Institute, pp. 27–32.

Vance, David B. "Groundwater Remediation by No Action: The Role of Natural Attenuation." *The National Environmental Journal.* July/August 1995, pp. 23–4.

—— . "Redox Reactions in Remediation." *Environmental Technology.* July/August 1996, pp. 24–5.

Wadden, Richard A. and Peter A. Scheff. *Engineering Design for The Control of Workplace Hazards.* New York: McGraw–Hill Book Co., 1987.

Walker Katherine D. and Christopher Hagger. "Practical Use of Risk Assessment in The Selection of A Remedial Action." *Hazardous Materials Control Monograph Series: Risk Assessment Volume I.* Silver Spring, MD: Hazardous Materials Control Research Institute, pp. 40–4.

Winklehaus, Charles, Barbara Turnham, S. Thomas Golojuch and Jeffery Sepesi. "Application of U.S. EPA's Superfund Health Assessment Methodology to Several Major Superfund Sites." *Hazardous Material Control Monograph Series: Health Assessment.* Silver Spring, MD: Hazardous Material Control Research Institute, pp. 125–30.

10 Fundamentals of Receptor Assessment

We have determined how a contaminant is generated, how much we can expect to enter a particular medium as a pathway, and how it may travel across the pathway. Let's turn our attention to the receptor. For each COC, what is its toxicity with respect to humans and wildlife? Who is exposed? Or will be? How many people are possibly affected? How many live where the exposure is? Work there? Play, shop, or go to school there? What ecological habitats are affected? What kinds of wildlife are exposed? Domestic animals? Crops? What is the magnitude of the dose received, particularly by humans, directly, or indirectly?

Evaluation of health consequences, caused by exposure to chemicals, is conducted in two parts. First, we must determine the likelihood that toxic injury will occur or otherwise adversely affect life, given the conditions of exposure to the COC. Typically, a search is made of literature to find relevant toxicity data, epidemiological studies, and perhaps clinical case studies. If necessary, data is extrapolated from animal toxicity studies, but this introduces uncertainty. Table 10.1 summarizes the topics and data considered in this phase of the exposure assessment. Once this information has been assembled and sifted, the exposure assessor must make a judgment based on what he or she knows about the conditions of exposure and the potential health effects as outlined in the various studies. The nature of the population at risk is a critical factor in this decision. Hopefully, the assessor will be able to conclude that the likelihood of any adverse effects is sufficiently small as to provide passive protection to public health. In other words, we hope not to have to take any action.

CHEMICAL INTAKE — GENERAL

The way we normally determine chemical intake is to determine by measurement, or calculation, the dose received — hence the term receptor. The dose-response assessment part of the exposure assessment is the determination of the relationship that exists between the amount received, and the probability of health effects occurring. A dose is generally calculated for the most exposed individual (MEI), if the exposure is environmental, or for the individual worker, if the exposure is in the workplace. Dose is rarely measured; rather, in practice, exposure is measured. The difference between exposure and dose is the difference between what can be received and the actual amount taken into the body, per unit weight per unit time (mg/kg-day). Two particular doses, ADI and UCR, are determined based on the assumption that the general population is at risk. When individuals are at risk, such as in a workplace, it is not necessary to collect demographic and other information, which is needed to study potential health effects in the general population. This is not to

---

TABLE 10.1
Summary of Receptor Toxicity Data

Pharmacokinetic Properties
Adsorption
Distribution and Storage
Metabolism (Transformation)
Excretion
Acute Toxicity
IDLH
LD_{50}
LC_{50}
Organ systems affected
Subacute and Subchronic Toxicity
Short-term Toxicity Studies
Target Organs
Chronic Toxicity
Long-term Studies in Lab. Animals
Long-term Human Exposure Reports
Epidemiology
Clinical Studies and Case Reports
Mutagenicity
Teratogenicity
Include human data where available
Carcinogenicity

---

say that unusual properties of certain materials would not put certain subpopulations at particular risk. For instance, one of the things that makes lead paint so harmful, in low income housing projects, is that the dry paint flakes, containing the lead, may have a sweet taste due to residual paint solvents, a taste not unlike candy. This makes the flakes all the more tempting for a child regardless whether hunger is a factor.

How the dose-response assessment is conducted depends on whether the health effects in question have a threshold or not. Most contaminants do not cause the exhibition of signs and symptoms until some threshold dose has been accumulated, or more properly, until some threshold exposure is measured. A threshold dose is the minimum amount of the contaminant necessary to produce adverse effects. Exposure limits such as PEL, REL, and TLV are based on this scenario, and for these kinds of contaminants the dose-response assessment estimates the exposure below which no significant adverse effect is expected statistically. Data from literature about the effects of such exposure are likely to be presented in terms of No-Observable Adverse Effects Level (NOAEL) for noncarcinogenic effects. NOAEL is the maximum experimental dose at which no toxic effects were observed. An experimental NOAEL is never taken directly as a human threshold dose. Thresholds vary not only species to species (interspecies), such as rat to human, but, even more so, they vary among individuals within a species (intraspecies). The same dose, administered under identical conditions, to two different species may present entirely different effects or vast differences in the degree of effect. A lethal dose administered to a human, may not affect a horse or dog, hence the use of strychnine and arsenic in many veterinary preparations. However, the toxicological effects of COCs on humans is typically determined by animal testing, so it behooves the experimenter to select a test species that most closely approximates the physiological processes in the human body.

EPA uses the term Average Daily Dose (ADD) for noncancerous effects and which is averaged over the period of dosing.

$$ADD_{pot} = \frac{D_{tot}}{M_B T} \tag{10.1}$$

where: ADD_{pot} = average daily dose, mg/kg-day
        D_{tot} = total dose, mg
        M_B = body weight, kg, and
        T = averaging time, days.

For cancer, biological response is in terms of lifetime probabilities and dose is described as a Lifetime Average Daily Dose (LADD) even if the exposure does not occur over the lifetime. The dose is

$$D_{tot} = C_{COC} \times R_1 \times D_E \tag{10.2}$$

where: C_{COC} = contaminant concentration, mg/kg,
        R_1 = intake rate, kg/day, and
        D_E = duration of exposure, days.

These units are used when soil or contaminated foodstuff is ingested. When the media is water, C is given in mg/l and R in l/day. For air, C is in mg/m^3 and R in m^3/day.

These doses are related to the exposure as follows. The duration of exposure, D_E, is a length of time that may be expressed as t_2-t_1. The concentration of the exposure may vary during this interval and is expressed as

$$E = \int_{t_1}^{t_2} C(t)dt \tag{10.3}$$

Potential dose is the amount of the COC that is ingested, inhaled, or applied to the skin. *Applied dose* is the amount of the COC at the absorption barrier, whether that is the skin. Lung, or gastrointestinal lining, and which is available for absorption. The amount of COC that is absorbed and available for interaction with biologically significant receptors is called the *internal dose*.

A potential exists for more variation, in response to a particular dose, among individual humans than between responses observed among a small population of test rats. Clearly, not all the members of a given population of a species will respond alike to an identical dose . Some sensitive individuals will have a response to the COC at a lower dose than more resistant individuals do. Eight reasons for intraspecies variation are: age and maturity; gender and hormonal status; genetic makeup; state of health; environmental factors; synergism; potentiation; and antagonism.

The very young and the very old are often more susceptible to chemical exposure than older children and adults. Obviously, infants and children are more susceptible to a COC than are adults. However, elderly individuals have reached a state where their diminished physiological capabilities make them unable to deal with a COC as effectively as when they were younger. Therefore, very young and very old age groups, in general, are more susceptible to lower doses of COCs.

In some cases, COCs have been know to affect either males or females but not the other gender. Reproductive toxins are the easiest examples. Women may be affected, but not men, or, women and men may be affected in different ways within the reproductive system of the body. Women have a naturally higher percent body fat than men do per pound of total weight, so they can accumulate more lipophilic (fat-soluble) COCs. Other physiological difference between males and females, such as strength and metabolism, account for some dose-response variation. Anabolic steroids build up proteins, while catabolic steroids break them down.

Some variation of dose-response is attributable to genetics. Clearly, where genetic factors diminish or impair physiological processes, especially the natural bodily defenses, variation is accounted for. Some people, for instance, due to a hereditary abnormality, lack a full complement of the G6PD (glucose-6-phosphodehydrogenase) enzyme, leading to damage to their red blood cells when they are dosed with aspirin or certain antibiotics. Persons with the normal form of the enzyme use these medicines safely.

Not least in importance, the state of health of the receptor can cause variation in response. Individuals with poor health, in general, do not process and excrete COCs as well as well persons do. Ill persons are more susceptible to the effects of COCs. Their bodies have decreased capacity for COC insult.

Environmental factors that may contribute to the response for a COC include workplace conditions, living conditions, and personal habits. Air pollution may complicate or conceal a response. Previous exposures to the COC may work in combination with other toxic mechanisms to alter the response to the COC.

Synergism is the manifestation of an enhanced response when two or more toxic COCs act in a combined fashion. Painters who are alcohol, or at least heavy drinkers, generally develop hepatotoxicity at a faster rate than someone who is merely exposed to paint vapors.

A potentiator is not toxic, but increases the toxicity of some other chemical. For instance, isopropanol is not hepatotoxic, but when a person is exposed to both isopropanol and carbon tetrachloride, the hepatotoxicity of the carbon tetrachloride is increased.

Antagonists are chemicals that lessen the predicted effect when combined. Four types of antagonists are functional, chemical, dispositional, and receptors. Functional antagonists produce opposite effects on the same physiological function. Phosphates reduce lead absorption in the GI tract, by forming insoluble lead phosphate, for instance. Chemical antagonism involves the reaction of the antagonist with the toxic compound to form a less toxic product. Toxic heavy metals, such as lead, arsenic, and mercury, can be bound up with chelating agents. Dispositional antagonists alter absorption, metabolism, distribution, or excretion, such as the way some alcohols use enzymes in their metabolism. Methanol breaks down to formaldehyde, which breaks down to formic acid. Ethanol breaks down to acetaldehyde, which breaks down to acetic acid. The aldehydes cause hangover, and even blindness. Since ethanol is more readily metabolized than methanol, methanol is not metabolized and gets excreted before it breaks down to formaldehyde. When Antabuse is administered to alcoholics, the metabolism of acetaldehyde is inhibited, and the patient gets a severe, prolonged hangover if ethanol is consumed. Receptor antagonists either bind to the same tissue receptor as the toxic chemical or blocks the action of the receptor, reducing the toxic effect. The toxic effects of organophosphate pesticides on receptor cells are blocked by atropine.

Typically, a safety factor is applied to the NOAEL, to derive a tolerable dose, called the Acceptable Daily Intake (ADI) in mg/kg-day for humans, by dividing the experimental NOAEL by an uncertainty factor. The ADI is considered the dose tolerable for the general human population. When a dose can be received by more than one pathway, the dose obtained by each route of exposure, regardless of the pathway, is first calculated. Total dose is determined by adding the route-specific doses together.

When calculated from animal study data, the acceptable daily intake must be based on NOEL, NOAEL, LOEL, or LOAEL, but with a safety factor built in. The factor F is determined:

$$F = F_1 F_2 F_3 \qquad (10.4)$$

where: F_1 = interspecies variation safety factor,
F_2 = intraspecies variation safety factor, and
F_3 = study sensitivity safety factor.

The interspecies variation factor is used to account for differences in responses between species such as human and rat. If the study produced human data, $F_1 = 1$. The value ranges from one to ten for animal data, depending on how analogous the study is for biokinetics and mechanism of toxicity between humans and the test species. F_2 accounts for intraspecies variation. Not all humans have the same sensitivities. If the human cohort of the study match the humans to be protected, use one as the value. An example would be if data from a study of iron workers were used to develop an occupational exposure limit then $F_2 = 1$, but if the data were to be used to develop an acceptable limit for the general population, the $F_2 = 10$ might be used instead. F_3 is used when a lifetime limit is desired but the data is from a short term test. If so, $F_3 = 10$. In most cases, the test data is divided by either one hundred ($F_1 = F_2 = 10$; $F_3 = 1$) or one thousand ($F_1 = F_2 = F_3 = 10$). The test data of interest are either NOEL, NOAEL, LOEL, or LOAEL, which is adjusted as follows:

$$NOAEL_{adj} = \frac{NOAEL}{F} \qquad (10.5)$$

The NOEL, LOAEL, and LOEL can be adjusted in the same manner. Except that an additional factor of ten, $F_3 = 100$, where LOEL or LOAEL is derived from short-term study data.

Pathway affects the dose calculation methodology. To calculate a dose received from ingestion of contaminated drinking water we need to know:

- concentration of COC in water
- volume of water consumed/day
- fraction COC absorbed through gastrointestinal tract
- weight of person drinking the water

Different factors determine the dose received when a COC is inhaled (such as breathing rate) or absorbed through the skin (such as skin penetration rate). Natural barriers to a COC, such as human skin, impede the intake (dose) and distribution once in the body (effective dose). Barrier effectiveness is partially dependent upon route of exposure. The information necessary can be deduced from the units of available data. DHD is specified in milligrams of COC per kilograms of body weight per day (mg/kg-day), whereas environmental concentrations are given as:

- concentration in water/liquids   mg/l
- concentration in air mg/m^3
- concentration in soil   mg/kg
- concentration in food   mg/kg

CARCINOGENICITY

Carcinogens are usually not examined for other chronic effects as these other effects typically require such a large dose compared to the doses that cause cancer. Cancer-causing agents, in other words, have no threshold, and to the best of our knowledge, any level of exposure may yield cancer after some amount of time. For carcinogenicity, the assessor's task is to extrapolate response data from doses in the experimental range to a response estimate in the dose range typical of an environmental exposure.

The most serious misunderstanding about cancer is that when we speak of cancer, we are not talking about one disease, but any of many diseases. A single disease called cancer is a myth. Even the mechanisms for cancer are complicated. Two carcinogenic mechanisms have been identified by researchers: genotoxic and epigenetic.

Genotoxic cancer is caused by electrophilic carcinogens that alter genes by means of interaction with DNA. If we all lived long enough we would all develop genotoxic cancer, sooner or later, or, more accurately, we would all develop benign tumors that were predisposed to become malignant. Three types of genotoxins are direct carcinogens, procarcinogens, and inorganic carcinogens. Direct, or primary, carcinogens act without any bioactivation. Procarcinogens require biotransformation to activate them. Inorganic carcinogens have been categorized preliminarily as genotoxic because they have the capacity to damage DNA.

Epigenetic carcinogens do not act directly with genetic material. Cocarcinogens are epigenetic substances that increase the overall response of a genotoxic substance when they are present together at the same time. Promoters increase the response of a genotoxic carcinogen when present after the genotoxic was present first. Promoters do not cause cancer by themselves. The solid-state carcinogens, such as asbestos and metal foils, depend on their physical form to cause cancer, though the exact mechanism is unknown. Hormones are not typically genotoxic, but they alter endocrine balance, and often act as promoters. Immunosuppressors stimulate virally induced, transplanted, or metastatic neoplasms by weakening the host immune system, and may be classified as epigenetic carcinogens.

Genotoxic carcinogens may be effective after a single exposure, or act in a cumulative manner, or act with other genotoxic carcinogens that affect the same target organ. Some epigenetic carcinogens only cause cancer when concentrations are high and exposure is long.

DNA is a critical target for carcinogens. Many carcinogens are metabolized so that they react with DNA. Defects in DNA repair predispose to cancer development. Many carcinogens are also mutagens, so inhibitors and inducers of carcinogens affect mutagenic activity. Several inheritable or chromosomal abnormalities predispose to cancer development. Initiated dominant tumor cells persist, consistent with changes in DNA. Cancer is inheritable at the cellular level, which condition may be the result of an alteration of DNA. All cancers display chromosomal abnormalities. Neely expresses the sequence of events leading to cancer using the symbols of a chemical equation:

$$DNA \underset{k_2}{\overset{k_1}{\longleftrightarrow}} DNA - x \overset{k_3}{\longrightarrow} C \qquad (10.6)$$

where: DNA   = deoxyribonucleic acid in the gene,
        DNA-x = altered DNA,
        C   = cancerous cell,
        k_1 = rate constant for production of DNA-x,
        k_2 = rate constant for repair of DNA-x, and
        k_3 = rate constant for production of cancerous cells.

Cancer happens when the cell's control center is disrupted with the subsequent alteration of reproduction and differentiation.

The rate constant k_1 represents the initiation process, which is driven to the right by three chief ways. Chemical interaction with DNA may produce a new structure, here labeled DNA-x, where the x represents a covalent bond. High energy, such as radiation exposure, can alter normal DNA to DNA-r, to distinguish this cell structure from DNA-x. Some DNA is randomly altered during the normal process of cell division and the divergent cell structure created by this process can lead to cancer, or at the very least to a benign tumor or neoplasm.

The process governed by k_2 is the operation of the defense mechanisms of the body, which work cooperatively to fend off and remove the aberrant cells before

TABLE 10.2
Dose-Response Assessment Requirements for Carcinogens

Human epidemiological data are preferred over animal data.

Data from animal species which respond most like humans should be used in the absence of suitable human data.

Emphasize biologically acceptable data from long-term animal studies showing the greatest sensitivity.

Data from exposure route of concern is preferred over other routes.

Describe carefully any considerations made in making route- to-route extrapolations if route of concern is not used.

Count all animals having neoplasms even if they have multiple tumor sites or multiple types (this is called pooling).

Benign tumors are counted with malignant tumors.

they can establish themselves in a host system. For instance, the body has an immunological mechanism that recognizes and neutralizes abnormal molecules before they begin to multiply. Unfortunately, other mechanisms work to help the aberrant molecules to get established. Repair mechanisms in the body use enzymes to convert altered DNA to normal DNA. These mechanisms can also be easily overwhelmed.

The first step in exposure assessment is to make an explicit determination of risk of cancer as a function of dose. Risk is the probability that cancer will occur. Risk has values between zero and one, expressed as $P(0)$ and $P(1)$. The EPA includes the requirements listed in Table 10.2 for a dose-response assessment of carcinogens.

The risk determined for each unit of dose is labeled the unit carcinogenic risk (UCR). UCR in risk per 1 mg/kg-day is determined by applying one of several mathematical models to carcinogenicity dose-response data. The risk of a specific dose is found by multiplying UCR by the dose.

The method of evaluating the receptor is summarized in Table 10.3 for discussion. A maximum lifetime risk of 1 in 1 million is always the goal. Action decisions must be aimed at preserving the $P(0.000001)$ risk level. Table 10.3 also shows that the minimum goal for noncarcinogenic health effects is DHD $\leq$ SADI. The SADI is derived by dividing ADI by various safety factors which are determined by the conditions present at the specific site of contamination.

Can the presence or absence of effects observed in a human population be used to predict effects of exposure in another exposed human population? When cancer effects in exposed humans are attributed to exposure to an exogenous agent, EPA makes the default assumption that such data predicts cancer in any other human population. This does not err on the side of public health conservatism, as you might think, as most studies linking cancer in humans to exogenous agents or reporting no effects are studies of occupational exposures. The sex, age, and general health of workers in an occupational setting in no way represents the general population in an environmental setting. How would infants and children, people whose health is not good, and males and females who are not represented in the workforce respond differently? We do not know and hence the assumption is not rigorously conservative.

TABLE 10.3
Evaluation of Receptor

1. Determine the daily human dose (DHD) for the COC.
2. Identify ADI and UCR from toxicity and carcinogenicity data.
3. Adjust ADI to suit site-specific circumstances and call this value SADI (site-specific ADI).
4. Compare DHD to SADI.
5. For carcinogens only, DHD x UCR = risk.
6. Action is required if DHD > SADI.
7. Action required if cancer risk > 1 in 1,000,000.
8. Best action makes DHD <<< SADI.
9. Best action makes cancer risk <<< 10^{-6}.

Contrarily, when cancer effects are not found in an exposed human population, an inference may not be drawn about the absence of carcinogenic effects in another exposed population. Unfortunately, epidemiologic studies have low power to detect and attribute responses, particularly when extrapolating null results from a healthy, worker population to other potentially sensitive exposed humans.

What about the presence or absence of effects observed in an animal population? Can these results be used to predict effects in exposed humans? EPA's default assumption is that positive effects in animal cancer studies indicate that the agent under study have potential for carcinogenic effects in humans. If no adequate human data are available, positive effects in animals may be used as a basis for assessing carcinogenic hazard to humans. When well-conducted animal studies do not detect cancer effects in two or more appropriate species and other available information does not support the carcinogenic potential of the agent under study, including the absence of human data to the contrary, one can conclude that the agent is not likely to possess human carcinogenic potential.

Target organs of carcinogenesis for agents that cause cancer in both animals and humans are most often concordant at one or more sites. Concordance by site is not uniform, however. The default assumption is that target organ concordance is not a prerequisite for evaluating implications of animal studies for humans. The mechanisms of control of cell growth anddifferentiation are concordant among species but with marked differences among species in the way control is managed in various tissues. In humans, the mutation of tumor suppressor gene p53 is perhaps the most frequently observed change in the genetics of tumors. This tumor suppressor is also observed in some rodent tissues, but other growth control mechanisms predominate in rodents. An animal response may very well be due to changes in a control that are relevant to human physiology, but may appear in animals in a different way. The National Toxicology Program (NTP) and the International Agency for Research on Cancer (IARC) both include benign tumors as carcinogenic effects in animal studies, if the tumors have the capacity to progress to malignancies with which they are associated. Of course, an animal study that reports more malignant than benign tumors should receive more weight than one that reports more benign than malignant tumors.

TABLE 10.4
Causes of Congenital Malformations

Heredity
Maternal diseases
Maternal malnutrition
Physical injury
Radiation
Chemical exposure

Benign tumors that do not progress to malignancy have to be assessed on a case-by-case basis due to the wide range of possibilities concerning the significance of their presence. Even if they do not turn malignant, benign tumors are serious health effects in their own right. A benign tumor on the brain is not harmless. Benign tumors may indicate the need for further studies, especially if observed in a short term study. Also, knowledge of the mode of action associated with the benign tumor response sometimes aids researchers in interpreting other tumor responses associated with the same agent.

How do metabolic pathways relate across species? Assume that a similarity exists in the basic pathways of metabolism between any two species and that a metabolite occurring in one species will be detected in the other. If a comparative metabolism study shows otherwise, you may use the data appropriately.

TERATOGENICITY

Teratogenicity is always examined since it cannot be predicted quantitatively. Teratogenicity is the capacity of a chemical to produce congenital malformation. Monsters is the Latin word for which teratogenicity gets its name. Table 10.4 lists causes of congenital malformations. Structural abnormalities are caused during the embryonic period, the first five to six weeks after conception. Once the structure of the body has been set, no exposure to any chemical can change the structure. Physiological and minor defects are caused during the fetal period, from eight to thirty-six weeks. Before or after this time, exposures have little effect. A previous exposure may still be flushing from the body, however, so remove newly pregnant women from exposure situations as soon as possible after learning of the conception. If a work area contains known teratogens, a good industrial hygiene practice is to prohibit women capable of child-bearing from working in the area.

Teratology is a precise science. Studies on laboratory animals have demonstrated that the exact day during the pregnancy is vitally important, so test exposures have to be evaluated daily. For example, thalidomide causes birth defects in rats only when administered to the mother on the twelfth day of gestation. Table 10.5 summarize the conditions on which the degree and nature of birth defects depend. Table 10.6 lists various factors with teratogenic potential. Chemicals that have been identified as teratogens are listed in Table 10.7.

TABLE 10.5
Factors Affecting the Degree and Nature of Birth Defects

Developmental state of embryo or fetus when the chemical is administered.
Dose, route of exposure, and exposure interval.
Transplacental absorption of COC and levels in tissues of embryo/fetus.
Ability of maternal liver to metabolize and detoxify COC.
Biological half-life COC or its metabolites.
State of cell cycle when COC reaches toxic concentration.
Capacity of embryonic/fetal tissues to detoxify or biotransform COC.

TABLE 10.6
Factors with Teratogenic Potential

Dietary deficiency
Hormonal deficiency
Hormonal excess
Hormonal and vitamin antagonists
Vitamin excess
Antibiotics
Heavy metals
Azo dyes
Anoxia producers
Certain chemicals (listed in Table 10.7)
Hypthermia, hyperthermia
Radiation
Infections

The two best known human teratogens have nothing to do with chemical exposures: ionizing radiation and German measles.

MUTAGENICITY

Mutagens cause mutations in the genetic code, altering the DNA of the exposed individual. Mutations may consist of chromosomal breaks, rearrangement of chromosome pieces, gain or loss of entire chromosomes, or changes within the gene structure. Table 10.8 lists known human mutagens.

Mutagens can pass effects into the human gene pool in the form of germinal or reproductive cell mutations that would take generations to breed out, except that humans do not select mates based on breeding needs. [The author does not intend to even suggest such a thing.] Another concern about mutagens is the possibility that somatic cell mutations may produce carcinogenic or teratogenic responses.

TABLE 10.7
Chemicals Identified as Teratogens

Acetamides (animals)
Anesthetic gases (humans)
Benzene (animals)
Boric acid (animals)
Caffeine (animals)
Carbon tetrachloride (animals)
Chloroform (animals)
Cyclohexanone (animals)
Dioxin [2,3,7,8-TCDD] (animals)
DMSO (animals)
Formamides (animals)
Hydoxyurea (animals)
Insecticides [in general] (animals)
Nitrosamines (animals)
Organic mercury compounds (humans)
Pesticides [in general] (animals)
Propylene glycol (animals)
Quinine (animals)
Salicylate (animals)
Sulfonamides (animals)
Thalidomide (humans)
Thiadiazole (animals)
Xylene (animals)

TABLE 10.8
Known Human Mutagens

Benzene
Ethylene oxide
Ethyleneimine
Hydrazine
Hydrogen peroxide
Ionizing radiation

THRESHOLD EXPOSURES

Other than cancer, chronic health effects are observed only after the exposure has reached a threshold. Exposures lower than the threshold are assumed to produce no effects; exposures greater than the threshold produce a response. Thresholds are determined by animal testing, epidemiological studies and reviews of clinical and case histories. Where the value is taken solely from animal studies it must be converted to a human ADI.

TABLE 10.9
Sources of Exposure Limits

PEL (Permissible Exposure Limit)	OSHA
TLV (Threshold Limit Value)	ACGIH
REL (Recommended Exposure Limit)	NIOSH
WEEL (Workplace Environmental Exposure Level)	AIHA
ERPG (Emergency Response Planning Guideline)	AIHA
IDLH (Immediately Dangerous to Life & Health)	NIOSH
OEL (Occupational Exposure Limit)	generic EPA
ORG (Occupational Reproductive Guideline)	ORNL
NCEL (New Chemical Exposure Limit)	EPA

Organizational abbreviations in Table 10.9:

ACGIH American Conference of Governmental Industrial Hygienists
AIHA   American Industrial Hygiene Association
EPA   Environmental Protection Agency
NIOSH National Institute of Occupational Safety & Health
ORNL   Oak Ridge National Laboratory
OSHA   Occupational Safety & Health Administration

How do people get exposed to hazardous and toxic chemicals in the environment? Mostly by eating. In certain localities or during emergencies by inhalation downstream from the source. Occasionally people contact these chemicals with their skin.

Chemical intoxication is traditionally diagnosed by the patient's history of exposure and the presentation of appropriate symptoms and finding of clinical signs. History of exposure is nearly "obvious" in a workplace setting. If trichloroethylene (TCE) is being used to as a vapor degreaser for metal parts, it should come as no surprise to us to find traces of TCE metabolites in the liver. In the workplace or in the outdoor environment the question we all want to have an answer to is, what is a safe level of exposure?

Threshold limit values (TLV's are recommended exposure limits made by the American Conference of Governmental Industrial Hygienists and PEL's are mandatory limits issued by OSHA) whether recommended or mandatory are based on known effects to exposure to one chemical over known time periods such as the work week. Table 10.9 shows some sources of exposure limits. These limits do not account for all symptoms and signs presented to and found by medical doctors. By the time the chemical shows up in a pathology report, it may be too late to recommend a more restrictive exposure limit.

Whenever a noncarcinogenic contaminant of air, water or soil has a TLV/PEL assigned to it the common practice in environmental exposure assessment has been to add a factor to account for a biological lifetime rather than a working lifetime and to account for exposure to those segments of the population which are more susceptible to the contaminant than the adult worker.

Remember that recommended or regulated exposure limits such as the TLV and PEL respectively are based on the assumption that there is a concentration below which no adverse effects (the no-effect-level) will be observed and as long as we keep a working adult exposed to less than the no-effect-level during the occupational lifetime (40 hours/week × 50 weeks/year and 40 years = 80,000 hours) his or her better-than-average health will be maintained. Therefore, in environmental practice we calculate the environmental exposure level (EEL):

$$EEL = \frac{TLV \times D_{af} \times M_{af}}{S_f} \times 10^3 \tag{10.7}$$

where: EEL is expressed in $\mu g/m^3$ and
          TLV = ACGIH-TLV or OSHA-PEL as 8-hr TWA, mg/m^3
          D_{af} = duration of exposure adjustment factor = 0.12
          M_{af} = magnitude of exposure adjustment factor = 0.72
          S_f = safety factor, 10 to 1,000

The duration of exposure factor is the ratio of the working lifetime (80,000 hours) divided by the biological lifetime for women (24 hours/day × 7 days/week × 52 weeks/year × 77.8 years = 6.8×10^5 hours).

Reasons exist to use other duration of exposure factors however. For instance, an assessment of exposure to residual pesticides on a farm was conducted. The chief route of exposure was inhalation of dust raised during tilling of the soil in the Spring each year. Therefore, the exposure rate was taken as ten days/year for forty years.

The magnitude of exposure adjustment factor is a more sophisticated adjustment as it is based on the increased risk faced by a ten year old child as compared with an adult breathing the same airborne contaminant. The child has a greater ventilation rate per unit body weight than an adult has. The daily air volume exchanged by an adult male (16 lpm × 60 min/hr × 24 hr/day/70 kg = 328 l/kg) is 0.72 that of a 10-year-old child (8 lpm × 60 min/hr × 24 hr/day/25 kg = 454 l/kg).

The safety factor (S_f) is based on the quality of data about the toxicity of a contaminant. If the information is used to derive the EEL is based on extensive experimental data a safety factor close to 10 should be used. If data is sparse or practically nonexistent a factor closer to a 1,000 ought to be used.

Therefore, we must ask ourselves as industrial hygienists and scientists what do *no-effect-level, safety factor* and *acceptable daily intake* really mean? Is it feasible that changes to the neurological system and perhaps other organs too happen stealthily even before irreversible tissue damage is noticed? Many people complain of chemical exposure without being taken seriously or in some cases are pacified with "we looked into it but found nothing for you to worry about." Take for instance the soldiers, airmen, and guardsmen returning from the "Desert War" with the "Saudi syndrome." They present symptoms, such as fatigue, malaise, slowed motor reactions, appetite impairment, headache, irritability, and rashes) that can be attributed to other common maladies. Essentially these kinds of symptoms are related to stress, a "bug," other syndromes such as chronic fatigue, or anything but chemical exposure. Recently, reports have indicated that these men and women may have received a dose of chemical or biological agents after all. In the broader context of chemical exposure, Root, Katzin and Schnare claim that such symptoms are likely to be "sensitive indicators of general organ system toxicity" and refer to them as "subclinical" or "subliminal" signs of chemical toxicity.

Health effects are easier to detect in the workplace where exposure levels are elevated and constant, or nearly so. Chronic low levels of exposure yield the kinds of symptoms that could be anything with respect to pathological cause. After many years of such exposure a pathological diagnosis may be possible but by then physiological changes are irreversible and permanent. Damage to the central and peripheral nervous systems can be devastating. Not to mention the "C" word, cancer.

Root, Katzin and Schnare devised an interesting table from reviewing studies of exposure to polyhalogenated biphenyls (PHBs, specifically PBBs and PCBs). They divided the studies into six exposure cohorts if that term may be used in such a way: ambient, low-level, occupational, extended occupational, lifelong ambient and massive. They found that the accumulated PHB in blood and body fat increased respectively as I have listed the exposure cohorts. The ambient exposure cohort had no observed health indications and biological functions were normal with ppb level PHB accumulated in the blood and fat for less than 35 years. The low-level exposure cohort, in the less than 1 ppm BHP accumulation range for more than 35 years, presented subtle symptoms such as fatigue, weakness, nervousness, pain in the joints and headaches among others. There is a wide overlap of these symptoms with other syndromes and diseases. Symptoms noted with lipophilic chemicals such as PHBs are not different from those reported for hydrophilic chemicals indicating they may relate to psychiatric or psychological conditions. Root, Katzin and Schnare are convinced however that differences between exposed groups and control groups cannot be explained without considering exposure to toxic substances in the etiology.

Continuing with our review of Root, Katzin and Schnare's article, it is interesting that at the occupational exposure level, about 25 ppm of BHP accumulated mostly in fat, subclinical and clinical signs are observed. They cite immune dysfunction, elevated CEA titer and elevated SGOT–SGPT. The extended occupational exposure cohort accumulated levels of BHP in fat approaching 200 ppm and presented readily identifiable signs and symptoms. Dermal abnormalities such as chlor–acne were presented. Abdominal pain was a frequent complaint as was eye irritation. Common complaints maybe, except the chlor–acne, but easily verifiable and traceable as to etiological origin.

A massive exposure to PCB caused a death by major system failure. The pathological report noted 13 ppm of PCB in body fat. After accumulating only ten ppm of PCB in fat, a cancer death was attributed to a lifetime ambient exposure. Unfortunately the toxicology of very few chemicals have been thoroughly researched. Most studies refer to blood level accumulations yet the levels stored in fat may be up to 500 times as much as that stored in blood. What does blood level accumulations have to do with chronic or subclinical conditions then? Table 10.10 summarizes symptoms associated with environmental exposures. The prevalence of symptoms is given in Table 10.11 in order of most often at the top to least presented at the bottom.

Mobilization of lipophilic chemicals from fat cells may not be desirable, as some medical researchers believe this could lead to the "flashback" phenomena wherein a symptom which has not presented itself in some time may recur. Other medical

TABLE 10.10
Summary of Environmental Exposure Symptoms

Associative Symptoms
  Memory Impairment
  Confusion
  Slowed Adolescent Development
Physiological Response Symptoms
  Headaches
  Sleeplessness
  Sleepiness
Ophthalmic Symptoms
  Eye Irritation
  Dimness of Sight
  Blurred Vision
  Oscillation of Eye
  Pupil Reactions
Gastrointestinal Symptoms
  Weight Loss (greater than 10 lbs.)
  Nausea
  Vomiting
  Abdominal Pain
  Abdominal Cramps
  Diarrhea
Musculoskeletal Symptoms
  Joint Pains
  Joint Swelling
  Muscular Aches and Pains
Motor Symptoms
  Speech Impairment
  Muscle Weakness
  Tremors
  Difficulty Walking
  Seizures
  Incoordination
  Dizziness
  Fatigue
Central Nervous Symptoms
  Depression
  Nervousness
  Irritability
  Emotional Instability
Sensory Nervous Symptoms
  Vision Impairment
  Hearing Impairment
  Loss of Smell
  Burning Sensation
  Paresthesias
  Hallucinations

TABLE 10.10 (continued)
Summary of Environmental Exposure Symptoms

Skin
  Rash
  Acne
  Sun Sensitivity
  Skin Darkening
  Skin Thickening
  Discoloration of Nails
  Deformation of Nails
  Dryness of Skin
  Sweating
  Slow or Poor Healing of Cuts

TABLE 10.11
Prevalence of Chemical Exposure Symptoms

Fatigue
Joint/Muscle Pain
Headaches
Abdominal Pain
Nervousness
Rash
Paresthesia
Constipation
Weakness
Dizziness/Incoordination
Acne
Skin Thickening
Disorientation

researchers believe that the risk of flashback is less than that of continuing to store the chemical in the body and that therefore one should be detoxified.

CHEMICAL INTAKE — INTERSPECIES

How are human equivalent doses extrapolated from animal studies? One way to determine the human oral dose is to adjust the animal oral dose by a scaling factor of body weights, M_B, raised to the 0.75 power. In other words, the human dose D_H

$$D_{H-oral} = D_{A-oral} \left(\frac{M_{B-human}}{M_{B-animal}} \right)^{0.75} \tag{10.8}$$

The adjustment factor represents the scaling of metabolic rates across animals of different sizes. If more sophisticated data is available concerning the relative differences in metabolism, the 0.75 power can be refined.

The human equivalent inhalation dose is estimated from lung deposition and internal dose data. Toxicokinetic data and metabolic parameters on the specific agent can be used to refine the equivalent dose. Toxicological modeling may be used to estimate from applied, to internal, to delivered dose.

For extrapolating from route-to-route, an agent that causes internal tumors by one route of exposure is assumed to be carcinogenic, if it is absorbed by the secondary route to give an internal dose. For internal tumors, internal dose is significant no matter what the route of exposure. The metabolism of the agent will be the same for a given internal dose. Evidence from metabolism studies is required to demonstrate that an agent will behave differently by one route of exposure as opposed to another.

A curve-fitting model is used to correlate dose responses outside the range of data available. When mode of action information supports linearity, or is insufficient to support nonlinearity, a linear approach is used. Direct action of the agent on DNA gives a linear response. Rate-limiting steps in metabolic processes are typically linear.

CHEMICAL INTAKE — CARCINOGEN

Scientists use a two-stage model for predicting cancers based on the mechanisms of carcinogenesis. First there is a proliferation of initiated, first-stage cells. Environmental agents affect cell transition rates, first-stage cell proliferation rates and cell death rates. Generally the model is written as:

$$P(x,t) = 1 - e^{-M[1+(1+S)x]}(1+Sx)\frac{\left(e^{[G(x)t]} - 1 - G(x)t\right)}{G(x)^2} \tag{10.9}$$

where: S = the smallest relative transition rate,
$S+1$ = the largest relative transition rate,
M = the scaling factor,
x = a constant lifetime exposure of the reactive form of the agent at the molecular or cellular site of action,
t = the age at which the risk is evaluated, and
$G(x)$ = the exposure-dependent growth rate of the first-stage, the growth of preneoplastic cells.

If we can assume that both transition rates are the same linear function of dose x, and that the growth rate of preneoplastic cells is independent of the exposure level the model can be simplified. In this case,

$$P(x,t) = 1 - e^{-M(1+Sx)^2}\frac{\left[e^{(Gt)} - 1 - Gt\right]}{G^2} \tag{10.10}$$

which reduces to the form

$$P(x) = I - e^{\left[-A(1+Sx)^2\right]}$$ (10.11)

at steady-state conditions, that is, if the time of observation is constant across all exposure groups.

Thorslund, Charnley and Anderson point out five strengths of this form of the model:

1. at low doses the model converges to a linear, no-threshold form;
2. it can be used to adjust for different lengths of observation among exposure groups;
3. the time-dependent form of the model has only two parameters that have to be estimated (therefore goodness-of-fit tests can be run on data from a standard bioassay with one control and two exposure groups;
4. a stable point estimate of risk can be obtained directly; and
5. the mathematical form of the model follows directly from the most widely accepted hypothesis of cancer induction.

Potency is the measure of the strength of any substance to cause cancer in humans. They are calculated for substances in Carcinogen Groups A, B and C which have suitable dose-response data but not for substances in Groups D and E since these are not potential carcinogens properly called. When considering the relative potency of j carcinogens compared to another carcinogen at response level:

$$R_j(p) = \frac{x(p)}{y_j(p)}$$ (10.12)

where $x(p)$ is the number of exposure units of the test carcinogen, and $y_j(p)$ is the number of exposure units of the jth carcinogen required to a total carcinogenic response rate of p in the test system. If the mechanism of the test carcinogen is the same as the jth carcinogen:

$$R_j(p) = R_j$$ (10.13)

In other words, the relative potency is independent of the response level. This assumption is the same as the hypothesis of simple action, which is used to estimate the joint response to multiple agents for various biological endpoints. The joint response to a species of carcinogenic chemicals is dependent upon total species exposure, T:

$$T = \sum_{j=1}^{m} R_j y_j + x$$ (10.14)

TABLE 10.12
Observed Rates

Agent	Exposure	Response
Control	0	r_o/n_o
First Member	x	r/n
jth Member	y_j	r_j/n_j

After Thorslund, Charnley and Anderson.

where: m = total number of chemical species members less one,
       y_j = exposure to the jth species member,
       x = exposure to the species indicator, and
      R_j = relative potency of the jth family member compared to the indicator
         (first) member.

The probability of a cancer response given exposure T is P(T) where the function P is the dose-response relationship for the lead member of the chemical species. Using a two-stage identical transition rate model, the dose-response for the test system is:

$$P(x) = 1 - e^{-A(1+Sx)^2} \qquad (10.15)$$

for the lead species member and

$$P(y_j) = 1 - e^{-A(1+SR_j y_j)^2} \qquad (10.16)$$

for j = 1,2,...,m carcinogenic species members.
    By equating observed rates in Table 10.12 with the function response we can estimate relative potencies, which are:

$$r_0/n_0 = 1 - e^{-A} \qquad (10.17)$$

$$r/n = 1 - e^{-A(1+Sx)^2} \qquad (10.18)$$

$$r_j/n_j = 1 - e^{-A(1+R_j y_j)^2} \qquad (10.19)$$

where: j = 1,2,...,m. Solving algebraically:

TABLE 10.13
Relative Potencies of Some PAHs

Benzo[a]pyrene	1.0
Benz[a]anthracene	0.145
Benzo[b]fluoranthene	0.140
Benzo[k]fluoranthene	0.066
Benzo[ghi]perylene	0.022
Chrysene	0.0044
Dibenz[ah]anthracene	2.82
Indeno[1,2,3-cd]pyrene	0.232

$$R_j = \frac{x\left[\ln\left(1 - \frac{r_j}{n}\right) \middle/ \ln\left(1 - \frac{r_0}{n_0}\right)\right]^{1/2} - 1}{y_j\left[\ln\left(1 - \frac{r}{n}\right) \middle/ \ln\left(1 - \frac{r_0}{n_0}\right)\right]^{1/2} - 1} \qquad (10.20)$$

Table 10.13 is a summary of Relative Potency estimates derived from indicator PAHs by Thorslund, Charnley, and Anderson.

For a multistage carcinogen with k stages, the lifetime increased cancer risk per kg body weight per day is:

$$P(D) = 1 - e^{-\left(q_1 D + q_2 D^2 + \dots + q_k D^k\right)} \qquad (10.21)$$

Crump and Howe calculate the cumulative lifetime risk of developing cancer due to exposure to a constant dose of a carcinogen beginning at age S_1 continuing until age S_2. To Crump and Howe risk is dependent on the number of stages k in the multi-stage process and particularly upon that stage r which is dose-dependent. If stage 1 is dose-related then

$$H(t) = dc \qquad (10.22)$$

when t is less than s_1

$$H(t) = dc = 0$$

when the value of t is greater than or equal to s_1, but less than s_2

$$H(t) = dc = \left(t - s_1\right)^k$$

and, when t is greater than or equal to s_2

$$H(t) = dc = (t - s_1)^k - (t - s_2)^k$$

where: $H(t)$ = cumulative incidence due to exposure
        t = age
        d = dose
        c = constant derived from carcinogenic potency
        k = stages
        s_1 = age at onset of exposure
        s_2 = age at termination of exposure.

The potency constant c is

$$c = \frac{b_1 \pi a_i}{k! a_1} \tag{10.23}$$

Dividing the cumulative incidence due to some short-term environmental exposure H_e by the cumulative incidence due to a lifetime exposure H_L at the same dose level yields the expected risk as a fraction of the lifetime the lifetime risk estimate F which is computed:

$$F = \frac{\displaystyle\sum_{t=s_1}^{70} (t - s_1)^k - \sum_{t=s_2}^{70} (t - s_2)^k}{\displaystyle\sum_{t=0}^{70} (t - s_1) k} \tag{10.24}$$

People over a certain age at the onset of exposure to a carcinogen are at much lower risk than those exposed persons who are younger. For instance, take a chemical or combination of chemicals with a thirty year interim period between onset of exposure and first detection of cancer symptoms. Everyone who was forty-five or older at the time of onset of exposure has much less of a risk of contracting cancer than someone who was younger than forty-five at the time, simply because in the interim thirty years those in the older group who have survived are seventy-five or older.

INGESTION

Taking hazardous substances into the body, by eating contaminated food, or drinking contaminated water, is the most common nonoccupational way of being exposed. The human body absorbs nearly one hundred percent of ingested COCs. The gastrointestinal (GI) tract is typically unharmed by the invasive COC. The primary exception to GI tract immunity involves irritating or corrosive chemicals. Those

COCs that are insoluble in the acidic fluids of the GI tract, consisting of the stomach, small intestines, and large intestines, are excreted. Soluble chemicals are absorbed through the lining of the GI tract, and transported by the circulatory system to whichever internal organ can be damaged.

The COC in the environment is the exposure, which ends at the opening of the mouth. The potential dose is received when contaminated food or water or accidentally ingested materials enter the mouth. The applied dose comes into contact with the membranes of the gastrointestinal tract. That amount of COC that passes through a G.I. membrane is the internal dose, and after metabolism becomes the biologically effective dose.

Fish bioaccumulate pollutants in the water and we consume them. Plants uptake contaminants from the soil and we either eat them directly or eat animals which eat those plants. Foraging animals transfer the hazardous materials to milk and meat. We consume a certain amount of soil directly (about 0.1 g/day for an adult) despite our penchant for cleanliness. A farmer tilling soil on an open tractor ingests about 1 mg/day of soil. Inhalation of fugitive dust for residents next to a dusty farm or mine can be twenty-five days per month, ten months per year for seventy years. The calculation is

$$I = \frac{C_W R_I F_E D_E}{M_B T} \tag{10.25}$$

where: I = intake by ingestion, mg/kg-day,
  C_W = concentration of the COC in water, mg/l,
  R_I = rate of ingestion, l/day,
  F_E = frequency of exposure, days/year,
  D_e = duration of exposure, years,
  M_B = body weight, kg, and
  T = average time, days.

$F_E D_E/T = 1$ when $F_E = 365$ days, $D_E = 70$ years, and $T = 25,550$ days. So, Equation 10.25 often simplifies to

$$I = \frac{0.75 C_W R_I}{M_B} \tag{10.26}$$

since seventy-five percent of water consumption is at home. If R_I is given in terms of total daily ingestion instead of ingestion of the contaminated food only, then a diet fraction representing the contaminated food must be determined and applied to the numerator of the right-side of Equations 10.25 or .26 appropriately. The daily intake of liquids is assumed to be two liters for adults and one for infants with ten kilograms body mass or less. Drinking rate is assumed to include coffee, tea, other beverages prepared with tapwater, and fruit and vegetable juices. Level of physical activity, temperature, and humidity affect these levels.

TABLE 10.14
Data Requirements for Human
Groundwater Receptors

Mean GW travel time to nearest down gradient well
GW use of uppermost aquifer
Populations served by affected aquifers
(A: 3 miles or 4.8 km downgradient)
(B: also 3 miles, period)
Mean GW travel time to nearest downgradient SW
(Supplying domestic users or food-chain agriculture)
Population within 1,000 ft of source
Distance to nearest installation boundary

The average adult male body weight is seventy kilograms and the rate of intake of water is two l/day. Therefore, typically, Equation 10.26 can be further reduced to

$$I = 0.0214 C_w \qquad (10.27)$$

Keep in mind that any simplified equation should be used judiciously.

GROUNDWATER

A major source of drinking water in the United States is groundwater which can be a route of exposure when contaminated groundwater is consumed. Review the uses of groundwater in the vicinity of hazardous substance releases as shown in Table 10.14, Data Requirements. Can people be exposed by groundwater? Also evaluate the relationship between surface water and groundwater in the area.

Concentration of a COC in groundwater to be used as drinking water (this applies to surface water for drinking as well) can be used to calculate health hazard quotient representing a ratio of the estimated intake rates of the detected COC to the toxicity benchmark for the COC.

$$Q = \frac{C_w \times F_w}{B_h} \qquad (10.28)$$

where: Q = human-health hazard quotient for COC
C_w = COC concentration in ground- or surface water, μg/l
F_w = individual drinking water intake, 2 l/day

The intake of a COC by ingestion of groundwater is determined by

$$I_w = \frac{C_w F_w F_E D_E}{M_B T} \qquad (10.29)$$

TABLE 10.15
**Data Requirements for Surface
Water Reception**

Population served by SW supplies
   (3 miles or 4.8 km downstream)
Water quality classification of nearest SW
Population within 1,000 ft of source
Distance to nearest installation boundary
Land use and zoning within 1 mile of source

---

where: I_w = intake by ingesting water, mg/kg-day, and
         C_w = concentration of COC in water, mg/L.

Lactating women have the highest mean total fluid intake (2.24 l/day) compared to pregnant women (2.04 l/day) and control women (1.94 l/day). Rural women tend to consume more total water (1.99 l/day) than urban (1.93 l/day) and suburban women (1.13 l/day) do. Some variability of drinking habits is also due to region of the country where one lives. Higher rates of consumption are required for higher activity levels and higher air temperatures, also.

SURFACE WATER

Many cities and towns in the U.S. take their domestic water directly from a river, lake or reservoir. People are potentially exposed to contaminants in surface water by drinking or otherwise coming into contact with it. Examples of other types of contact are swimming, fishing, boating, skiing, etc. Data requirements for assessment of reception of a COC by surface water includes information outlined in Table 10.15. The calculation for the human-health hazard exposure to a COC in drinking water taken from surface water is:

$$Q = \frac{\left(C_w F_w\right)+\left(C_f F_f\right)}{B_h}$$
(10.30)

where: Q   = human-health hazard quotient for COC
         C_w = contaminant concentration in surface water, µg/l,
         F_w = adult drinking water intake, 2 l/day,
         C_f = contaminant concentration in fish, µg/g,
         F_f = mean daily fish consumption, 6.5 g/day, and
         B_h = health effects benchmark for COC, µg/day.

Remember that

$$C_f = C_w B_b$$
(10.31)

where B_b is the bioaccumulation factor (l/g) for the COC. With more than one ÇOC present,

$$Q_{total} = \sum_{i=1}^{n} Q_i \tag{10.32}$$

SEAFOOD

People who eat fish, which have been exposed to contaminated water, sediments, and/or organisms, can likewise be exposed. Commercial fisheries make exposure to contaminated seafood a real possibility hundreds of miles away from the ocean. Visitors to popular sports fishing can be exposed to contaminants they otherwise would not have been exposed to. Ingestion by way of eating fish is calculated by determining the concentration of contaminant in fish, C_f, in ppb. Generally this is accomplished for any edible organism by finding the concentration of the pollutant in the environment and multiplying by a chemical specific bioconcentration factor.

$$Q = \frac{C_f \times F_f}{B_h} \tag{10.33}$$

where F_f mean daily individual fish consumption, 6.5 g/day and

$$C_f = C_w \times B_b \tag{10.34}$$

where C_f is expressed in $\mu g/g$ and B_b is the bioaccumulation factor for COC in species, l/g. The bioconcentration factor (B_b) is the equilibrium ratio of the concentration of a substance in fish to the concentration in water. What fraction is bioaccumulated by the fish? Schaffer provides some B_b's for freshwater fish (vertebrates) which are summarized in Table 10.16. Of these, TCE has the lowest B_b as expected since it is rapidly metabolized in the body of most animals. Its typical half-life in almost any animal is one day or less.

Species of interest in water columns are pelagic and demersal fish while the bottom sediments are home to benthic organisms. Pelagic species occur in the open sea, the ocean. Demersal species occur near the bottom and benthic species occur on the bottom.

An effective average individual consumption of seafood per day of 14.3 grams is provided by the National Marine Fisheries Service. EPA uses an average individual consumption of 6.5 grams per day of freshwater and estaurine fish.

The average individual is assumed to weigh 65 kg (143 lb). Therefore, the dose of contaminated seafood ingested daily is

$$d = \frac{14.3}{65,000} C_f \tag{10.35}$$

TABLE 10.16
Bioconcentration Factor for Vertebrate Freshwater Fish

Substance	Median B_b (B_b Range)
Benzene	5.2
Benzo(a)pyrene (BaP)	1×10^2 (1 to 10^4)
Chlorobenzene	10.3
Chloroform	3.8
1,2-Dichlorethane	1.2
1,1-Dichloroethylene	5.6
Ethyl Benzene	37.5
Methylene Chloride	0.91
Polychlorinated Biphenyl (PCB)	3×10^5
Tetrachlorinated Dibenzodioxin (TCDD or dioxin)	1×10^4 (10^{-2} to 10^5)
Tetrachloroethylene	30.6
Toluene	10.7
1,1,1-Trichloroethane	5.6
Trichloroethylene (TCE)	20 (10^{-2} to 10^2)

or, for a larger person eating his alloted average individual consumption of freshwater fish

$$d = (6.5/126,000)\, C_f$$

Here, d is dose expressed in mg of contaminant per kg of body weight per day. C_f is the contaminant concentration in fish expressed in mg of contaminant per kg of fish tissue.

The National Marine Fisheries Service can provide the total annual catch of seafood by pelagic, demersal and benthic fish.

DOMESTIC ANIMALS AND FOWLS

Domestic animals would be suspected of being contaminated if their food and water supplies are contaminated. Evaluation of uptake and bioconcentration factors is important. The most current Nationwide Food Consumption Surveys by the U.S. Department of Agriculture can be used to estimate consumption. When EPA finalizes its *Exposure Factors Handbook*, the three most recent surveys will be cited.

WILD ANIMALS AND BIRDS

As with domestic animals, wild animals are contaminated by drinking contaminated water or eating contaminated food. Concentrate on uptake and biocencentration. Data requirements for groundwater and surface water are given in Tables 10.17 and 10.18 respectively.

TABLE 10.17
Data Requirements for Wildlife Reception of Groundwater

Mean GW travel time to downgradient habitat or natural area
Importance/sensitivity of downgradient habitat/natural area
Presence of "critical" environments within 1 mile

TABLE 10.18
Data Requirements for Wildlife
Reception of Surface Water

Importance/sensitivity of biota/habitats affected
Presence of "critical" environments within 1 mile

FOOD CROPS

Pollutants in soil are taken up by plants, and, if these plants or fruits or vegetables from them are consumed by people or animals, the effects can be expected to show up, after some amount of time. The next section will deal with forage crops, and how contaminants reach humans through livestock. In this section, we will look at the direct ingestion of crops by humans. The soil to plant uptake factor (CR) is the equilibrium ratio of the concentration of a substance in plants to the concentration of the substance in soil. Typically, the pollutant species is accumulated in the plant so the concentration in the plant will be several times that in soil. This is expressed by:

$$C_P = C_S \times CR \qquad (10.36)$$

However, the concentration in the plant depends on morphological structure. For instance,

$$C_{roots} > C_{leaves/stems} > C_{tubers} > C_{reproductiveparts}$$

While metals and inorganic pollutants are taken up by the roots, evidence seems to indicate that organic substances are taken up by above ground plant structures which are exposed to the volatilized portion of the contaminant as it escapes the soil. Therefore, CR in literature should specify the plant structure for which they apply.

Researchers have found CR values for PCBs in carrots to range from 3×10^{-3} for peeled roots to 0.125 for whole unpeeled roots. Some researchers report that the lower chlorinated isomers of PCBs are more readily accumulated than other isomers. CR values for PCBs in marsh grass range from 0.56 for all isomers found in the plants to 4.2 for all isomers found in aerial parts of the plant to 10 for the single

TABLE 10.19
Soil to Plant Uptake Factor

Substance	Median CR (CR Range)
Benzo(a)pyrene (BaP)	0.1
	$(7 \times 10^{-6}$ to 40)
Polychlorinated Biphenyl (PCB)	1.0
	$(3 \times 10^{-3}$ to 10)
Tetrachlorinated Dibenzodioxin	1×10^{-3}
(TCDD or dioxin)	$(3 \times 10^{-5}$ to 1.0)
Trichloroethylene (TCE)	0.1
	(0.01 to 100)

isomer found in greatest quantity in the aerial part. Table 10.19 lists some common CRs.

Seveso, Italy, is the site of the world's worst dioxin release to date. Field studies there have shown that dioxins are not taken up into roots in detectable quantities. However, the peels and other parts of carrots contained dioxin and seem to show that dioxin gets into the plant through airborne routes.

We have shown in a previous section how a dose is calculated from a given intake rate, R_I. At the time we did not concern ourselves with whether the rate was for dry food or moist food. Presumably, if the COC is with the solid foodstuff, the moisture content dilutes the concentration and vice versa. To calculate a dry intake rate

$$R_{I-dry} = R_{I-consumed}\left[\frac{(100 - W)}{100}\right] \qquad (10.37)$$

where W is the percent water content.

FORAGE CROPS

Transfer of contaminants from forage crops to the milk and meat of grazing animals is another pathway of interest in exposure studies. The forage to milk transfer coefficient (Fm) is used to predict the concentration of a contaminant in milk. Fm represents the equilibrium fraction of a cow's daily ingestion of a contaminant that is found in a liter of milk. The forage to meat transfer coefficient (Fr) is the equilibrium fraction of a cow's daily intake of a contaminant that is found in a pound of its meat.

The contaminant level in forage crops for milk cows is estimated or measured as C_{fi}, lb/lb or oz/lb. The concentration in milk is then calculated:

$$C_{milk} = C_{fi} \times R_{fi} \times Fm \qquad (10.38)$$

TABLE 10.20
Forage to Milk Transfer Factor

Substance	Median Fm (Fm Range)
Benzo(a)pyrene (BaP)	10^{-4}
	(10^{-6} to 10^{-2})
Polychlorinated Biphenyl (PCB)	0.01
	(10^{-4} to 0.1)
Tetrachlorinated Dibenzodioxin (TCDD or dioxin)	0.01
	(10^{-5} to 1.0)
Trichloroethylene (TCE)	10^{-6}

TABLE 10.21
Forage to Meat Transfer Factor

Substance	Median Fr (Fr Range)
Benzo(a)pyrene (BaP)	10^{-5}
	(10^{-7} to 10^{-3})
Polychlorinated Biphenyl (PCB)	0.005
	(10^{-4} to 0.1)
Tetrachlorinated Dibenzodioxin (TCDD or dioxin)	0.0001
Trichloroethylene (TCE)	10^{-6}
	(10^{-8} to 10^{-6})

The forage ingestion rate, R_{FI}, if not known by measurement is taken as 22 lb/day (10 kg/day). The units of Fm is in day/gallon and Table 10.20 gives some common Fm values.

$$Fm = \frac{C_{milk}}{C_{feed}I_{feed}} \qquad (10.39)$$

The B_b for milk has been assumed to be 0.7 for ingestion (Hubbard, et al.).

Another equilibrium transfer factor (Fr) is used to estimate the concentration of a hazardous chemical in beef or other meat producing animals. For instance the Fr for PCB for swine is 9×10^{-3} day/kg and for sheep it is 7×10^{-3}. Generally less substance is transferred from feed to meat than from feed to milk and Fr for any given substance can be estimated as one tenth of Fm as shown in Table 10.21. Lower chlorinated isomers of TCDD show less transfer than do higher chlorinated isomers.

SOIL ROOT ZONES

Over a period of time, the amount of contaminant in the soil root zone will deplete faster than we can account for by plant uptake. To account for this loss, soil root

TABLE 10.22
Removal Rate from Soil Root Zone

Substance	Median λ_s (λ_s Range)
Benzo(a)pyrene (BaP)	3.8×10^{-3}
	(1.9×10^{-4} to 0.7)
Polychlorinated Biphenyl (PCB)	1.9×10^{-4}
	(1.9×10^{-5} to 0.7)
Tetrachlorinated Dibenzodioxin (TCDD or dioxin)	3.8×10^{-4}
	(1.9×10^{-5} to 0.7)
Trichloroethylene (TCE)	0.1
	(7.9×10^{-3} to 0.7)

zone removal is assumed to be exponentially related to the contaminant's half-life. Many physical, chemical and biological processes are occurring in the soil to deplete the contaminants. Some of these processes are erosion, hydrolysis and biodegradation to name three. There is a complex interrelationship between these processes so total contaminant loss must be measured to avoid having to account for each complexity and to arrive at a reasonable removal rate.

The half-life of PCB can range from weeks to decades in soils and sediments. Ten years is considered to be a conservative half-life to use for PCB in soil. The maximum and minimum half-lives are assumed to be 100 years and 7 days respectively. TCDD on soil surface has a measured half-life of 10 hours. However, this chemical does not follow the exponential half-life and as time increases so does the half-life of TCDD. The assumed best estimate of half-life of TCDD in soil is 5 years but the range of probable values is from 1 day to 100 years. A chemical such as benzo(a)pyrene is metabolized rapidly by plants and the principle soil removal process is biodegradation. Therefore, the best estimate BaP half-life is 0.5 years, the minimum is 1 day and the maximum is 10 years. TCE has both high volatility and biodegradability. The best estimate for its half-life in the soil root zone is 1 week. Probable half-lives range from 1 day to 1 year. Table 10.22 lists some values for λ.

DIRECT INGESTION OF SOIL

The intake of a COC from contaminated soil is due to some ingestion of soils from foods, but also from the transfer of soils from hand to mouth in activities such as gardening, cleaning, and recreating. Intake by ingestion is calculated:

$$I = \frac{10^{-6} C_{soil} R_I f_I F_c D_E}{M_B T}$$
(10.40)

where: I   = intake via ingestion, mg/kg-day,
C_{soil} = concentration of COC in soil, mg/kg,
R_I = ingestion rate, mg soil/day,

TABLE 10.23
Lifetime Soil Ingestion (I_{soo}) Rate

Age Group	I_{soo} Rate (g/day)
0–9 months	0
9–18 months	1
1.5–3.5 years	10
3.5–5 years	1
5–18 years	0.1
over 18 years	0.1

After Ford and Gurba.

f_I = fraction ingested, dimensionless,
F_e = frequency of exposure, days/year,
D_E = duration of exposure, years,
T = period over which exposure is averaged, days, and
M_B = body weight, kg.

Although we tend to think only of toddlers and younger children eating dirt all people do. Table 10.23 lists lifetime soil ingestion (I_{soo}) by age groups. The interesting thing about I_{soo} (the soil ingestion rate) given in Table 10.23, which is that the toddler group, 1.5 to 3.5 years, consumes a disproportionate amount of soil yet the health effects of soil consumption is typically examined over a lifetime of seventy years. As you can see, sixty-eight years are weighted in favor of a no-effect I_{soo}. Clearly the toddlers are the most sensitive of our population when it comes to exposure to soil contamination by means of direct ingestion.

Ford and Gurba provide a table of values to derive a lifetime average soil ingestion ($I_{soo}|_{avg}$). They take the number of years spent in each age group, multiple that by 365 days per year and the I to get a cumulative quantity of soil ingested. This yields 10,495 grams of soil over a 70 year lifetime. Then they take this number and divide it by 70 kg (the average weight of the adult male) and by the number of days in 70 years to arrive at the $I_{soo}|_{avg} = 0.006$g/kg day. Their point is that this number is meaningless for protecting the public. The proper number to use is 10 g/day of ingested soil to account for the worst case.

The FAO/WHO Expert Committee on Food Additives devised the allowable daily intake (ADI) factor as a more useful health criterion for threshold contaminants. The ADI is intended to represent the amount of a contaminant which an adult could receive daily over a lifetime without showing signs or symptoms of health effects. The ADI is no guarantee of absolute safety any more than the LASI is.

Another useful factor is called the soil criterion (SC). This factor uses a 10 kg body weight and a 10 g/day SI.

$$SC = 14.29(ADI) = mg/kg = \mu g/g$$

TABLE 10.24
Unit Carcinogenic Risks

Compound	UCR (mg/kg day)$^{-1}$
Arsenic	15
DDT	8.42
Dieldrin	30.4

where ADI is in mg/day. The generic equation to calculate SC for other SI's and body weights is:

$$SC = 1,000 ADI \frac{BW}{SI} \qquad (10.41)$$

where SI is in g/day and BW is a body weight ratio in kg/70 kg.

This procedure does not apply if the soil contaminant is carcinogenic. For one thing the 10 g/day SI is no longer appropriate since it is for a sensitive age group with regard to threshold chemicals. Carcinogens do not have a threshold, however they do not act overnight with merely one exposure or even with short duration exposure. Cancers are typically caused by low-level exposures of long term duration. A few exceptions to this statement exist, mainly mesothelioma from asbestos exposure and Burkett's lymphoma from Epstein–Barr virus, however, for chemicals in soil, let us assume, for the present discussion,that we can rely on the need for a long exposure to initiate neoplasms.

Since cancer is considered to be cumulative over the lifetime the soil exposure should be assessed in a cumulative manner. Ford and Gurba calculated a LASI of 0.006 g/kg day. If a lifetime average body weight is used instead 70 kg, a slightly higher LASI is obtained. For discussion let us use the lower figure. The biggest factor when dealing with carcinogens in soil, since lifetime exposure is the concern, is to account for degradation of the contaminant in the soil over 70 years. The environmental half-life, if known, allows us to account for biodegradation, hydrolysis and photolysis. Biodegradation data is not readily available but there is some data on the hydrolysis and photolysis of organics. Absorption through membranes is assumed to be one hundred percent unless proved otherwise.

The half-life, $t_{1/2}$ in years, is adjusted for the human lifespan by dividing by seventy. Assume linear decay unless the data source states differently. Do not apply this correction of half-life to metals in soil as they do not decay with time.

EPA has a Carcinogen Assessment Group which publishes unit carcinogenic risks (UCR) expressed as excess cancer risk from a lifetime ingestion of 1 mg/kg day of a carcinogen. Table 10.24 lists some UCR's.

This permits calculation of lifetime allowable daily intake (LADI) in mg/kg day by selecting an appropriate risk level.

TABLE 10.25
Environmental Half-Life

Compound	$t_{1/2}$ years
DDT	14.6
Dieldrin	0.14

TABLE 10.26
Soil Criteria for Carcinogens Ingested

Compound	SC mg/kg day
Arsenic	0.01
DDT	0.1
Dieldrin	2.74

$$LADI = \frac{Risk}{UCR} \qquad (10.42)$$

The risk is an excess case in a certain number of exposures. For carcinogens the number most used is 1 in a million or 1:1,000,000 or 1×10^{-6}. The LADI for arsenic, DDT and dieldrin are 6.67×10^{-8}, 1.19×10^{-7} and 3.29×10^{-8} mg/kg day respectively.

In order to convert this acceptable dose into a soil criterion, SC (mg/kg), we calculate as follows:

$$SC = \frac{1,000 LADI}{\left(\dfrac{t_{1/2}}{70} \times LASI \right)} \qquad (10.43)$$

Recall that LASI was 0.006 mg/kg day. Some half-lives are given in Table 10.25 and Table 10.26 gives some SC. Metals have no half-lives. In the case of arsenic the derived SC is less than the background level as the global mean concentration of this carcinogen is 6 ppm or 600 times the calculated allowable! Obviously, the background must be taken as the acceptable but that may be one explanation why cancer is so prevalent.

DERMAL CONTACT

In terms of weight, the skin is the largest, or rather the heaviest, organ. Only, the lungs have a greater surface area. The skin is a fairly effective barrier between the environment and the other organs of the body. The eyes and the lungs are the only other organs, besides the skin, that are in direct contact with the environment. Of these three, the skin provides the best defense against chemicals.

The skin consists of two principle layers, the outer layer is the epidermis, and the inner layer is the dermis. Besides the tough outer covering we call the skin, the epidermis contains hair follicles and sweat glands that penetrate to the dermis. The dermis contains sweat ducts, sebaceous glands, connective tissue, fat, and blood vessels as well as hair follicles and sweat glands. Chemicals attacking the skin can penetrate through the sweat glands, hair follicles, or sebaceous glands, though this amount is relatively small. Most of any chemical is absorbed directly through the epidermis, if not washed away quickly with copious amounts of uncontaminated water. The top layer of the skin is the stratum corneum, a thin cohesive membrane of dead surface skin that is sloughed off every two weeks through a process of skin dehydration and polymerization of intracellular substance. Skin permeability depends almost entirely on the epidermis, how healthy it is, and how intact it is. The dermis is a collection of porous, watery, nonselective diffusion media.

Intact skin has several natural defense functions. Healthy, intact epidermis minimizes absorption of chemicals and provides an effective physical barrier to bacteria. The sweat glands regulate the heat of the body. The sebaceous glands secrete fatty acids, which are bacteriostatic and fungistatic. The skin pigment, melanocytes, minimize damage from ultraviolet radiation from the sun. The connective tissue in the dermis provides elasticity against trauma. From the many blood vessels in the dermis, the lymph-blood system provides immunologic responses to infection by cleaning the watery portion in the lymph nodes, and introducing antibodies and phagocytes in the blood stream.

How readily the skin absorbs chemicals depends mainly on the health and intactness of the skin, as mentioned before, and on the chemical properties of the substance attacking. Absorption is enhanced by breaking the epidermis by abrasions or other wounds. Absorption is directly proportional to skin hydration. This means that more will be absorbed when the person is hot and sweaty. Also, when the person is hot, the sweat cells open up to secrete sweat, allowing the chemical to enter directly, but also providing a means, through the sweat, to dissolve solids. Absorption is also directly proportional to increasing concentration of the COC, increasing contact time, and the surface area affected. More absorption takes place if the natural pH of the skin (pH = 5) is raised or lowered. Absorption is inversely proportional to the particle size of particulate substances. Sometimes COCs are components of mixtures. If a mixture contains an agent that will damage skin, or otherwise render it more susceptible to penetration, absorption of the COC is enhanced. Surface active agents, such as DMSO, are carriers for COCs attacking the skin. A slight electrical charge on the skin can induce ion movement across membranes.

Direct contact of the skin, eyes, and mucous membranes with contaminated soils, or exposed hazardous substances, on a site, may be a significant route of exposure, if people and animals have access to the site. Skin absorption of a COC is assumed to be one hundred percent, unless a study shows some other value to be the case. One study used an absorption factor of 0.10 for dermal contact (Hubbard, et al.). The surface area of skin is taken to be 457 in^2 (2,948 cm^2) or about 3.2 ft^2 (0.3 m^2). A total body surface area, A_S, in m^2 can be calculated

$$A_S = \frac{4W + 7}{W + 90} \tag{10.44}$$

where W is weight in kg. Generically, the Gehan and George formula is used and although criticized it gives a reasonably good estimate of surface area:

$$A_S = KW^{2/3} \tag{10.45}$$

where K is a constant. The DuBois and DuBois formula is:

$$A_S = a_0 H^{a_1} W^{a_2} \tag{10.46}$$

where: $a_0 = 0.007182$
            $a_1 = 0.725$
            $a_2 = 0.425$

which has been refined over the years and is now expressed

$$A_S = 0.0235 H^{0.42246} W^{0.51456} \tag{10.47}$$

or

$$\ln A_S = -3.7508 + 0.42246 \ln H + 0.51456 \ln W \tag{10.48}$$

Health effects caused by skin contact may seem innocuous. For instance, reddening of the skin, or mild dermatitis, is rarely taken seriously. Many workers do not even report these cases to their employers, though they should. More severe effects, such as destruction of skin, and debilitating dermatitis, are more often reported to employers. However, many chemicals cross the boundary of the skin, and are absorbed into the blood stream, without the victim even being fully aware of what is occurring. Chemicals that are absorbed through the skin may be causing systemic damage, without the victim being aware of the damage until it progresses to a certain point.

Primary irritants act directly on skin at the attack site. Acetone, benzyl chloride, carbon disulfide, chloroform, chromic acid, ethylene oxide, hydrogen chloride, iodine, mercury, methyl ethyl ketone, phenol, phosgene, picric acid, styrene, sulfur dioxide, toluene, and xylene are primary irritants. Photosensitizers increase sensitivity to light, producing redness and irritation of skin. Acridine, creosote, furfural, naphtha, pyridine, and tetracyclines are photosynthesizers. Allergic sensitizers include ammonia, benzoyl peroxide, chromates, chromic acid, cobalt, formaldehyde, mercury, nitrobenzene, phthalic anhydride, and toluene diisocyanate. These produce allergic reactions after repeated exposures.

The eyes are more sensitive to chemical exposure than the skin is, and even a brief exposure to the eyes may cause severe health effects directly to the eyes. Any COC that can be absorbed through the skin may also be absorbed into the blood stream through the eyes, and be transported to a target organ, elsewhere in the body. Acids damage the eyes depending on pH and the protein combining capacity of the acid. Whatever an acid burn is like, the first few hours after contact, is pretty much how the long-term damage will be. Sulfuric acid not only burns, but removes water from the eyes and generates heat of dilution. Hydrochloric acid will produce severe damage up to pH 1, but has almost no effect at pH 3 or greater. Picric and tannic acids produce damage over the entire acidic pH range (1–7).

However, unlike an acid burn, an alkali burn may continue to grow worse as time passes. In fact, the victim and inexperienced first medical responders may be fooled into thinking that the damage is mild, when, in fact, a serious burn is in progress. Flush alkali burns for a long time. A fifteen minute minimum flush time is widely recommended. Chemical burns produce ulceration, perforation, and clouding of the cornea or lens. In the case of alkalis, pH and length of exposure have more weight than the exact alkali.

Organic solvents dissolve fats in the eyes, causing immediate pain, and dulling of the cornea. Unless the solvent is hot upon contact, the damage is typically slight. Acetone, ethanol, and toluene are such solvents. Lacrimators immediately cause the eye to tear up, even at low concentrations. Lacrimators induce an instant reaction without causing damage, whereas irritants cause damage. However, very concentrated solutions of lacrimators can cause chemical burns and destruction of corneal material. Chloroacetophenone (tear gas) and MACE are lacrimators.

Other COCs form cataracts, damage the optic nerve, or damage the retina when they contact eye tissue. They may reach the eye via the blood system rather than by direct contact. Naphthalene and thallium cause cataracts. Naphthalene and phenothiazone cause retina damage. Methanol and thallium cause optic nerve damage, including permanent blindness.

Contact with contaminated soil provides the opportunity for exposure via skin absorption. At the outer surface of the skin is the applied dose. After uptake through the skin it is an internal dose. After undergoing metabolism on its way to the target organ, the absorbed, or biologically effective, dose produces an unwanted effect.

$$D_{pot} \int_{t_1}^{t_2} C(t) R_I dt \qquad (10.49)$$

For a series of events in which the skin is dosed:

$$D_{pot} \int_{i=1}^{n} C_i R_{Ii} D_E \qquad (10.50)$$

If D_E is very brief,

$$D_{pot} = C_{avg}R_I\big|_{avg}D_E \qquad (10.51)$$

The dermal absorbed dose is

$$D_{abs} = \frac{10^{-6}C_{soil}A_{skin}f_{adh}f_{abs}F_eT}{M_Bt_{avg}} \qquad (10.52)$$

where: D_{abs} = absorbed dose, mg/kg-day,
         A_{skin} = affected surface area of the skin, cm^2,
         f_{adh} = soil-to-skin adherence factor, mg/cm^2-event, and
         f_{abs} = absorption factor, dimensionless.

For the typical adult, A_{skin} is taken as 18,150 cm^2.

The adherence of soil to the skin is vital when the exposure scenario includes dermal contact with contaminated soil. Significant parameters of soils are cation exchange capacity, organic content, clay mineralogy, and particle size distribution. Adherence increases with decreasing particle size. Adherence is directly proportional to moisture content. Clay and organic content have little effect on adherence. However, adherence levels vary on different parts of the body. The hands, knees, and elbows have the highest adherence levels, while the face provides the least opportunity for adherence. Outdoor work, outdoor recreation, and gardening activities yield high levels of soil adherence, with the highest levels seen among those who contact wet soils.

The amount of chemical absorbed through the skin per contact event is estimated using a skin permeability coefficient, K_p. The coefficient is expressed in grams per square meter per hour, or in terms the rate at which the chemical permeates the skin, cm^2/h. As a molecule encounters a membrane it will be held back or else the molecule partitions into the membrane and migrate across the full thickness of the membrane. K_p is a function of the path length of chemical diffusion, the membrane partition coefficient for the chemical, and the diffusion coefficient of the molecule in the membrane. *Lag time* consists of an unsteady state period after the initial contact before steady state diffusion is established.

For organic compounds that exhibit octanol–water partitioning, the skin permeability coefficient, in cm/h, is

$$\log K_p = -2.72 + 0.71\log K_{O/W} - 0.0061MW \qquad (10.53)$$

where: $K_{O/W}$ = octanol–water coefficient, dimensionless, and
         MW = molecular weight, g/g-mole.

Table 10.27 lists the skin permeability of several commercially available solvents. Goldblum, Clegg, and Erving have suggested using 4.34 g/m^2-hr as a typical number.

TABLE 10.27
Skin Permeability of Commercial Solvents

Solvent	Permeability Constants	
	g/m²h	cm/h
Butyl acetate	1.9	0.0002
Dimethyl acetamide	217	0.0220
Dimethyl formamide	101	0.0107
Dimethyl sulfoxide	284	0.0258
5% DMSO	7.4	0.0008
10% DMSO	14.8	0.0016
20% DMSO	83	0.0087
40% DMSO	153	0.0161
Ethanol	12.6	0.0016
Gamma-butyrolactone	1.4	0.0001
Methyl ethyl ketone	128	0.0159
Methylene chloride	46	0.0034
N-methyl-2-pyrrolidone	323	0.0296
Propylene carbonate	1.7	0.0001
Sulfolane	0.2	0.00002
Toluene	2.6	0.0003
Water, heavy	15.1	0.0004

Skin acts as a semipermeable membrane and a safety factor must used to account for sensitive individuals.

The distribution of K_p in the stratum corneum and the viable epidermis is estimated using EPA's parameter B:

$$B = \frac{K_{O/W}}{10^4} \qquad (10.54)$$

The skin diffusivity constant, D_{SC}, in cm²/h, is determined as

$$\log \frac{D_{SC}}{l_{SC}} = -2.72 - 0.0061 MW \qquad (10.55)$$

where l_{SC} is the length of stratum corneum, cm, and is typically assumed to be 10^{-3} cm. The lag time, τ, is

$$\tau = \frac{l_{SC}^2}{6D_{SC}} \qquad (10.56)$$

To estimate the time to reach steady state, t^*, EPA uses the value of B. If $B \leq 0.1$, then $t^* = 2.4\tau$. Between 0.1 and 1.17,

$$t^* = (8.4 + 6\log B)\tau \qquad (10.57)$$

If B is greater than 1.17,

$$t^* = 6\left(b - \sqrt{b^2 - c^2}\right)\tau \qquad (10.58)$$

where

$$b = \frac{2}{\pi}(1 + B)^2 - c \qquad (10.59)$$

and

$$c = \frac{1 + 3B}{3} \qquad (10.60)$$

Now we can calculate the dermal absorption per event, DA_{event}:

$$DA_{event} = 2K_P C_v \sqrt{\frac{6\tau t_{event}}{\pi}} \qquad (10.61)$$

in mg/cm^2, if t_{event} is less than τ. If not, then

$$DA_{event} = K_P C_v \left[\frac{t_{event}}{1 + B} + 2\tau\left(\frac{1 + 3B}{1 + B}\right)\right] \qquad (10.62)$$

where C_v is the concentration of organic (mg/ml) in both equations.

But what if the organic is not in an aqueous solution, rather in soil? The EPA has considered this case and has determined that the skin permeability coefficient for chemicals in soil, cm/h, is

$$K_{P,s}^{soil} = \frac{K_P}{K_{soil/W}} \qquad (10.63)$$

where K_P is the skin permeability coefficient, cm/h, determined earlier for chemicals in water, and

$$K_{soil/W} = K_D \rho_{soil} \qquad (10.64)$$

where: K_D = soil/water partition coefficient, l/kg
         ρ_{soil} = density of soil, g/cm^3.

$K_{soil/w}$ is in units of either 10^{-3} kg/g or 10^3 cm/l. The dermal absorption for each event is

$$DA_{event} = C_{soil} f_{adh} f_{abs} \qquad (10.65)$$

where: C_{soil} = COC concentration in soil, mg/kg, or 10^{-6} kg/mg,
             f_{adh} = adherence factor of soil to skin, mg/cm^2, and
             f_{abs} = absorption factor.

C_{soil} is a decay of the original concentration in soil on skin:

$$C_{soil} = C_{soil}^0 e^{-(K_{soil}+K_{vol})t} \qquad (10.66)$$

where

$$K_{soil} = 1000 K_{P,s}^{soil} \rho_{soil} / f_{adh} \qquad (10.67)$$

and

$$K_{vol} = 3600 K_h D_{air} / f_{adh} K_D l \qquad (10.68)$$

where: t   = time, h,
             K_h = Henry's law constant, dimensionless, and
             l   = thickness of boundary layer at air–soil interface, cm.

K_{soil} and K_{vol} are expressed in terms of inverse time, h^{-1}.
    The absorption fraction is estimated in terms of parameters we are already familiar with

$$f_{abs} = \frac{\rho_{soil} K_{P,s}^{soil}}{f_{adh}(K_{soil}+K_{vol})}\left[1 - e^{-(K_{soil}+K_{vol})t_{event}}\right] \qquad (10.69)$$

EPA has determined that f_{adh} ranges from 0.2 mg/cm^2 to 1.5 mg/cm^2 per event. Due to the study methods, EPA uses 0.2 mg/cm^2 as an average weight of soil adhered to skin and 1.0 mg/cm^2 as a reasonable upper limit.

INHALATION

Exposure to most chemicals will be in the physical form of gases, vapors, mists, or particulate matter. Therefore, inhalation is the major route of entry for many, though not all, exposure assessments. Exposure ends at the openings of the mouth and nose where the potential intake is the potential dose. After inhalation, the COC is either exhaled or deposited in the respiratory tract, where it is an applied dose that either damages tissue through direct contact, or diffuses into the blood stream through the alveoli–blood interface, which ranges from 30–70 m³ surface area. This internal dose is subject to metabolism and becomes the biologically effective dose when it acts on an organ. As usual, since very little internal data is available, the biologically effective dose, or any upstream dose for that matter, is assumed to equal the exposure in the worst case.

The respiratory tract has vital function elements in constant, direct contact with the environment. The lungs have seventy to one hundred square meters of surface area, as opposed to two square meters of skin, and ten square meters of gastrointestinal tract. Three regions make up the respiratory tract: the nasopharyngeal region, the tracheobronchial region, and the pulmonary acinus.

The nasopharyngeal region extends from nose to larynx and the passages are lined with ciliated epithelium and mucous glands. In this region large particles are filtered out of incoming air, but another function of the nasopharyngeal passages is to increase the relative humidity and moderate the temperature of the inhaled air. In nasopharyngeal region, also called the upper respiratory tract (URT), the water-soluble, or larger particle size, COC may damage tissues in a variety of ways, ranging from irritation, to severe corrosive destruction of tissue. Substances that are less water soluble, and the finer particle sizes, make it into the deep lungs, where, in the alveoli, they are absorbed into the blood and can be circulated to organs that have an affinity for the COC, and where health effects occur. Inhaled particles settle in the respiratory tract according to the scheme in Table 10.28. The inhalability I of an aerosol or particle is defined as:

$$I = 0.5\left[1 + e^{-0.06d_{ae}}\right]$$ (10.70)

for diameters d_{ae} up to 100 µm.

The tracheobronchial region consists of trachea, bronchi, and bronchioles. The region is an airway that connects the nasopharyngeal region and alveoli. The passageways here are lined with ciliated epithelium coated with mucous. The cilia whip back and forth similar to anemones in water. This motion acts like an escalator to move particles from the deep lungs to the nasopharyngeal passages where they are expectorated or swallowed. Smoke and cough suppressants paralyze the cilia, though the paralysis wears off after a few hours of no dose. However, long-term use of burning tobacco or cough suppressants can permanently paralyze the cilia, thereby destroying a natural defense capability of the body.

The pulmonary acinus is the basic functional unit of the lungs, where the action is. Gas exchange from the atmosphere to blood back to the atmosphere takes place in this region, which consists of small bronchioles, and connecting alveoli. Humans have approximately one hundred million alveoli contact the pulmonary capillaries where the gas transfer actually occurs.

The amount of a COC available for inhalation is

$$I_D = Cbh$$ (10.71)

TABLE 10.28
Particle Deposition in the Respiratory Tract

5–30 micron	Nasopharyngeal region
1–5 micron	Tracheobronchial region
0.01–1 micron	Alveolar region

TABLE 10.29
NIOSH Inhalation Rates

Activity Level	b, m³/hr
Rest	0.56
Light work	1.18
Medium work	1.75
Medium-heavy work	2.63
Heavy work	3.60
Maximum work	7.90
NIOSH 1976	

where: I_D = daily inhalation exposure, mg/day
C = airborne concentration of COC, mg/m³
b = inhalation rate, m³/hr
h = duration, hr/day.

The inhalation rate, b, increases with physical activity. NIOSH found that workers typically breathe about ten cubic meters of air in an eight hour work shift, or 1.25 m³/hr on the average. Table 10.29 lists inhalation rates for various levels of activity. The inhalation rate will vary over the period of a shift, depending on the activity level at the time of observation. Hence, exposure measurements are taken over a period of time and time-weight averaged. Breathing rates can be estimated based on energy expenditures according to the following expression:

$$V_E = EHV_Q \qquad (10.72)$$

where: V_E = ventilation rate, l/min or m³/hr,
E = energy expenditure rate, KJ/min or MJ/hr,
H = oxygen consumed in energy production, l/KJ or m³/MJ,
V_Q = ventilatory equivalent, dimensionless.

Weighted average oxygen uptake was 0.05 l/KJ based on studies cited by EPA. Ventilatory equivalent was 27 based on a geometric mean of data from studies reviewed by EPA.

The duration, h, is determined by subjective knowledge of the exposure situation. For instance, for a worker being exposed during the entire work shift, h = 8 hours/day. For a toddler at home, being exposed to radon gas, h = 24 hours/day. The airborne concentration is either provided from monitoring data, or estimated.

The average daily inhalation exposure for an industrial worker is

$$I_D = 1.25Ch \qquad (10.73)$$

and, if the duration is an eight hour workday, $I_D = 10C$.

The general expression for intake by inhalation is

$$I = \frac{C_A R_I t_E F_E D_E}{M_B T} \qquad (10.74)$$

where: I   = intake by inhalation, mg/kg-day,
         C_m = modeled concentration, mg/m³,
         R_{insp} = inspiration rate, m³/hour, and
         t_c = exposure time, hours/day.

An expression that accounts for materials volatilizing from groundwater while showering is

$$C_A = \frac{C_W V_W}{V_{shower}} \qquad (10.75)$$

and

$$C_A = \frac{C_W V_W}{V_{bathroom}} \qquad (10.76)$$

which can be combined as

$$I = \frac{C_W R_I V_W}{M_B}\left(\frac{D_{shower}}{V_{shower}} + \frac{D_{bathroom}}{V_{bathroom}}\right) \qquad (10.77)$$

where R_I = 20 m³/day for the average adult. The time spent showering and in the bathroom after the shower is generally taken as ten minutes (1/6 hour) each. The volumes of the shower and bathroom are generally assumed to be ten cubic meters each.

The lifetime average daily dose (LADD) in mg/kg day is calculated

$$LADD = \frac{C \times I \times E}{BW \times L} \qquad (10.78)$$

where: BW = body weight, 70 kg
        C   = concentration, mg/m^3
        E   = cumulative chemical exposure, hr
        I   = inhalation rate, 20 m^3/day
        L   = lifetime, 25,550 days (70 yr)

From the LADD, a risk can be calculated:

$$R = CPF \times LADD \qquad (10.79)$$

where: CPF = cancer potency factor
        R   = risk

Absorption for inhalation is assumed to be one hundred percent unless a study demonstrates otherwise. The values of 0.75 have been used for inhalation of dust from tilling and inhalation of fugitive dust from neighboring farming or mining operations (Hubbard, et al.).

Asphyxiants are gases that deprive body tissues of oxygen. Simple asphyxiants are physiologically inert gases, such as nitrogen, helium, methane, argon, or neon, that displace oxygen, if their partial pressure is great enough. As the partial pressure of oxygen is lowered, the point of suffocation is eventually reached. Chemical asphyxiants prevent the tissues of the body from using available oxygen. Carbon monoxide binds to hemoglobin two hundred times more readily than oxygen does. Cyanide prevents the transfer of oxygen from blood to tissues by inhibiting transfer enzymes that are necessary.

Irritants are chemicals that constrict and irritate airways. Irritation may lead to edema, the formation of fluid, and infection, thus complicating what, at first, was not necessarily a serious matter. Some severe irritants are ammonia, chlorine, hydrogen chloride, and hydrogen fluoride.

COCs that produce cell death (necrosis) and edema are ozone and nitrogen dioxide. Fibrotic tissue produced by chemical exposure can become massive enough to block airways, and decrease lung capacity. Asbestos, beryllium, and silicates produce fibrosis. Allergens are chemicals that induce allergic responses. Isocyanates and sulfur dioxide are allergens. Allergic reactions are never presented on the first exposure, rather on subsequent exposures, after the immune system has made antibodies for the COC. Biosynosis is an allergic reaction to cotton fiber. Carcinogens cause lung cancer from which about three thousand people die daily. Arsenic, asbestos, cigarette smoke, and coke oven emissions are carcinogenic. Benz-alpha-pyrine is a newly identified lung carcinogen.

The respiratory tract can also be a route for COCs to reach other organs of the body. Many hepatotoxic solvents are inhaled and absorbed into the circulatory system, where the solvent molecules or their metabolites end up in the liver.

How can we know if the receptor has received an estimated dose? Some contaminants remain intact in the body such as the heavy metals. Such contaminants can be tracked by biological monitoring. That is by measuring their concentration in urine, blood, hair, nails, perspiration, tears, milk, saliva, feces and exhaled air. Collection and analysis of blood and urine and other body fluids is far less invasive than biopsy methods on target organs. So far though such biological monitoring methods have had limited use in the control workplace chemical hazards and even less use in environmental studies of human exposures. Yu and Sherwood describe one such study in which a relationship was established between urinary elimination and radioactive hand contamination from an airborne concentration of uranium. The study found a linear correlation between urinary uranium and airborne uranium: 0.67 to 1.27 µg U/l urine per µg U/m^3 air. Urinary uranium, E_U in µg/l, can be estimated

$$E_U = 227 + 1.04 C_{U,air} \tag{10.80}$$

Also, some chemicals change identity in the body's metabolism process. Detection of these contaminants in body fluids requires the measure of a metabolite, the new, transformed compound of the contaminant. For instance two metabolites of DDT are DDD and DDE.

A special case is the exposure to acid contaminated fog. Several deaths were attributed to acid fog — smog — in London, the Meuse River Valley in Belgium, and the Monangahela River Valley in Pennsylvania, earlier this century. Acid fog is traced to sulfur dioxide (SO_2) and nitrous oxides (NO_x) emissions from industrial combustion processes. Although volatile organic acids are a minor contributor the mineral acids predominate. In the Northeast, sulfate ($SO^=_4$) from heavy use of coal as power plant fuel dominates the acid pollutant species. In Southern California, where the automobile rules, nitrate (NO^-_3) is ubiquitous.

A fog is an unstable aerosol, which is highly dependent on water concentration, droplet size, and pH, which are in turn variable over distance and time. A fog is formed when the relative humidity reaches one hundred percent and water starts condensing onto pre-existing suspended particulate matter, which become the nuclei of the fog droplets. As the formation of fog begins, the droplets are relatively small and strongly acidic. As the fog continues to develop, droplets increase in size, diluting the hydrogen ion concentration [H^+]. Thus, pH increases from the inception of fog to some later point in time. When evaporation starts, particle size of the droplets decrease and the fog becomes more acidic again. During a fog which persists for longer than early to midmorning, perhaps for several days, changes in droplet size may occur several times with the pH of the fog decreasing or increasing with size. The droplet size in the early stages of fog, when pH is acidic, ranges from eight to twelve microns in diameter. The range of droplet diameters is eight to sixteen microns during the late acidic stage of a fog as evaporation is decreasing droplet size.

Tissue fluid osmolarity typically measures around 300 mOsm. Osmolarity is the concentration of an osmotic solution, measured in milliosmols. Hypoosmolar aerosols are believed to cause bronchoconstriction. The dose of acid fog inhaled is expressed as total hydrogen ion (H^{+TOTAL}):

$$H^{+=}_{TOTAL} [H^+] \times \left\{ \left(t_{QB} \right) \left(\dot{V}_{E(QB)} \right) + \left(t_{Ex} \right) \left(\dot{V}_{E(Ex)} \right) \right\} \tag{10.81}$$

where: t_{QB} = period of breathing at rest, min, and
         t_{Ex} = period of breathing under exertion, min.

INJECTION

COCs may also enter the body when the skin is penetrated mechanically, or when contaminated objects puncture it. As the COC circulates with the blood, and gets deposited in target organs, health effects present themselves. This route of exposure is not found much in industry, but is not unheard of, either. Take for instance, a mechanic is working on a hydraulic line, which is still under pressure. A leak develops in the pressurized line. The hydraulic fluid may be ejected in a fine stream, barely visible to the naked eye, yet, capable of cutting like a knife. Injection has occurred. If the hydraulic fluid is also toxic, toxic injection has occurred.

INTERNAL DOSE

The internal dose is that which is absorbed through some barrier of the body: skin, lung membrane, or G.I. membrane. Expressions for D_{int} are analogous to other dose terms. For uptake processes, notably skin absorption:

$$D_{int} = \int_{t_1}^{t_2} C(t)K_pA_{skin}(t)dt \qquad (10.82)$$

where: K_p = permeability coefficient, and
         A_{skin} = surface area exposed.

The carrier medium may or may not cross the intake boundary, but some of the COC must, or we have no hazard. Wouldn't that be ideal? Therefore, expressions of transport internal to the body do not rely on the movement of the carrier (air, water, solid), but on the COC itself. The flow of the COC across the barrier is called the flux, J, and is almost never measurable. Flux depends on such factors as the nature of the COC, the nature of the barrier, active transport, passive diffusion processes, and the concentration of the COC contacting the exterior of the barrier. The permeability constant referred to above defines the relationship between the flux and the exposure concentration. Fortunately, K_p can be determined experimentally. So, internal dose becomes

$$D_{int} = C_{avg}K_pA_{skin}\big|_{avg}D_E \qquad (10.83)$$

Average daily internal dose is

$$ADD_{int} = \frac{C_{avg}K_p\overline{A}_{skin}D_E}{M_BT} \qquad (10.84)$$

Potential dose is the COC in the total amount of medium contacting the skin, whether or not all the COC actually contacts the skin:

$$D_{pot} = C_{avg}M_{medium} = C_{avg}F_{adh}\overline{A}_{skin}D_E \qquad (10.85)$$

Applied dose is the amount of COC that actually contacts and

$$D_{int} = D_{app}\int_{t_1}^{t_2} f(t)dt \qquad (10.86)$$

is the relationship between internal dose and applied dose. The term f(t) is a nonlinear absorption function that is typically unmeasurable. The dimensions of f(t) are mass absorbed per mass applied per unit time. This function is dependent upon concentration gradient of the COC, carrier medium, type of skin, and skin moisture. Taken over time, from first exposure until some time t, f(t) becomes f_{abs}, which is the fraction of the applied dose that is absorbed after t. The absorption factor is a cumulative number with the potential to increase to 1 (100% absorption) but typically reaches some steady-state level well below that. Thus, Equation 10.41 becomes

$$D_{int} = D_{app}f_{abs} \qquad (10.87)$$

But, if all the bulk carrier eventually contacts the skin, then

$$D_{int} = D_{pot}f_{abs} \qquad (10.88)$$

which leads to

$$ADD_{int} = \frac{C_{avg}M_{medium}f_{abs}}{M_BT} \qquad (10.89)$$

For intake processes, such as inhalation and ingestion, the potential dose is the amount inhaled or ingested. The potential and applied doses are thought to be about equal, so

$$D_{int} = D_{app}f_{abs} = D_{pot}f_{abs} = C_{avg}\overline{R}_ID_Ef_{abs} \qquad (10.90)$$

Average daily dose for two-step intake-uptake processes is

$$ADD_{int} = ADD_{pot}f_{abs} = \frac{C_{avg}\overline{R}_ID_Ef_{abs}}{M_BT} \qquad (10.91)$$

SYSTEMIC PROCESSES

Once a COC has been absorbed into the body, three mechanisms determine the fate of the chemical and the health effects that may come from it being in the body: metabolism, storage, and excretion.

Chemicals that are metabolized are transformed by means of chemical reactions in the body. The products of metabolic reactions are *metabolites*. Other chemicals, and metabolites, are stored, after distribution, to specific target organs. Storage may slow down metabolism, thus, making a chemical more persistent in the body. Excretory mechanisms include exhaled breath, perspiration, urine, feces, or detoxification. Excretion rids the body of the COC over a period of time, which may be days or months in some cases, and a lifetime, for those chemicals that have such a low elimination rate that they persist, causing deleterious effects for a long time.

CENTRAL NERVOUS SYSTEM

Nerve cells, or neurons, have a high metabolic rate, but no tolerance for anaerobic metabolism. Therefore, inadequate oxygen flow, anoxia, to the brain, kills neurons immediately. Some neurons may die before oxygen or glucose transport stops completely. Neurons are easily affected by both simple and chemical asphyxiants. Also, COCs that reduce the oxygen content of blood by reducing respiration rate restricts the flow of oxygen to neurons. Barbiturates and the narcotic substances lower respiration rate. COCs, such as aniline, arsine, benzene, ethylene chlorohydrin, nickel, and tetraethyl lead, produce cardiac arrest, hypotension, hemorrhaging, or thrombosis, any of which reduce blood pressure, and starve neurons for oxygen.

Some COCs directly damage neurons, or else inhibit neuron function through specific action on parts of the cell, producing such symptoms as dullness, restlessness, muscle tremor, convulsions, loss of memory, epilepsy, idiocy, loss of muscle contraction, and abnormal sensations. Other neuron damaging COCs additionally produce weakness of the lower extremities, and abnormal sensations. Some may even prevent the neurons from producing the proper muscle contraction, which may lead to death due to respiratory paralysis. Personality disorders and madness are an additional possibility from nerve damage.

LIVER

The effect of chemicals on the liver is well known and drinking alcohol is the best known chemical exposure that is a hepatotoxin. Chemicals produce lesions on the liver, the degree depending on the chemical itself, and the duration of the exposure. Hepatotoxin responses are acute, chronic, or 0biotransformation. The acute response amounts to cell death. Chronic responses are two: cirrhosis and/or carcinoma. Cirrhosis is a progressive fibrotic disease associated with liver dysfuntion and jaundice. Malignant tumors can also be initiated by hepatotoxins. The liver, being the principal organ that chemically alters COCs entering the body, biotransforms toxicants. Here is an example of how whisky is transformed.

ethanol → acetaldehyde → acetic acid → water + carbon dioxide

Biotransformation may lead to toxic metabolites:

carbon tetrachloride → chloroform

Diet, hormone activity, and alcohol consumption affect the metabolic capacity of the liver to biotransform COCs, not to mention the fact that the liver can also be simply overcome by the exposure.

KIDNEYS

The kidneys are susceptible to nephrotoxins. Although the kidneys are merely one percent of body weight, twenty to twenty-five percent of the blood is contained in the kidneys during rest. This means that large quantities of a COC can reach the kidneys relatively quickly. Because of their tremendous work load, the kidneys need a lot of oxygen and nutrients. One-third of the blood plasma that reaches the kidneys is filtered there. About 98–99 percent of salt and water from the blood stream is reabsorbed through the kidneys, thereby concentrating the salt. Passive diffusion with the kidneys is increased by pH changes, thus cellular concentrations of the COC accumulates. The active secretion process of the kidneys may also concentrate the COC. The biotransformation rate is high in the kidneys, but that can work against the body in some cases. Heavy metals denature proteins, and produce cell toxicity, and accumulate in the kidneys, making them sensitive. The heavy metals include arsenic, cadmium, chromium, gold, mercury, lead, and silver. Gold is introduced to the body medically when arthritis is being treated. Metabolism of halogenated organic compounds produces toxic metabolites. These COCs include alkanes and alkenes that contain chlorine, fluorine, bromine, or iodine ions.

CIRCULATORY SYSTEM

The blood system is damaged by agents that affect blood cell production, hematopoietic toxins, agents that affect the components of blood, or the oxygen-carrying capacity of red blood cells, asphyxiants. Bone marrow produces most of the components of blood, so hematopoietic agents can be devastating. Hematopoietic toxins include arsenic, benzene, bromine, and methyl chloride. Though not a chemical, ionizing radiation also damages bone marrow.

Thrombocytes, or platelets, help prevent blood loss by forming blood clots. Aspirin inhibits clotting. Benzene decreases the number of thrombocytes. Tetrachloroethylene, or perchloroethylene, increases platelet count. Leukocytes, or white blood cells, defend the body against foreign organisms by engulfing and destroying the invading organisms, or by producing antibodies. Steroids reduce the white blood cell count dramatically. Boron hydrides, magnesium oxide, naphthalene, and tetrachloroethylene increase leukocyte count. Benzene and phosphorus decrease the leukocyte count. Erythrocytes, red blood cells, transport oxygen in the blood and are therefore, very important immediately to many systems of the body. The iron compound hemoglobin picks up an oxygen molecule as it passes through the fine capillaries near the alveoli in the lungs. The normal, reduced state of hemoglobin is called oxyhemoglobin because it readily accepts oxygen. Carbon monoxide combines two hundred times more readily with hemoglobin than oxygen does, forming carboxyhemoglobin. This condition causes asphyxiation. However, carboxyhemoglobin is reversible to oxyhemoglobin despite the greater affinity for carbon monoxide. Other chemicals form irreversible methemoglobin, a fully oxidized state that refuses oxygen or to be changed. Aniline, hydrogen cyanide, dinitrobenzene, mercaptans, nitrobenzene, 2-nitropropane, sodium nitrite, and trinitrotoluene (TNT) cause methemoglobinemia.

The spleen filters bacteria and particulate matter, such as deteriorated red blood cells, from the blood. Iron is recovered from hemoglobin in the spleen and recycled. Adult spleens produce leukocytes. Embryonic spleens produce all types of blood cells, assisting the bone marrow production of these cells. Chlorobenzene and nitrobenzene directly damage the spleen.

REPRODUCTIVE SYSTEM

OSHA PELs, ACGIH TLVs, NIOSH RELs, and other recognized exposure limits fail to consistently consider reproductive toxicity. Over five thousand chemicals, drugs, and natural substances listed in the Registry of Toxic Effects of Chemical Substances (RTECS) have positive reproductive effects in toxicological studies. Jankovic and Drake screened the studies on these compounds and developed a short-list of 213 compounds, defining a reproductive toxin as one that interferes with the reproductive or sexual functioning of either male or female from puberty through adulthood. Developmental toxins are those that produce effects in offspring from conception through puberty with four principle manifestations: death of the conceptus, structural abnormality, altered growth, and functional deficiency.

Using animal studies, the inhaled absorbed dose for reproductive end points is

$$D_{inh} = NOAEL \times \frac{R_1}{W} \times \frac{T}{24} \qquad (10.92)$$

where: D_{inh} = inhaled absorbed dose, animal, mg/kg-day,
$NOAEL$ = No Observed Adverse Effect Level, mg/m^3,
R_I = inhalation rate, m^3/day,
W = body weight, kg, and
T = exposure duration, hrs.

The equivalent dose for humans, based on the average female body weight (60 kg) and 10 m^3 inhalation volume for eight hours is

$$D_I = 6D_{inh} \qquad (10.93)$$

where D_I is in mg/m^3. To calculate the human reproductive equivalent based on animal dermal or oral data:

$$D = 6D_{animal} \qquad (10.94)$$

where: D   = human dose, mg/m^3, and
         D_{animal} = animal dose, mg/kg-day.

REFERENCES

Ayers, Kenneth W., *et al. Environmental Science and Technology Handbook*. Rockville, MD: Government Institutes, 1994.

Bowes, Stephen M.,III, Marcie Francis, Beth L. Laube and Robert Frank. "Acute Exposure to Acid Fog: Influence of Breathing Pattern on Effective Dose." *American Industrial Hygiene Association Journal*. February 1995, pp. 143–150.

Cockerham, Lorris G. and Barbara S. Shane. *Basic Environmental Toxicology*. Boca Raton, FL: CRC Press, 1994.

Cogliano, Vincent James. "The U.S. EPA's Methodology for Adjusting The Reportable Quantities of Potential Carcinogens." *Hazardous Materials Control Monograph Series: Health Assessment*. Silver Spring, MD: Hazardous Materials Control Research Institute, pp. 114–7.

De Serres, Frederick J. and Arthur D. Bloom, eds. *Ecotoxicity and Human Health: A Biological Approach to Environmental Remediation*. Boca Raton, FL: CRC Press/Lewis Publishers, 1996.

Desmarais, Anne Marie C. and Paul J. Exner. "The Importance of The Danger Assessment in Superfund Feasibility Studies." *Hazardous Materials Control Monograph Series: Health Assessment*. Silver Spring, MD: Hazardous Materials Control Research Institute, pp. 59–62.

EPA. *Dermal Exposure Assessment: Principles and Applications*. Washington, D.C.: Exposure Assessment Group, Office of Health and Environmental Assessment, January 1992.

EPA. "EPA Guidelines on Exposure Assessment." *57 FR 22890*, May 29, 1992.

EPA. *Exposure Factors Handbook*. Rockville, MD: Government Institutes, Inc., 1996.

EPA Office of Research and Development. *Proposed Guidelines for Carcinogen Risk Assessment*. EPA/600/P-92/003C, April 1996.

Ford, Karl L. and Paul Gurba. "Health Risk Assessments for Contaminated Soils." *Hazardous Materials Control Monograph Series: Health Assessment*. Silver Spring, MD: Hazardous Materials Control Research Institute, pp. 63–4.

Goldbloom, David K., John M. Clegg, and John D. Erving. "Use of Risk Assessment Groundwater Model in Installation Restoration Program (IRP) Site Decisions." *Environmental Progress*. May 1992, pp. 91–7.

Hallenbeck, William H. *Quantitative Risk Assessment for Environmental and Occupational Health*. 2nd Ed. Boca Raton, FL: Lewis Publishers, 1993.

Hartman, Catherine E., George J. Schewe, Leslie J. Ungers and Michael J. Petruska. "Waste Oil Risk Assessment Study." *Hazardous Materials Control Monograph Series: Health Assessment*. Silver Spring, MD: Hazardous Materials Control Research Institute, pp. 35–8.

Hubbard, Amy E., Robert J. Hubbard, John A. George and William A. Hagel. "Quantitative Risk Assessment as The Basis for Definition of Extent of Remedial Action at The Leetown Pesticide Superfund Site." *Hazardous Materials Control Monograph Series: Risk Assessment Volume I*. Silver Spring, MD: Hazardous Materials Control Research Institute, pp. 33–9.

Karam, Joseph G. and Martha J. Otto. "Ocean Disposal Risk Assessment Model." *Hazardous Materials Control Monograph Series: Health Assessment*. Silver Spring, MD: Hazardous Materials Control Research Institute, pp. 1–8.

Jankovic, John and Frances Drake. "A Screening Method for Occupational Reproductive Health Risk." *AIHA Journal.* July 1996.

Neely, W. Brock. *Introduction to Chemical Exposure and Risk Assessment.* Boca Raton, FL: CRC Press/Lewis Publishers, 1994.

Partridge, Lawrence J. "The Application of Quantitative Risk Assessment in Selecting Cost-Effective Remedial Alternatives." *Hazardous Materials Control Monograph Series: Risk Assessment Volume I.* Silver Spring, MD: Hazardous Materials Control Research Institute, pp. 64–73.

Partridge, Lawrence J. and Arthur D. Schatz. "Application of Quantitative Risk Assessment to Remedial Measures Evaluation at Abandoned Sites." *Hazardous Materials Control Monograph Series: Risk Assessment Volume I.* Silver Spring, MD: Hazardous Materials Control Research Institute, pp. 1–5.

Rodricks, Joseph V. "Comparative Risk Assessment: A Tool for Remedial Action Planning." *Hazardous Materials Control Monograph Series: Risk Assessment Volume I.* Silver Spring, MD: Hazardous Materials Control Research Institute, pp. 45–8.

Root, David E., David B. Katzin and David W. Schnare. "Diagnosis and Treatment of Patients Presenting Subclinical Signs and Symptoms of Exposure to Chemicals which Bioaccumulate in Human Tissue." *Hazardous Materials Control Monograph Series: Health Assessment.* Silver Spring, MD: Hazardous Materials Control Research Institute, pp. 31–4.

Schaffer, Steven A. "Environmental Transfer and Loss Parameters for Four Selected Organic Priority Pollutants." *Hazardous Materials Control Monograph Series: Health Assessment.* Silver Spring, MD: Hazardous Materials Control Research Institute, pp. 26–30.

Shade, Bill and Michael Jayjock, private correspondence on spreadsheet models for estimating air concentrations and dermal exposure based on EPA–OTS methodologies, April 9, 1993.

Smith, Ellen D., Lawrence W. Barnthouse, Glenn W. Suter II, James E. Breck, Troyce D. Jones and Dee Ann Sanders. "Improving The Risk Relevance of Systems for Assessing The Relative Hazard of Contaminated Sites." *Hazardous Materials Control Monograph Series: Risk Assessment Volume I.* Silver Spring, MD: Hazardous Materials Control Research Institute, pp. 27–32.

Thorslund, Todd W., Gail Charnley and Elizabeth L. Anderson. "Innovative Use of Toxicological Data to Improve Cost-Effectiveness of Waste Cleanup." *Hazardous Materials Control Monograph Series: Health Assessment.* Silver Spring, MD: Hazardous Materials Control Research Institute, pp. 77–83.

Ursin, Christian, Charles M. Hansen, John W. Van Dyk, Peter O. Jensen, Ib J. Christensen, and Joergen Ebbehoej. "Permeability of Commercial Solvents through Living Human Skin." *American Industrial Hygiene Journal.* July 1995, pp. 651–660.

Wilsey, P.W., J.W. Vincent, M.J. Bishop, L.M. Brousseau, and I.A. Greaves. "Exposures to Inhalable and "Total" Oil Mist Aerosol by Metal Machining Shop Workers." *AIHA Journal.* December 1996, pp. 1149–53.

Yu, Rong Chun and R. Jerry Sherwood. "The Relationships between Urinary Elimination, Airborne Concentration, and Radioactive Hand Contamination for Workers Exposed to Uranium." *AIHA Journal.* July 1996.

11 EPA and OSHA Guidelines

EPA and OSHA guidelines are excellent reference tools for the exposure assessor, but they differ on one point. The EPA guidelines are just that and have to be integrated with regulations where appropriate. The OSHA guidelines are included in standards (or regulations) on specific substances on in broader standards, such as hazardous materials handling.

The following is a paraphrase of the EPA guidelines on ecological exposure assessment. These guidelines were chosen because they are comprehensive and, while dealing with the outdoor environment, can serve as a basis for assessing the indoor environment. The guidelines are paraphrased to make them a teaching tool and to focus them on the goals and objectives of this book on calculating exposures. In no way is the intention of EPA questioned or second guessed or meant to be changed by this paraphrase. However, the chief risk in paraphrasing such a document is precisely that meanings can be changed and intentions misinterpreted. Therefore, take warning. Before embarking on an actual exposure assessment, avail yourself of a copy of the complete guidelines.

EPA GUIDELINES ON ECOLOGICAL
EXPOSURE ASSESSMENT

The U.S. Environmental Protection Agency (EPA) initiated a program in 1984 to ensure "scientific quality and technical consistency" of risk assessments with the goal to develop guidelines to be used internally within EPA. The Guidelines for Estimating Exposures (1986 Guidelines) were finalized in 1986; the Proposed Guidelines for Exposure-Related Measurements (1988 Proposed Guidelines) were issued in 1988. The latter document was intended to be a companion and supplement to the former document. After receiving comments on these documents EPA issued Guidelines for Exposure Assessment in 1992 which combines, reformats and updates the earlier two guidelines substantially.

The proposed guidelines were EPA's first agency-wide ecological risk assessment guidelines. Broad in scope, the describe general principles and provide numerous examples to show how ecological risk assessment can be applied to a wide range of systems, stressors, and biological/spatial/temporal scales. A general approach provides flexibility, allowing regional EPA offices to develop more specific guidance, suited to their particular needs. Because of their broad scope, the guidelines do not provide detailed guidance in specific areas nor are they highly prescriptive. Frequently, rather than requiring that certain procedures always be followed, the guidelines describe the strengths and limitations of alternate approaches. EPA expresses preferences where possible, but declined to be specific because ecological risk assessment is a relatively new, rapidly evolving discipline, and requirements for specific approaches could soon become outdated. EPA is also expand the references in the proposed guidelines to include additional review articles or key publications that provide a window to the literature. EPA has stated that it intends to develop a series of shorter, more detailed guidance documents on specific ecological risk assessment topics after the proposed guidelines have been finalized.

Since the National Research Council proposed an ecological risk paradigm in 1993, a marked increase in discussion of ecological risk assessment issues has occurred at meetings of professional organizations, and numerous articles and books on the subject have been published. EPA work proceeded in a step-wise fashion. Preliminary work began in 1989 and included a series of colloquia sponsored by EPA's Risk Assessment Forum to identify and discuss significant issues in ecological risk assessment. Based on this early work and on a consultation with the Science Advisory Board (SAB), EPA decided to produce ecological risk assessment guidance sequentially, beginning with basic terms and concepts and continuing with the development of source materials for guidelines. The first product of this effort was the Risk Assessment Forum report, *Framework for Ecological Risk Assessment*, which proposed principles and terminology for the ecological risk assessment process. Since then, EPA solicited suggestions for ecological risk assessment guidelines structuring and sponsored the development of other peer-reviewed materials, including ecological assessment case studies, and a set of issue papers that highlight important principles and approaches that EPA scientists should consider in preparing these guidelines.

The nature and content of the guidelines were shaped by the documents mentioned above, as well as numerous meetings and discussions with individuals both within and outside of EPA. By early 1995, EPA had begun to solicit responses to the planned nature and structure of the proposed guidelines at three colloquia with Agency program offices and regions, other Federal agencies, and the public. Draft guidelines were discussed at an external peer review workshop in December, 1995. Subsequent reviews included EPA's Risk Assessment Forum and the Regulatory and Policy Development Committee, and interagency comment by members of subcommittees of the Committee on the Environment and Natural Resources of the Office of Science and Technology Policy.

INTRODUCTION

EPA defines ecological risk assessment as a process for organizing and analyzing data, information, assumptions, and uncertainties to evaluate the likelihood of adverse ecological effects. Ecological risk assessment provides a critical element for environmental decisionmaking. The full definition of ecological risk assessment is:

> The process that evaluates the likelihood that adverse ecological effects may occur or are occurring as a result of exposure to one or more stressors.

> — (U.S. EPA, 1992a)

Several terms within this definition require further explanation.

Likelihood: Descriptions of risk may range from qualitative judgments to quantitative probabilities. While risk assessments may include quantitative risk estimates, the present state of the science often may not support such quantitation. EPA would prefer to convey qualitatively the relative magnitude of uncertainties to a decision maker than to ignore them because they may not be easily understood or estimated.

Adverse ecological effects: Ecological risk assessments deal with anthropogenic changes that are considered undesirable because they alter valued structural or functional characteristics of ecological systems. An evaluation of adversity may consider the type, intensity, and scale of the effect as well as the potential for recovery.

May occur or are occurring: Ecological risk assessments may be prospective or retrospective. Retrospective ecological risk assessments evaluate the likelihood that observed ecological effects are associated with previous or current exposures to stressors. Many of the same methods and approaches are used for both prospective and retrospective assessments, and in the best case, even retrospective assessments contain predictive elements linking sources, stressors and effects.

One or more stressors: Ecological risk assessments may address single or multiple chemical, physical, or biological stressors. Because risk assessments are conducted to provide input to management decisions, this guidance focuses on stressors generated or influenced by anthropogenic activity.

The overall ecological risk assessment process is shown in Figure 11-1. Problem formulation is the first phase of the process where the assessment purpose is stated, the problem defined, and the plan for analyzing and characterizing risk determined. In the analysis phase, data on potential effects of and exposures to stressors identified during problem formulation are technically evaluated and summarized as exposure and stressor-response profiles. These profiles are integrated in risk characterization to estimate the likelihood of adverse ecological effects. Major uncertainties, assumptions, and strengths and limitations of the assessment are summarized during this phase. While discussions between risk assessors and risk managers are emphasized both at risk assessment initiation (planning) and completion (communicating results), these guidelines maintain a distinction between risk assessment and risk management. Risk assessment focuses on evaluating the likelihood of adverse effects, and risk management involves the selection of a course of action in response to an identified risk that is based on many factors (e.g., social, legal, political, or economic) in addition to the risk assessment results.

The bar along the right side of Figure 11-1 shows several activities associated with risk assessments: data acquisition, iteration, and monitoring. While the risk assessment may focus on data analysis and interpretation, acquiring the appropriate quantity and quality of data for use in the process is critical. If such data are lacking, the risk assessment may stop until the necessary data are acquired. The process is more frequently iterative than linear, since the evaluation of new data or information may require revisiting a part of the process or conducting a new assessment.

Monitoring data provides important input to all phases of the risk assessment process. For example, monitoring provides the impetus for initiating a risk assessment by identifying changes in ecological condition. In addition, monitoring data is used to evaluate the results predicted by the risk assessment. For example, follow-up studies are used to determine whether techniques used to mitigate pesticide exposures in field situations in fact reduce exposure and effects as predicted by the risk assessment. Or, for a hazardous waste site, monitoring helps verify whether source reduction resulted in anticipated ecological changes. Monitoring is also critical for determining the extent and nature of any ecological recovery that may occur. The experience gained by comparing monitoring results to evaluate risk assessment predictions helps improve the risk assessment process and is encouraged.

ECOLOGICAL RISK ASSESSMENT IN A MANAGEMENT CONTEXT

Ecological risk assessment is important for environmental decisionmaking because of the high cost of eliminating environmental risks associated with human activities and the necessity of making regulatory decisions in the face of uncertainty. Even so, ecological risk assessment provides only a portion of the information required to make risk management decisions.

Contributions of Ecological Risk Assessment to Environmental Decision-making. Ecological risk assessments provide EPA with input to a diverse set of environmental decisionmaking processes, such as the regulation of hazardous waste sites, industrial chemicals, and pesticides, or the management of watersheds affected by multiple nonchemical and chemical stressors. The ecological risk assessment process has several features that contribute to managing ecological risks.

In a risk assessment, changes in ecological effects can be expressed as a function of changes in exposure to a stressor. This inherently predictive aspect of risk assessment is particularly useful to the decision maker who must evaluate tradeoffs and examine different alternatives. Risk assessments include an explicit evaluation of uncertainties. Uncertainty analysis lends credibility and a degree of confidence to the assessment that strengthens its use in decisionmaking and helps the risk manager focus research on those areas that will lead to the greatest reductions in uncertainty. Risk assessments provide a basis for comparing, ranking, and prioritizing risks. The risk manager uses such information to help decide among several management alternatives. Risk assessments emphasize consistent use of well-defined and relevant endpoints. This is especially important for ensuring that the results of the risk assessment will be expressed in a way that the risk manager can use.

Risk Management Considerations. Although risk assessors and risk managers interact both at the initiation and completion of an ecological risk assessment, risk managers decide how to use the results of an assessment and whether a risk assessment should be conducted. While a detailed review of management issues is beyond the scope of the guidelines, key areas are highlighted.

A risk assessment is not always required for management action. When faced with compelling ecological risks and an immediate need to make a decision, a risk manager might proceed without an assessment, depending on professional judgment and statutory requirements.

Initial management decisions or statutory requirements significantly affect the scope of an assessment. So, it is important for risk managers to consider a broader scope or alternative actions for a risk assessment. A particular statute may require the assessment to focus on one type of stressor, such as chemicals, when other, perhaps more important, stressors, such as habitat alteration, are in the system. In other situations, it may be feasible to evaluate a range of options. For example, before requesting an ecological risk assessment of alternative sites for the construction and operation of a dam for hydroelectric power, risk managers may consider larger issues, such as the need for the additional power and the feasibility of using other power-generating options.

Risk managers consider many factors in making regulatory decisions. Legal mandates require the risk manager to take certain courses of action. Political and social considerations lead the risk manager to make decisions that are either more or less ecologically protective. Economic factors are also critical. While ecological risk assessment provides critical information to risk managers, it is only part of the whole environmental decisionmaking process.

PLANNING RISK ASSESSMENT: DIALOGUE BETWEEN RISK MANAGERS AND ASSESSORS

The purpose for an ecological risk assessment is to produce a scientific evaluation of ecological risk that enables managers to make informed environmental decisions. To ensure that ecological risk assessments meet risk managers' needs, a planning dialogue between risk managers and risk assessors is a critical first step toward initiating problem formulation and plays a continuing role during the conduct of the risk assessment. Planning is the beginning of a necessary interface between risk managers and risk assessors. The planning process is distinct from the scientific conduct of an ecological risk assessment itself, and political and social issues, while helping to define the objectives for the risk assessment, must not bias the scientific evaluation of risk.

During the planning dialogue, risk managers and risk assessors each have important perspectives. Generally, risk managers are charged with protecting societal values, such as human health and the environment, and must ensure that the risk assessment provides information relevant to a decision. To meet this charge, risk managers describe why the risk assessment is needed, what decisions it will support, and what they want to receive from the risk assessor. Risk managers also consider what problems they have encountered in the past when trying to use risk assessments for decisionmaking. The ecological risk assessors' role is to ensure that science is effectively used to address ecological concerns. Risk assessors describe what they can provide to the risk manager, where problems are likely to occur, and where uncertainty may be problematic. Both evaluate the potential value of conducting a risk assessment to address identified problems.

Both risk managers and risk assessors are responsible for coming to agreement on the goals, scope, and timing of a risk assessment and the resources that are available and necessary to achieve the goals. Together they use information on the area's ecosystems, regulatory endpoints, and publicly perceived environmental values to interpret the goals for use in the ecological risk assessment.

The first step in planning is to determine if a risk assessment is the best option for making the decision required. Questions concerning what is known about the degree of risk, what management options are available to mitigate or prevent it, and the value of conducting a risk assessment compared with other ways of learning about and addressing environmental concerns are asked during these discussions.

Sometimes, a risk assessment may add little value to the decision process. The risk manager and risk assessor must explore alternatives for addressing potential risk before continuing to the next planning stage.

When the decision is made to conduct a risk assessment, planning focuses on (1) establishing management goals that are agreed on, clearly articulated, and contain a way to measure success; (2) defining the decisions to be made within the context of the management goals; and (3) agreeing on the scope, complexity, and focus of the risk assessment, including the expected output and the technical and financial support available to complete it. To achieve these objectives, risk managers and risk assessors each play an active role in planning the risk assessment.

Establishing Management Goals

Management goals for a risk assessment are established by risk managers but are derived in a variety of ways. Many EPA risk assessments are conducted based on legally established management goals. For example, national regulatory programs generally have management goals written into the law governing the program. In this case, goal setting was previously completed through public debate in establishing the law. In most cases, legally established management goals do not provide sufficient guidance to the risk assessor. For example, the objectives under the Clean Water Act to protect and maintain the chemical, physical and biological integrity of the nation's waters are open to considerable interpretation. The EPA staff interprets the law in regulations and guidance. Significant interaction between the risk assessor and risk manager is required to translate the law into management goals for a particular location or set of circumstances.

EPA has increasingly emphasized place-based or community-based management of ecological resources. Management goals are taking on new significance for the ecological risk assessor. Management goals for places such as watersheds are formed as a consensus based on diverse values reflected in Federal, state, and local regulations; constituency group agendas; and public concerns. Significant interactions among a variety of interested parties are required to generate agreed-on management goals for the resource. Public meetings, constituency group meetings, evaluation of resource management organization charters, and other means of looking for management goals shared by these diverse groups are necessary unless EPA provides specific guidance. Diverse risk management teams may elect to use social scientists trained in consensus-building methods to help establish management goals. While management goals derived in this way may require further definition, these goals are generally supported by the audience for the risk assessment.

Regardless of how management goals are established, goals that explicitly define which ecological values are to be protected are more easily used to design a risk assessment for decisionmaking than general management goals. Whenever goals are general, risk assessors must interpret the goals into ecological values that can be measured or estimated and ensure that the managers agree with their interpretation. Legally mandated goals generally are interpreted by EPA. This interpretation may be performed once and then applied to the multiple similar assessments. For other risk assessments, the interpretation is unique to the ecosystem being assessed and must be done on a case-by-case basis as part of the planning process.

MANAGEMENT DECISIONS

A risk assessment is shaped by the kind of decision it will support. When a management decision is explicitly stated and closely aligned to management actions, the scope, focus, and conduct of the risk assessment are well defined by the specificity of the decision to be made. Some risk assessments are used to help establish national policy applied consistently across the country. Other risk assessments are designed for a specific site. When options are known prior to the risk assessment, a number of assumptions inherent in those options need to be explicitly stated during planning to ensure that the decision criteria are not altering the scientific validity of the risk assessment by inappropriately applying assumptions or unnecessarily limiting the variables. For many risk assessments, a range of options for managing risk are possible. When different management options have been identified, risk assessment can predict potential risk across the range of these options.

SCOPE AND COMPLEXITY OF THE RISK ASSESSMENT

Although the purpose for the risk assessment determines whether it is national, regional, or local, the resources available for conducting the assessment determines how extensive and complex it can be within this framework and the level of uncertainty that can be expected. Each risk assessment is constrained by the availability of data, scientific understanding, expertise, and financial resources. Within these constraints, designing a risk assessment may be complex. Risk managers and risk assessors must discuss in detail the nature of the decision, available resources, opportunities for increasing the resource base, and the output that will best provide the information for decisions required.

Part of the agreement on scope and complexity is based on the maximum uncertainty acceptable in whatever decision the risk assessment supports. The lower the tolerance for uncertainty, the greater the scope and complexity needed in the risk assessment. Risk assessments completed in response to legal mandates and likely to be challenged in court often require rigorous attention to acceptable levels of uncertainty to ensure that the assessment will be used in a decision. A frank discussion between the risk manager and risk assessor must address sources of uncertainty in the risk assessment and ways uncertainty can be reduced through selective investment of resources. Where appropriate, planning could account for the iterative nature of risk assessment and include explicitly defined steps. These steps, or tiers, represent increasing levels of complexity and investment, with each tier designed to reduce uncertainty. The plan may include an explicit definition of iterative steps with a description of levels of investment and decision criteria for each tier. Guidance on addressing the interplay of management decisions, study boundaries, data needs, uncertainty, and specifying limits on decision errors may be found in EPA's guidance on data quality objectives.

PLANNING OUTCOME

The planning phase is complete when agreements are reached on the management goals, assessment objectives, the focus and scope of the risk assessment, resource availability, and the type of decisions the risk assessment is to support. Agreements may encompass the technical approach to be taken in a risk assessment as determined by the regulatory or management context and reason for initiating the risk assessment, the spatial and temporal scales.

In mandated risk assessments, planning agreements are often codified in regulations, and little documentation of agreements is warranted. In other risk assessments, a summary of planning agreements may be important for ensuring that the risk assessment remains consistent with early agreements. A summary provides a point of reference for determining if early decisions may need to be changed in response to new information. However, a summary is recommended to help ensure quality communication between and among risk managers and risk assessors and to document the decisions that have been agreed upon.

Once planning is complete, the formal process of risk assessment begins through the initiation of problem formulation. During problem formulation, risk assessors should continue the dialogue with risk managers following assessment endpoint selection and once the analysis plan is completed. At these points, potential problems can be identified before the risk assessment proceeds.

PROBLEM FORMULATION PHASE

Problem formulation is a formal process for generating and evaluating preliminary hypotheses about why ecological effects have occurred, or may occur, from human activities. As the first stage of an ecological risk assessment, problem formulation provides the foundation on which the entire assessment depends. During problem formulation, management goals developed during planning are evaluated to establish objectives for the risk assessment, the problem is defined, and the plan for analyzing data and characterizing risk is determined. Deficiencies in problem formulation compromise all subsequent work.

Products of Problem Formulation. Successful problem formulation depends on the quality of three work products: (1) assessment endpoints that adequately reflect management goals and the ecosystem they represent, (2) conceptual models that describe key relationships between a stressor and assessment endpoint or among several stressors and assessment endpoints, and (3) an analysis plan. Essential to the development of these products are the effective integration and evaluation of available information.

Problem formulation is not necessarily completed in the order presented here. First, the order in which products are produced is directly related to why the ecological risk assessment is initiated. Second, problem formulation is inherently interactive and iterative, not linear. Substantial reevaluation is expected to occur within and among all products of problem formulation.

Integration of Available Information. The foundation for problem formulation is the integration of available information on the sources of stressors and stressor characteristics, exposure, the ecosystems potentially at risk, and ecological effects. When key information is of the appropriate type and sufficient quality and quantity, problem formulation proceeds effectively. When key information is unavailable in one or more areas, the risk assessment may be temporarily suspended while new data are collected. If new data cannot be collected, then the risk assessment depends on what is known and what can be extrapolated from that information. Complete information is typically not available at the beginning of risk assessments. When this is the case, the process of problem formulation assists in identifying where key data are missing and provides the framework for further research where more data are needed. Where data are few, a clear articulation of the limitations of conclusions, or uncertainty, from the risk assessment becomes increasingly critical in risk characterization.

The reason why an ecological risk assessment is initiated directly influences what information is available at the outset, and what information must be found. A risk assessment can be initiated because a known or potential stressor may be released into the environment, an adverse effect or change in condition is observed, or better management of an important ecological value is desired. Risk assessments are sometimes initiated for more than one reason.

Risk assessors beginning with information about the source or stressor seek available information on the effects the stressor is associated with and the ecosystems that it is found in. Risk assessors beginning with information about an observed effect or change in condition need to seek information about potential stressors and sources. Risk assessors starting with concern over a particular ecological value need additional information on the specific condition or effect of interest, the ecosystems potentially at risk, and potential stressors and sources.

The initial use of available information is a scoping process similar to that used to develop environmental impact statements. During this process, data and information, whether actual, inferred, or estimated, are considered to ensure that nothing important is overlooked. A comprehensive evaluation of all information provides the framework for generating a large array of risk hypotheses to consider. After the initial scoping process, information quality and applicability to the particular problem of concern are increasingly scrutinized as the risk assessor proceeds through problem formulation. When analysis plans are formed, data validity becomes a significant factor to consider.

As the complexity and spatial scale of a risk assessment increase, information needs escalate. Ecosystems characteristics directly influence when, how, and why particular ecological entities may become exposed and exhibit adverse effects due to particular stressors. Predicting risks from multiple chemical, physical, and biological stressors requires an understanding of their interactions. Risk assessments for a region or watershed, where multiple stressors are the rule, require consideration of ecological processes operating at larger spatial scales.

Problem formulation proceeds with the identification of assessment endpoints, and the development of conceptual models and the analysis plan. However, the order in which these task are done is influenced by the reason for initiating the assessment. Early recognition that initiation effects the order of product generation will help facilitate the development of problem formulation.

Selecting Assessment Endpoints. Assessment endpoints are explicit expressions of the actual environmental value that is to be protected. Assessment endpoints are critical to problem formulation because they link the risk assessment to management concerns and are central to conceptual model development. Their relevance to ecological risk assessment is determined by how well they target susceptible ecological entities. Their ability to support risk management decisions depends on how well they represent measurable characteristics of the ecosystem that adequately represent management goals. The selection of ecological concerns and assessment endpoints in EPA has traditionally been done by individual EPA program offices. Recently, EPA activities such as the watershed protection approach and community-based environmental protection used contributions by interested parties in the selection of ecological concerns and assessment endpoints.

Selecting What to Protect. The ecological resources selected to represent management goals for environmental protection are reflected in the assessment endpoints that drive ecological risk assessments. Assessment endpoints often reflect environmental values that are protected by law, provide critical resources, or provide an ecological function that would be significantly impaired, or that society would perceive as having been impaired, if the resource were altered.

Although any number of assessment endpoints may be identified, consider the practicality of using particular assessment endpoints helps refine selections. For example, when the attributes of an assessment endpoint can be measured directly, extrapolation is unnecessary; therefore this uncertainty is not introduced into the results. Assessment endpoints that cannot be measured directly but can be represented by measures that are easily monitored and modeled still provide a good foundation for the risk assessment. Assessment endpoints that cannot be linked with measurable attributes are not appropriate for a risk assessment.

Use three principal criteria when selecting assessment endpoints: (1) their ecological relevance, (2) their susceptibility to the known or potential stressors, and (3) whether they represent management goals. Of these, ecological relevance and susceptibility are essential for selecting scientifically valid assessment endpoints. Rigorous selection based on these criteria must be maintained. However, to increase the likelihood that the risk assessment will be useful in management decisions, assessment endpoints that represent societal values and management goals are more effective. Given the complex functioning of ecosystems and the interdependence of ecological entities, it is likely that assessment endpoints can be selected that are responsive to management goals while meeting scientific criteria. This provides a way to address changes that may occur over time in the public's perception of ecological value. For example, wetlands viewed as infested swamps thirty years ago are now considered prime wildlife habitat. Assessment endpoints that meet all three criteria provide the best foundation for an effective risk assessment.

Ecologically relevant endpoints reflect important characteristics of the system and are functionally related to other endpoints that help sustain the natural structure, function, and biodiversity of an ecosystem. For example, ecologically relevant endpoints may contribute to the food base (primary production), provide habitat, promote regeneration of critical resources (decomposition or nutrient cycling), or reflect the structure of the community, ecosystem, or landscape (species diversity or habitat mosaic). Changes in ecologically relevant endpoints result in unpredictable and widespread effects.

Ecological relevance is most important when risk assessors are identifying the potential cascade of adverse effects that result from the loss or reduction of one or more species or a change in ecosystem function. Careful selection of assessment endpoints that address both specific organisms of concern and landscape-level ecosystem processes is important in landscape-level risk assessments. In some cases, it may be possible to select one or more species and an ecosystem process to represent larger functional community or ecosystem processes.

Determining ecological relevance in specific cases requires expert judgment based on site-specific information, preliminary site surveys, or other available information. The less information available, the more critical it is to have informed expert judgment to ensure appropriate selections. If assessment endpoints in a risk assessment are not ecologically relevant, the results of the risk assessment may predict risk to the assessment endpoints selected but seriously misrepresent risk to the ecosystem of concern, which could lead to misguided management.

Ecological resources are considered susceptible when they are sensitive to a human-induced stressor to which they are exposed. *Sensitivity* refers to how readily an ecological entity is affected by a particular stressor and is directly related to the mode of action of the stressors. For example, chemical sensitivity is influenced by individual physiology and metabolic pathways. Sensitivity is also influenced by individual and community life-history characteristics. For example, species with long life cycles and low reproductive rates will be more vulnerable to extinction from increases in mortality than those with short life cycles and high reproductive rates. Species with large home ranges may be more sensitive to habitat fragmentation when the fragment is smaller than their required home range compared to those with smaller home ranges within a fragment. However, habitat fragmentation may also affect species with small home ranges where migration is a necessary part of their life history and fragmentation prevents exchange among subpopulations.

Sensitivity may be related to the life stage of an organism when exposed to a stressor. Frequently, young animals are more sensitive to stressors than adults. For example, Pacific salmon eggs and fry are very sensitive to sedimentation from forest logging practices and road building because they can be smothered. Age-dependent sensitivity, however, is not only in the young. In many species, special events like migration and molting represent significant energy investments that make organisms more vulnerable to an array of possible stressors. Finally, sensitivity may be increased by the presence of other stressors or natural disturbances. For example, the presence of insect pests and disease may make plants more sensitive to damage from ozone.

Measures of sensitivity include mortality or adverse reproductive effects from exposure to toxics, behavioral abnormalities, avoidance of significant food sources or nesting sites, or loss of offspring to predation because of the proximity of stressors such as noise, habitat alteration or loss, community structural changes, or other factors.

Exposure is the other key determinant in susceptibility. Exposure can mean co-occurrence, contact, or the absence of contact, depending on the stressor and assessment endpoint. The amount and conditions of exposure directly influence how an ecological entity will respond to a stressor. Thus, to determine what entities are susceptible, it is important to consider information on the proximity of an ecological resource to the stressor, the timing of exposure in terms of frequency and duration, and the intensity of exposure occurring during sensitive life stages of the organisms.

Adverse effects of a particular stressor may be important during one part of an organism's life cycle, such as early development or reproduction. Adverse effects may result from exposure to a stressor or to the absence of a necessary resource during a critical life stage. For example, if fish are unable to find suitable nesting sites during their reproductive phase, risk is significant even when water quality is high and food sources abundant. The interplay between life stage and stressors can be very complex.

Exposure may occur in one place or time, and effects may not occur until another place or time. Both life history characteristics and the circumstances of exposure influence susceptibility in this case. For example, the temperature of the incubation medium of marine turtle eggs affects the sex ratio of the offspring. But the population impacts of a change in incubation temperature may not be observable until years later when the cohort of affected turtles begins to reproduce. Delayed effects and multiple stressor exposures add complexity to evaluations of susceptibility. For example, although toxicity tests may determine receptor sensitivity to one stressor, the degree of susceptibility may depend on the co-occurrence of another stressor that significantly alters receptor response. Conceptual models need to reflect these factors. If a species is unlikely to be exposed to the stressor of concern, it is inappropriate as an assessment endpoint.

Ultimately, the value of a risk assessment depends on whether it can support quality management decisions. Risk managers are more willing to use a risk assessment for making decisions when the assessment is based on values and organisms that people care about. These values, interpreted from management goals into assessment endpoints, provide a defined and measurable entity for the risk assessment. Candidates for assessment endpoints might include entities such as endangered species, commercially or recreationally important species, functional attributes that support food sources or flood control, or aesthetic values, such as clean air in national parks or the existence of charismatic species like eagles or whales.

Selection of assessment endpoints based on public perceptions alone could lead to management decisions that do not consider important ecological information. While being responsive to the public is important, it does not obviate the requirement for scientific validity as represented by the sections on ecological relevance and susceptibility. Many ecological entities and attributes meet the necessary scientific rigor as assessment endpoints; some will be recognized as valuable by risk managers and the public, and others will not. Midges, for example, can represent the base of a complex food web that supports a popular sports fishery. They may also be considered pests. While both midges and fish are important ecological entities in this ecosystem and represent key components of the aquatic community, selecting the fishery as the assessment endpoint and using midges as a critical ecological entity to measure allow both entities to be used in the risk assessment. This choice maintains the scientific validity of the risk assessment and is responsive to management concerns. In those cases where the risk assessor identifies a critical assessment endpoint that is unpopular with the public, the risk assessor may find it necessary to present a persuasive case in its favor based on scientific arguments.

Defining Assessment Endpoints. Assessment endpoints provide a transition between broad management goals and the specific measures used in an assessment. They help assessors identify measurable attributes to quantify and predict change. Assessment endpoints also help the risk assessor determine whether management goals have been or can be achieved.

Two elements are required to define an assessment endpoint. The first is the valued ecological entity. This can be a species, a functional group of species, an ecosystem function or characteristic, a specific valued habitat, or a unique place. The second element is the characteristic about the entity of concern that is important to protect and potentially at risk. For example, it is necessary to define what is important. For piping plovers that may be nesting and feeding success. For eelgrass it may be areal extent and patch size. Wetlands require an endemic wet meadow community structure and function. For an assessment endpoint to provide a clear interpretation of the management goals and the basis for measurement in the risk assessment, both an entity and an attribute are required.

Assessment endpoints are distinct from management goals and do not represent what the managers or risk assessors want to achieve. As such they do not contain words like protect, maintain, or restore, nor do they indicate a direction for change such as loss or increase.

Defining assessment endpoints can be difficult. They may be too broad, vague, or narrow, or they may be inappropriate for the ecosystem requiring protection. Ecological integrity is a frequently cited, but vague, goal and an even more vague assessment endpoint. Integrity can only be used effectively when its meaning is explicitly characterized for a particular ecosystem, habitat, or entity. This may be done by selecting key entities and processes of an ecosystem and describing characteristics that best represent integrity for that system.

Assessment endpoints must be appropriate for the ecosystem of concern. Selecting a game fish that grows well in reservoirs may meet a feasible management goal, but would be inappropriate for evaluating risk from a new hydroelectric dam if the ecosystem of concern is a stream in which salmon spawn. Although the game fish will satisfy the fishable goal and may be highly desired by local fishermen, a reservoir species does not represent the ecosystem at risk. A vague viable fish populations assessment endpoint substituted by reproducing populations of indigenous salmonids could therefore prevent the development of an inappropriate risk assessment.

Clearly defined assessment endpoints provide direction and boundaries for the risk assessment and minimize miscommunication and reduce uncertainty. Assessment endpoints directly influence the type, characteristics, and interpretation of data and information used for analyses and the scale and character of the assessment. For example, an assessment endpoint such as egg production of pond invertebrates defines local population characteristics and requires very different types of data and ecosystem characterization compared with watershed aquatic community structure and function. If concerns are local, the assessment endpoints should not focus on landscape concerns. Where ecosystem processes and landscape mosaics are of concern, survival of a particular species would provide inadequate representation.

Assessment endpoints that are poorly defined, inappropriate, or at the incorrect scale can be very problematic.

The presence of multiple stressors influences the selection of assessment endpoints. When it is possible to select one assessment endpoint that is sensitive to many of the identified stressors, yet responds in different ways to different stressors, it is possible to consider the combined effects of multiple stressors while still discriminating among effects. For example, if recruitment of a fish population is the assessment endpoint, it is important to recognize that recruitment may be adversely affected at several life stages, in different habitats, through different ways, by different stressors. The measures of effect, exposure, and ecosystem and receptor characteristics chosen to evaluate recruitment provide a basis for discriminating among different stressors, individual effects, and their combined effect.

The assessment endpoint provides a basis for comparing a range of stressors if carefully selected. For example, the National Crop Loss Assessment Network selected crop yields as the assessment endpoint to evaluate the cumulative effects of multiple stressors. Although the primary stressor was ozone, the crop-yield endpoint allowed them to consider the effects of sulfur dioxide and soil moisture. An endpoint should be selected so that all the effects can be expressed in the same units. This is important when selecting assessment endpoints for addressing the combined effect of multiple stressors. However, in situations where multiple stressors act on the structure and function of aquatic and terrestrial communities in a watershed ecosystem, an array of assessment endpoints that represent the ecosystem community and processes is more effective than a single endpoint. When based on differing susceptibility to an array of stressors, the careful selection of assessment endpoints helps risk assessors distinguish among effects from diverse stressors. Exposure to multiple stressors may lead to effects at different levels of biological organization, for a cascade of adverse responses that should be considered.

Although assessment endpoints must be defined in terms of measurable attributes, selection does not depend on the ability to measure those attributes directly or on whether methods, models, and data are currently available. If the response of an assessment endpoint cannot be directly measured, it may be predicted from responses of surrogate or similar entities. Although for practical reasons it is helpful to use assessment endpoints that have well-developed test methods, field measurement techniques, and predictive models, it is not necessary for these methods to be established protocols. Measures that will be used to evaluate assessment endpoint response to exposures for the risk assessment are often identified during conceptual model development and specified in the analysis plan.

Risk assessors and risk managers must agree that selected assessment endpoints represent the management goals for the particular ecological value and the rationale for their selection should be clear to others.

Conceptual Models. A conceptual model, in problem formulation, is a written description and visual representation of predicted responses by ecological entities to stressors to which they are exposed, and the model includes ecosystem processes that influence these responses. Conceptual models represent relationships such as exposure scenarios, which may qualitatively link land-use activities to sources and their stressors, or may describe primary, secondary, and tertiary exposure pathways, or may describe co-occurrence between exposure pathways, ecological effects, and ecological receptors.

Conceptual models for ecological risk assessments are developed from information about stressors, potential exposure, and predicted effects on an ecological entity, the assessment endpoint. Depending on why a risk assessment is initiated, one or more of these categories of information is known at the outset. The process of creating conceptual models helps identify the unknown elements.

The complexity of the conceptual model depends on the complexity of the problem, number of stressors, number of assessment endpoints, nature of effects, and characteristics of the ecosystem. For single stressors and single assessment endpoints, conceptual models can be relatively simple relationships. In situations where conceptual models describe both the pathways of individual stressors and assessment endpoints and the interaction of multiple and diverse stressors and assessment endpoints, several subordinate models normally will be required to describe individual pathways. Other models may then be used to explore how these individual pathways interact.

Conceptual models consist of two principal products: a set of risk hypotheses that describe predicted relationships between stressor, exposure, and assessment endpoint response, along with the rationale for their selection; and a diagram that illustrates the relationships presented in the risk hypotheses.

Risk Hypotheses. Hypotheses are assumptions made in order to evaluate logical or empirical consequences. Risk hypotheses are statements of assumptions about risk based on available information and are formulated using a combination of expert judgment and information on the ecosystem at risk, potential sources of stressors, stressor characteristics, and observed or predicted ecological effects on selected or potential assessment endpoints. Hypotheses may predict the effects of a stressor event before it happens, or they may postulate why observed ecological effects occurred and ultimately what sources and stressors caused the effect. Depending on the scope of the risk assessment, the set of risk hypotheses may be very simple, predicting the potential effect of one stressor on one receptor, or extremely complex, as is typical in value-initiated risk assessments that often include prospective and retrospective hypotheses about the effects of multiple complexes of stressors on diverse ecological receptors.

Risk hypotheses should be developed even when information is incomplete, but the amount and quality of data affect the specificity and level of uncertainty associated with risk hypotheses and the conceptual models they form. When preliminary information is conflicting, risk hypotheses can be constructed specifically to differentiate among competing predictions. The predictions can then be evaluated systematically either by using available data during the analysis phase or by collecting new data before proceeding with the risk assessment. Hypotheses and predictions set a framework for using data to evaluate functional relationships such as stressor-response curves.

Early conceptual models are broad in scope, identifying as many potential relationships as possible. As more information is incorporated, the plausibility of specific risk hypotheses helps risk assessors sort through potentially large numbers of stressor-effect relationships and the ecosystem processes that influence them to identify those risk hypotheses most appropriate for the analysis phase. Justifications for selecting and omitting selecting hypotheses are documented.

Conceptual Model Diagrams. Conceptual model diagrams may be based on theory and logic, empirical data, mathematical models, or probability models. They are useful tools for communicating important pathways in a clear and concise way and can be used to ask new questions about relationships that help generate plausible risk hypotheses.

Conceptual model diagrams frequently contain boxes and arrows to illustrate relationships. When constructing these kinds of flow diagrams, use distinct and consistent shapes to distinguish stressors, assessment endpoints, responses, exposure routes, and ecosystem processes. Although flow diagrams are often used to illustrate conceptual models, no set configuration for conceptual model diagrams is required. Pictorial representations can also be effective. Regardless of the configuration, a significant part of the usefulness of a diagram is linked to the detailed written descriptions and justifications for the pathways and relationships shown. Without this, diagrams can misrepresent the processes illustrated.

When developing diagrams to represent a conceptual model, factors to consider include the number of relationships depicted, the comprehensiveness of the information, the certainty surrounding a pathway, and the potential for measurement. The number of relationships that can be depicted in one flow diagram depends on how comprehensive each relationship is. The more comprehensive, the fewer relationships that can be shown with clarity. Flow diagrams that highlight where data are abundant or scarce can provide insights on how the analyses should be approached and can be used to show the degree of confidence the risk assessor has in the relationship. Such flow diagrams can also help communicate why certain pathways were pursued and others were not.

Diagrams provide a working and dynamic representation of relationships and can be used to explore different ways of looking at a problem before selecting one or several to guide analysis. Once the risk hypotheses are selected and flow diagrams drawn, they set the framework for final planning for the analysis phase.

Uncertainty in Conceptual Models. Conceptual model development accounts for one of the more important sources of uncertainty in a risk assessment. If important relationships are missed or specified incorrectly, risks could be seriously under- or overestimated in the risk characterization phase. Uncertainty arises from lack of knowledge on how the ecosystem functions, failure to identify and interrelate temporal and spatial parameters, failure to describe a stressor or suite of stressors, or not recognizing secondary effects. In some cases, little may be known about how a stressor moves through the environment or causes adverse effects. In most cases, multiple stressors are the norm and a source of confounding variables, particularly for conceptual models that focus on a single stressor. Opinions of experts on the appropriate conceptual model configuration differ. While simplification and lack of knowledge may be unavoidable, risk assessors should document what is known, justify the model, and rank model components in terms of uncertainty. Uncertainty associated with conceptual models can be reduced by developing alternative conceptual models for a particular assessment to explore possible relationships. In cases where more than one conceptual model is plausible, the risk assessor must decide whether it is feasible to follow separate models through the analysis phase or whether the models can be combined into a better conceptual model. It is important to revisit, and if necessary revise, conceptual models during risk assessments to incorporate new information and recheck the rationale. Present conceptual models to risk managers to ensure the models communicate well and address key concerns the managers may have. This check for completeness and clarity provides an opportunity to assess the need for changes before analysis begins.

Throughout the process of problem formulation, ambiguities, errors, and disagreements occur, contributing to uncertainty. Wherever possible, eliminate sources of uncertainty through better planning. All uncertainty cannot be eliminated, so clearly summarize with a clear description of the nature of the uncertainties at the close of the problem formulation.

The hypotheses considered most likely to contribute to risk are pursued in the analysis phase. Provide the rationale for selecting and omitting risk hypotheses and acknowledge data gaps and uncertainties.

Analysis Plan. An analysis plan is a usual final stage of problem formulation, particularly in the case of complex assessments. Here, risk hypotheses are evaluated to determine how they will be assessed using available and new data. The analysis plan also delineates the assessment design, data needs, measures, and methods for conducting the analysis phase of the risk assessment.

The analysis plan includes the most important pathways and relationships identified during problem formulation that will be pursued in the analysis phase. Describe what will be done and, in particular, what will not be done. Address issues concerning the level of confidence needed for the management decision relative to the confidence that can be expected from an analysis in order to determine data needs and evaluate whether one analytical approach may be better than another. When new data are needed to conduct analyses, the feasibility of obtaining the data should be taken into account.

The selection of critical relationships in the conceptual model to pursue in analysis is based on: availability of information; strength of information about relationships between stressors and effects; the assessment endpoints and their relationship to ecosystem function; relative importance or influence and mode of action of stressors; and completeness of known exposure pathways. In situations where data are few and new data cannot be collected, combine existing data with extrapolation models so that alternative data sources may be used. This allows the use of data from other locations or on other organisms where similar problems exist and data are available. For example, the relationship between nutrient availability and algal growth is well established. Although differences will exist in how the relationship is manifested, based on the dynamics of a particular ecosystem, the relationship itself tends to be consistent. When using data that require extrapolation, identify the source of the data, justify the extrapolation method, and discuss major uncertainties apparent at this point.

Where data are not available, recommendations for new data collection should be part of problem formulation. An iterative, phased, or tiered approach to the risk assessment may provide an opportunity for early management decisions on issues that can be addressed using available data. A decision to conduct a new iteration is based on the results of any previous iteration and proceeds using new data collected as specified in the analysis plan. When new data cannot be collected, pathways that cannot be assessed are a source of uncertainty and should be described in the analysis plan.

Selecting Measures. In the analysis planning stage, measures are identified to evaluate the risk hypotheses. Three categories of measures are measures of effect, measures of exposure, and measures of ecosystem and receptor charcteristics. *Measures of effect* are used to evaluate the response of the assessment endpoint when exposed to a stressor. *Measures of exposure* point to how exposure may be occurring, including how a stressor moves through the environment and how it may co-occur with the assessment endpoint. *Measures of ecosystem and receptor characteristics* include ecosystem characteristics that influence the behavior and location of assessment endpoints, the distribution of a stressor, and life history characteristics of the assessment endpoint that may affect exposure or response to the stressor. These diverse measures increase in importance as the complexity of the assessment increases and are particularly important for risk assessments initiated to protect ecological values.

The analysis plan provides a synopsis of measures that will be used to evaluate risk hypotheses. Potential extrapolations, model characteristics, types and quality of data, and planned analyses, with specific tests for different types of data, are described. The plan discusses how results will be presented upon completion. The analysis plan provides the basis for making selections of data sets that will be used for the risk assessment.

The plan includes explanations of how data analyses will distinguish among hypotheses, an explicit expression of the approach to be used, and justifications for the elimination of some hypotheses and selection of others. It includes the measures selected, analytical methods planned, and the nature of the risk characterization options and considerations that will be generated: quotients, narrative discussion, stressor-response curve with probabilities. An analysis plan is enhanced if it contains explicit statements for how measures were selected, what they are intended to evaluate, and which analyses they support. During analysis planning, uncertainties associated with selected measures and analyses are articulated and, where possible, plans for addressing them are made.

Relating Analysis Plans to Decisions. The analysis plan is a risk manager-risk assessor checkpoint and an appropriate time for technical review. Discussions between the risk assessors and risk managers help ensure that the analyses will provide the type and extent of information that the manager can use for decision-making. Such discussions may also identify what can and cannot be done based on the preliminary evaluation of problem formulation, including which relationships to portray for the risk management decision. A reiteration of the planning discussion is important to ensure that the appropriate balance among the requirements for the decision, data availability, and resource constraints is established for the risk assessment.

ANALYSIS PHASE

The analysis phase consists of the technical evaluation of data to reach conclusions about ecological exposure and the relationships between the stressor and ecological effects. During analysis, risk assessors use measures of exposure, effects, and eco-system and receptor attributes to evaluate questions and issues that were identified in problem formulation. The products of analysis are summary profiles that describe exposure and the stressor-response relationship. When combined, these profiles provide the basis for reaching conclusions about risk during the risk characterization phase.

The conceptual model and analysis plan developed during problem formulation provide the basis for the analysis phase. By the start of analysis, the assessor should know which stressors and ecological effects are the focus of investigation and whether secondary exposures or effects will be considered. In the analysis plan, the assessor identified the information needed to perform the analysis phase. By the start of analysis, these data should be available.

The analysis phase is composed of two principal activities, the characterization of exposure and characterization of ecological effects. Both activities begin by evaluating data, such as the measures of exposure, ecosystem and receptor charac-teristics, and effects, in terms of their scientific credibility and relevance to the assessment endpoint and conceptual model. In exposure characterization, data are then analyzed to describe the source, the distribution of the stressor in the environ-ment, and the contact or co-occurrence of the stressor with ecological receptors. In ecological effects characterization, data are analyzed to describe the relationship between the stressor and response and to evaluate the evidence that exposure to the stressor causes the response, or stressor-response analyses. In many cases, extrapo-lation is necessary to link the measures of effect with the assessment endpoint.

Conclusions about exposure and the relationship between the stressor and response are summarized in profiles, which provide the opportunity to review what has been learned during the analysis phase and summarize this information in the most useful format for risk characterization. Depending on the risk assessment, the profiles may take the form of a written document or modules of a larger process model. Alternatively, documentation may be deferred until risk characterization. In any case, the purpose of the profiles is to ensure that the information needed for risk characterization has been collected and evaluated.

This process is flexible, with interaction between the ecological effects charac-terization and exposure characterization. When secondary stressors and effects are of concern, exposure and effects analyses are conducted iteratively for different ecological entities, and the analyses typically become so intertwined that they are difficult to differentiate. The distinction between the analysis phase and risk estima-tion can become blurred.

The nature of the stressor (chemical, physical, or biological) influences the types of analyses conducted and the details of implementation. Thus, the results of the analysis phase range from highly quantitative to qualitative, depending on the stres-sor and the scope of the assessment. The estimation of exposure to chemicals emphasizes contact and uptake into the organism, and the estimation of effects often entails extrapolation from test organisms to the organism of interest.

Evaluating Data And Models For Analysis. In problem formulation, the asses-sor identifies the information needed to perform the analysis phase and plans for collecting new data. The first step of the analysis phase is the critical evaluation of data and models to ensure that they can support the risk assessment.

Strengths and Limitations of Different Types of Data. The analysis phase relies on the measures identified in the analysis plan, which may come from laboratory or field studies or may be produced as output from a model. Data may have been developed for a specific risk assessment or for another purpose. A strategy that builds on the strengths of each type of data can improve confidence in the conclusions of a risk assessment.

Both laboratory and field studies, including field experiments and observational studies, can provide useful data for risk assessment. Because conditions can be controlled in laboratory studies, responses are typically less variable and smaller differences are easier to detect. However, the controls may limit the range of responses. For example, animals cannot seek alternate food sources, so they may not reflect responses in the environment. Field surveys are usually more representative of both exposures and effects, including secondary effects, found in natural systems than are estimates generated from laboratory studies or theoretical models. However, because conditions are not controlled, variability may be higher and differences are more difficult to detect. Field studies are most useful for linking stressors with effects when stressor and effect levels are measured concurrently. In addition, the presence of confounding stressors makes it difficult to attribute observed effects to specific stressors. Preferred field studies use designs that minimize effects of potentially confounding factors. Intermediate between laboratory and field are studies that use environmental media collected from the field to conduct studies of response in the laboratory. Such studies may improve the power to detect differences and may be designed to provide evidence of causality.

Most data are reported as measurements for single variables such as a chemical concentration or the number of dead organisms. In some cases, however, variables are combined into indices, and the index values are reported. Several indices are used to evaluate effects. For example, the rapid bioassessment protocols and the Index of Biotic Integrity, or IBI, have several advantages. They have the ability to provide an overall indication of biological condition by incorporating many attributes of system structure and function, from individual to ecosystem levels. They can also be used to evaluate responses from a broad range of anthropogenic stressors. Finally, indices minimize the limitations of individual metrics for detecting specific types of responses.

Although indices are very useful, they also have several drawbacks, many of which are associated with combining heterogeneous variables. For example, the final value may depend strongly on the function used to combine variables. Some indices, such as the IBI, combine only measures of effects. Differential sensitivity or other factors make it difficult to attribute causality when many response variables are combined. Such indices need to be separated into their components to investigate causality. Interpretation becomes even more difficult when an index combines measures of exposure and effects because double-counting may occur or changes in one variable can mask changes in another. Exposure and effects measures need to be separated in order to make appropriate conclusions. For these reasons, professional judgment plays a critical role in developing and applying indices.

Experience from similar situations provides another important data source that is particularly useful when predicting effects of stressors that have not yet been released. For example, lessons learned from past experiences with related organisms are often critical in trying to predict whether an organism will survive, reproduce, and disperse in a new environment. Another example is the evaluation of toxicity of new chemicals through the use of structure-activity relationships, or SARs. The simplest application of SARs is to identify a suitable analog for which data are available to estimate the toxicity of the compound for which data are lacking. More advanced applications involve the use of quantitative structure-activity relationships (QSARs), which describe the relationships between chemical structures and specific biological effects and are derived using information on sets of related chemicals. The use of analogous data without knowledge of the underlying processes may substantially increase the uncertainty in the risk assessment; however, these data may be the only option available.

While models are often developed and used as part of the risk assessment, sometimes the risk assessor relies on output of a previously developed model as input to the risk assessment. Models are particularly useful when measurements cannot be taken, for example when the assessment is predicting the effects of a chemical yet to be manufactured. Models can also provide estimates for times or locations that are impractical to measure and provide a basis for extrapolating beyond the range of observation. The quality of the model does not depend on how realistic it is, but on how well it performs in relation to the purpose for which it was built. Thus, the assessor must review the questions that need to be answered and then ensure that a model answers those questions. Models are simplifications of reality, so they may not include important processes for a particular system and may not reflect every condition in the real world. A model's output is only as good as the quality of its input variables, so critical evaluation of input data is important, as is comparing model outputs with measurements in the system of interest whenever possible.

Data and models for risk assessment are often developed in a tiered fashion. For example, simple models that err on the side of conservatism may be used first, followed by more elaborate models that provide more realistic estimates. Effects data may also be collected by using a tiered approach. Short-term tests designed to evaluate effects such as lethality and immobility may be conducted first. If the chemical exhibits high toxicity or a preliminary characterization indicates a risk, then more expensive, longer-term tests that measure sublethal effects such as changes to growth and reproduction can be conducted. Later tiers may employ multispecies tests or field experiments. Evaluate tiered data in light of the decision they are intended to support; data collected for early tiers may not be able to support more sophisticated needs.

Evaluating Measurement or Modeling Studies. Much of the information used in the analysis phase is available through published or unpublished studies that describe the purpose of the study, the methods used to collect data, and the results. Evaluating the utility of these studies relies on careful comparison of the objectives of the studies with the objectives of the risk assessment. In addition, study methods are examined to ensure that the intended objectives were met and that the data are of sufficient quality to support the risk assessment. Confidence in the information and the implications of using different studies is described during risk characterization, when the overall confidence in the assessment is discussed. In addition, the risk assessor should identify areas where existing data do not meet risk assessment needs. In these cases, EPA recommends collecting new data.

A study's documentation directly influences the ability to evaluate its utility for risk assessment. Studies should contain sufficient information so that results can be reproduced, or at least so the details of the author's work can be accessed and evaluated. An additional advantage is the ability to access findings in their entirety, providing the opportunity to conduct additional analyses of the data, if needed. For models, a number of factors increase the accessibility of methods and results. These begin with model code and documentation availability. Reports describing model results should include all important equations, tables of all parameter values, a description of any parameter estimation techniques, and tables or graphs of results.

Evaluating the Purpose and Scope of the Study. The assessor must often evaluate the utility of a study that was designed for a purpose other than risk assessment. In these cases, examine the objectives and scope of the original study to evaluate their compatibility with the objectives and needs of the current risk assessment. An examination of objectives can identify important uncertainties and ensure that the information is used appropriately in the assessment.

Similarly, a model may have been developed for purposes other than risk assessment. The model description should include the intended application, theoretical framework, underlying assumptions, and limiting conditions to help assessors identify important limitations in its application for risk assessment. For example, a model developed to evaluate chemical transport in the water column alone may have limited utility for a risk assessment of a chemical that partitions readily into sediments.

The variables and conditions examined by studies should also be compared with those variables and conditions identified during problem formulation. In addition, the range of variability explored in the study should be compared with the range of variability of interest for the risk assessment. In general, studies that minimize the amount of extrapolation needed are preferred and represent the measures identified in the analysis plan, the time frame of interest, considering seasonality and intermittent events, the ecosystem and location of interest, the environmental conditions of interest, and the exposure route of interest.

Evaluating the Design and Implementation of the Study. The design and implementation of the study are evaluated to ensure that the study objectives were met and that the information is of sufficient quality to support the purposes of the risk assessment. The study design provides insight into the sources and magnitude of uncertainty associated with the results. Among the most important design issues for studies of effects is whether a study had sufficient power to detect important differences or changes. Because this information is rarely reported, the assessor may need to calculate the magnitude of an effect that could be detected under the study conditions.

Risk assessors should evaluate evidence that the study was conducted properly. For laboratory studies, this may mean determining whether test conditions were properly controlled and control responses were within acceptable bounds. For field studies, issues include the identification and control of potentially confounding variables and the careful selection of reference sites. For models, issues include the program's structure and logic and the correct specification of algorithms in the model code.

Study evaluation is easier if a standard method or standard quality assurance/quality control (QA/QC) protocols are available and followed by the study. However, the assessor still needs to consider whether the precision and accuracy goals identified in the standard method were achieved and whether these goals are appropriate for the purposes of the risk assessment. For example, detection limits identified for one environmental matrix may not be achievable for another and may be higher than concentrations of interest for the risk assessment. Study results can still be useful even if a standard method was not used. However, an additional burden is placed on both the authors and the assessors to provide and evaluate evidence that the study was conducted properly.

Evaluating Uncertainty. Uncertainty evaluation is an ongoing theme throughout the analysis phase. The objective is to describe, and, where possible, quantify what is known and not known about exposure and effects in the system of interest. Uncertainty analyses increase credibility by explicitly describing the magnitude and direction of uncertainties, and provide the basis for efficient data collection of, or application of, refined methods. Sources of uncertainty discussed are relevant to characterizing both exposure and ecological effects shown in Table 11.1.

Sources of uncertainty when evaluating information include unclear communication of information to the assessor, unclear communication about how the assessor handled the information, and errors in the information (descriptive errors). These sources are characterized by critically examining sources of information and documenting rationales for decisions when handling it. The discussion should allow the reader to make an independent judgment about the validity of the decisions reached by the assessor.

Sources of uncertainty when estimating the value of a parameter include variability, uncertainty about a quantity's true value, and data gaps. The term *variability* is used to describe the true heterogeneity in a characteristic influencing exposure or effects. Examples include the variability in soil organic carbon, seasonal differences in animal diets, or differences in chemical sensitivity among different species. Heterogeneity is usually described during uncertainty analysis, although heterogeneity may not reflect a lack of knowledge and cannot usually be reduced by further measurement. Variability is described by presenting a distribution or specific percentiles from it, such as mean and 95th percentile.

Uncertainty about a quantity's true value includes uncertainty about its magnitude, location, or time of occurrence. This uncertainty is usually reduced by taking additional measurements. Uncertainty about a quantity's true magnitude is usually described by sampling error, or *variance* in experiments, or measurement error. When the quantity of interest is biological response, sampling error can greatly influence the ability of the study to detect effects. Properly designed studies specify sample sizes that are sufficiently large to detect important signals. Unfortunately, many studies have sample sizes that are too small to detect anything but gross changes. The discussion should highlight situations where the power to detect difference is low.

TABLE 11.1
Uncertainty Evaluation in the Analysis Phase

Source of Uncertainty	Example Analysis Phase Strategies	Specific Example
Unclear communication	Contact principal investigator or other study participants if objectives and methods of literature studies are unclear. Document decisions made during the course of the assessment.	Clarify whether the study was designed to characterize local populations or regional populations. Discuss rationale for selecting the critical toxicity study.
Descriptive errors	Verify that data sources followed appropriate QA/QC procedures.	Double-check calculations and data entry.
Variability	Describe heterogeneity using point estimates (e.g., central tendency and high end) or by constructing probability or frequency distributions. Differentiate from uncertainty due to lack of knowledge.	Display differences in species sensitivity using a cumulative distribution function.
Data gaps	Describe approaches used for bridging gaps and their rationales. Differentiate science-based judgments from policy-based judgments.	Discuss rationale for using a factor of 10 to extrapolate between a LOAEL and a NOAEL.
Uncertainty about a quantity's true value	Use standard statistical methods to construct probability distributions or point estimates (e.g., confidence limits). Evaluate power of designed experiments to detect differences. Consider taking additional data if sampling error is too large. Verify location of samples or other spatial features.	Present the upper confidence limit on the arithmetic mean soil concentration, in addition to the best estimate of the arithmetic mean. Ground-truth remote sensing data.
Model structure uncertainty (process models)	Discuss key aggregations and model simplifications. Compare model predictions with data collected in the system of interest.	Discuss combining different species into a group based on similar feeding habits.
Uncertainty about a model's form (empirical models)	Evaluate whether alternative models should be combined formally or treated separately. Compare model predictions with data collected in the system of interest.	Present results obtained using alternative models. Compare results of a plant uptake model with data collected in the field.

Nearly every assessment encounters situations where data are unavailable or where information is available on parameters that are different from those of interest for the assessment. Examples include using laboratory animal data to estimate a wild animal's response or using a bioaccumulation measurement from an ecosystem other than the one interest. These data gaps are usually bridged based on a combination of scientific data or analyses, scientific judgment, and policy judgment. For example, in deriving an ambient water quality criterion, data and analyses are used to construct distributions of species sensitivity for a particular chemical. Scientific judgment is used to infer that species selected for testing will adequately represent the range of sensitivity of species in the environment. Policy judgment is used to define the extent to which individual species should be protected. Differentiate among these elements when key assumptions and the approach used are documented.

Scientists may disagree on the best way to bridge data gaps, increasing uncertainty. Confidence can be increased through consensus building techniques such as peer reviews, workshops, and other methods to elicit expert opinion. Data gaps are often be filled by completing additional studies on the unknown parameter. Opportunities for reducing this source of uncertainty should be noted and carried through to risk characterization. Data gaps that preclude the analysis of exposure or ecological effects should also be noted and discussed in risk characterization.

An important objective of characterizing uncertainty in the analysis phase is to distinguish variability from uncertainties arising from lack of knowledge or uncertainty about a quantity's true value. This distinction facilitates the interpretation and communication of results. Thus, the assessor places error bounds on the distribution of exposure and on the estimate the proportion of the population that might exceed a toxicity threshold.

Sources of uncertainty surrounding the development and application of models include the structure of process models and the description of the relationship between two or more variables in empirical models. Process model description should include key assumptions, simplifications, and aggregations of variables. Empirical model descriptions should include the rationale for selection, and statistics on model performance, such as goodness of fit. Uncertainty in process or empirical models can be quantitatively evaluated by comparing model results to measurements taken in the system of interest or by comparing the results obtained using different model alternatives.

Methods for analyzing and describing uncertainty range from simple to complex. The calculation of one or more point estimates is one of the most common approaches to presenting analysis results. Point estimates that reflect different aspects of uncertainty have greater value if appropriately developed and communicated. Classical statistical methods, including confidence limits and percentiles, are readily applied to describing uncertainty in parameters. When a modeling approach is used, sensitivity analysis can be used to evaluate how model output changes with changes in input variables, and uncertainty propagation can be analyzed to examine how uncertainty in individual parameters affects the overall uncertainty of the assessment. The availability of software for Monte-Carlo analysis has greatly increased the use of probabilistic methods. EPA encourages assessors to follow best practices. Other methods like fuzzy mathematics and Bayesian methodologies are available, but have not yet been extensively applied to ecological risk assessment. EPA does not endorse the use of any one method over others and notes that the poor execution of any method can obscure rather than clarify the impact of uncertainty on an assessment's results. No matter what technique is used, the sources of uncertainty should be addressed.

Characterization of Exposure. Exposure characterization describes the contact or co-occurrence of stressors with ecological receptors based on measures of exposure and of ecosystem and receptor characteristics. These measures are used to analyze stressor sources, their distribution in the environment, and the extent and pattern of contact or co-occurrence. The objective is to produce a summary exposure profile that identifies the receptor, describes the course a stressor takes from the source to the receptor, and describes the intensity and spatial and temporal extent of co-occurrence or contact. The profile also describes the impact of variability and uncertainty on exposure estimates and reaches a conclusion about the likelihood that exposure will occur.

The exposure profile is combined with an effects profile to estimate risks. To be useful, the results must be compatible with the stressor-response relationship generated in the effects characterization.

Exposure Analyses. Exposure is analyzed by describing the source and releases, the distribution of the stressor in the environment, and the extent and pattern of contact or co-occurrence. The order of discussion of these topics is not necessarily the order in which they are evaluated in a particular assessment. For example, the assessor may start with information about tissue residues, and attempt to link these residues with a source.

Describe the Source. A source description identifies where the stressor originates, describes what stressors are generated, and considers other sources of the stressor. Exposure analyses may start with the source when it is known, but some analyses may begin with known exposures and attempt to link them to sources, while other analyses may start with known stressors and attempt to identify sources and quantify contact. The source is the first component of the exposure pathway and significantly influences where and when stressors eventually will be found. In addition, many management alternatives focus on modifying the source.

A source can be defined in several ways — as the place where the stressor is released or the management practice or action that produces stressors. In some assessments, the original source no longer exists and the source is defined as the current origin of the stressors.

In addition to identifying the source, the assessor describes the generation of stressors in terms of intensity, timing, and location. The location of the source and the environmental medium that first receives stressors are two attributes that deserve particular attention. In addition, the source characterization should consider whether other constituents emitted by the source influence transport, transformation, or bioavailability of the stressor of interest. In the best case, stressor generation is measured or modeled quantitatively, but sometimes can only be qualitatively described.

Many stressors have natural counterparts or multiple sources, and the characterization of these other sources can be an important component of the analysis phase. For example, many chemicals occur naturally, are generally widespread due to other sources, or may have significant sources outside the boundaries of the current assessment. Human activities may change the magnitude or frequency of natural disturbance cycles.

The way multiple sources are evaluated during the analysis phase depends on the objectives of the assessment articulated during problem formulation. One option is to focus only on the source under evaluation and calculate incremental risks attributable to that source. This strategy is common for assessments initiated with an identified source or stressor. You could also consider all sources of a stressor and calculate total risks attributable to that stressor. Relative source attribution is accomplished as a separate step. This is a common strategy for assessments initiated with an observed effect or an identified stressor. Finally, you could consider all stressors influencing an assessment endpoint and calculate cumulative risks to that endpoint. This is common for assessments initiated because of concern for an ecological value.

Source characterization can be particularly important for new biological stressors, since many of the strategies for reducing risks focus on preventing entry in the first place. Once the source is identified, the likelihood of entry may be characterized qualitatively. The description of the source can set the stage for the second objective of exposure analysis, which is describing the distribution of the stressor in the environment.

Describe the Distribution of the Stressor or Disturbed Environment. The second objective of exposure analyses is to describe the spatial and temporal distribution of the stressor in the environment. Because exposure occurs where receptors co-occur with or contact stressors in the environment, characterizing the spatial and temporal distribution of a stressor is a necessary precursor to estimating exposure. The stressor's distribution in the environment is described by evaluating the pathways that stressors take from the source as well as the formation and subsequent distribution of secondary stressors.

Stressors can be transported by many pathways in the environment. An evaluation of transport pathways ensures that measurements are taken in the appropriate media and locations and that models include the most important processes.

For chemical stressors, the evaluation of pathways usually begins by determining into which media a chemical will partition. Key considerations include physico-chemical properties such as solubility and vapor pressure. For example, lipophilic chemicals tend to be found in environmental compartments with higher proportions of organic carbon, such as soils, sediments, and biota. From there, the evaluation may examine the transport of the contaminated medium. Because constituents of chemical mixtures may have different properties, it is important to consider how the composition of a mixture may change over time or as it moves through the environment.

Ecosystem characteristics influence the transport of all types of stressors. The challenge is to determine the particular aspects of the ecosystem that are most important. In some cases, ecosystem characteristics that influence distribution are known. In other cases, much more professional judgment is needed.

The creation of secondary stressors can greatly alter conclusions about risk. Secondary stressors can be formed through biotic or abiotic transformation processes and may be of greater or lesser concern than the primary stressor. Evaluating the formation of secondary stressors is usually done as part of exposure characterization; however, coordination with the ecological effects characterization is important to ensure that all potentially important secondary stressors are evaluated. The evaluation of secondary stressors usually focuses on metabolites or degradation products or chemicals formed through abiotic processes. In addition, secondary stressors can be formed through ecosystem processes. While the possibility and rates of chemical transformation can be investigated in the laboratory, rates in the field may differ substantially, and some processes may be difficult or impossible to replicate in a laboratory. When evaluating field information, though, it may be difficult to distinguish between transformation processes and transport processes.

The distribution of stressors in the environment can be described using measurements, models, or a combination of the two. If stressors have already been released, direct measurements of environmental media or a combination of modeling and measurement is preferred. However, a modeling approach may be necessary if the assessment is intended to predict future scenarios or if measurements are not possible or practicable. For chemical stressors, refer to the exposure assessment guidelines (U.S. EPA, 1992d).

By the end of this step, the environmental distribution of the stressor or the disturbed environment should be described. This description can be an important precursor to the next objective of exposure analysis — estimating the contact or co-occurrence of the stressor with ecological entities. In cases where the extent of contact is known, describing the environmental distribution of the stressor can help identify potential sources, and ensure that all important exposures have been addressed. In addition, by identifying the pathways a stressor takes from a source, the second component of an exposure pathway is described.

Describe Contact or Co-occurrence. The third objective of the exposure analysis is to describe the extent and pattern of co-occurrence or contact between a stressor and a receptor — exposure. The objective of this step is to describe the intensity and temporal and spatial extent of exposure in a form that can be compared with the stressor-response profile generated in the effects assessment. The description of exposure is a critical element of estimating risk — if there is no exposure, there can be no risk.

Exposure can be described in terms of co-occurrence of the stressor with receptors, of the actual contact of a stressor with receptors, or of the uptake of a stressor into a receptor. The terms by which exposure is described depend on how the stressor causes adverse effects. Co-occurrence is particularly useful for evaluating stressors that can cause effects without actually contacting ecological receptors. Most stressors, however, must contact receptors to cause an effect. Finally, some stressors must not only be contacted, but also must be internally absorbed. For example, a toxicant that causes liver tumors in fish must be absorbed through the gills and reach the target organ to cause the effect.

Co-occurrence is evaluated by comparing the distribution of the stressor with the distribution of the ecological receptor. Maps of the stressor may be overlaid with maps of ecological receptors. The increased availability of geographic information systems (GIS) has provided new tools for evaluating co-occurrence.

Contact is a function of the amount of a stressor in an environmental medium and activities or behavior that brings receptors into contact with the stressor. Contact is often assumed to occur in areas where the two overlap. For chemicals, contact is quantified as the amount of a chemical ingested, inhaled, or in material applied to the skin — the potential dose. In its simplest form, it is quantified as an environmental concentration, with the assumptions that the chemical is well mixed and that the organism contacts a representative concentration. This approach is commonly used for respired media. For ingested media, another common approach combines modeled or measured concentrations of the contaminant with assumptions or parameters describing the contact rate.

Uptake is evaluated by considering the amount of stressor that is internally absorbed into an organism. Uptake is a function of the stressor, the medium, the biological membrane, and the organism. Because of interactions among these four factors, uptake varies on a situation-specific basis. Uptake is usually assessed by modifying an estimate of contact with a factor indicating the proportion of the stressor that is available for uptake (the bioavailable fraction) or actually absorbed. Absorption factors and bioavailability measured for the chemical, ecosystem, and organism of interest are preferred. Internal dose can also be evaluated by using a pharmacokinetic model or by measuring biomarkers or residues in receptors. Most stressor-response relationships express the amount of stressor in terms of media concentration or potential dose rather than internal dose. This practice limits the utility of using estimates of uptake for risk estimation. However, biomarkers and tissue residues can provide valuable confirmatory evidence that exposure has occurred, and tissue residues in prey organisms can be used for estimating risks to their predators.

The characteristics of the ecosystem and receptors must be considered to reach appropriate conclusions about exposure. Abiotic attributes may increase or decrease the amount of a stressor contacted by receptors. For example, the presence of naturally anoxic areas above contaminated sediments in an estuary may reduce the amount of time that bottom-feeding fish spend in contact with the contaminated sediments and thereby reduce exposure to the contamination. Biotic interactions can also influence exposure. For example, competition for high-quality resources may force some organisms to utilize disturbed areas. The interaction between exposure and receptor behavior can influence both the initial and subsequent exposures. For example, some chemicals reduce the prey's ability to escape predators and thereby may increase predator exposure to the chemical as well as the prey's risk of predation. Alternatively, organisms may avoid areas, food, or water with contamination they can detect. While avoidance reduces exposure to chemicals, it may increase other risks by altering habitat usage or other behavior.

Three dimensions must be considered when estimating exposure: intensity, time, and space. Intensity is the most familiar dimension for chemical and biological stressors and may be expressed as the amount of chemical contacted per day or the number of pathogenic organisms per unit area.

The temporal dimension of exposure has aspects of duration, frequency, and timing. Duration can be expressed as the time over which exposure occurs, exceeds some threshold intensity, or over which intensity is integrated. If exposure occurs as repeated, discrete events of about the same duration, frequency is the important temporal dimension of exposure. If repeated events have significant and variable duration, both duration and frequency must be considered. In addition, the timing of exposure, including the order or sequence of events, can be an important factor to describe.

In chemical assessments, the dimensions of intensity and time are often combined by averaging intensity over time. The duration over which intensity is averaged is determined by considering both the ecological effects of concern and the likely pattern of exposure. Because toxicological tests are usually conducted using constant exposures, the most realistic comparisons between exposure and effects are made when exposure in the real world does not vary substantially. In these cases, the arithmetic average exposure over the time period of toxicological significance is the appropriate statistic to use. However, as concentrations or contact rates become more episodic or variable, the arithmetic average may not reflect the toxicologically significant aspect of the exposure pattern. In extreme cases, averaging may not be appropriate at all, and assessors may need to use a toxic dynamic model to assess chronic effects.

Spatial extent is another dimension of exposure and is most commonly expressed in terms of area (hectares, square meters). At larger spatial scales, however, the shape or arrangement of exposure may be an important issue, and area alone may not be the appropriate descriptor of spatial extent for risk assessment. A general solution to the problem of incorporating pattern into ecological assessments has yet to be developed. However, the emerging field of landscape ecology and the increased availability of geographic information systems greatly expands the options for analyzing and presenting the spatial dimension of exposure.

This step completes exposure analysis. Exposure should be described in terms of intensity, space, and time, in units that can be combined with the effects assessment. In addition, the assessor should be able to trace the paths of stressors from the source to the receptors, completing the exposure pathway. The results of exposure analysis are summarized in the exposure profile.

Exposure Profile. The final product of exposure analysis is a summary profile of what has been learned. Depending on the risk assessment, the profile may be a written document, or a module of a larger process model. Alternatively, documentation may be deferred until risk characterization. In any case, the objective is to ensure that the information needed for risk characterization has been collected and evaluated. In addition, compiling the exposure profile provides an opportunity to verify that the important exposure pathways identified in the conceptual model were evaluated.

The exposure profile identifies the receptor and describes the exposure pathways and intensity and spatial and temporal extent of co-occurrence or contact. It also describes the impact of variability and uncertainty on exposure estimates and reaches a conclusion about the likelihood that exposure will occur.

The profile describes the relevant exposure pathways. If exposure can occur through many pathways, it is useful to rank them by contribution to total exposure. The profile describes the ecological entity that is exposed and represented by the exposure estimates. The assessor states how each of the three general dimensions of exposure (intensity, time, and space) was treated and why that treatment is necessary or appropriate. The profile should also describe how variability in receptor attributes or stressor levels can change exposure. Variability can be described by using a distribution or by describing where a point estimate is expected to fall on a distribution. Cumulative-distribution functions (CDFs) and probability-density functions (PDFs) are two common presentation formats. The point estimate/descriptor approach is used when information is too scarce to describe a distribution. EPA recommends using *central tendency* to refer to the mean or median of the distribution, *high end* to refer to exposure estimates that are expected to fall between the 90th and 99.9th percentile of the exposure distribution, and *bounding estimates* to refer to those higher than any actual exposure. The exposure profile summarizes important uncertainties such as lack of knowledge. In particular, the assessor should identify key assumptions and describe how they were handled, discuss, and quantify if possible, the magnitude of sampling and/or measurement error, identify the most sensitive variables influencing exposure, and identify which uncertainties can be reduced through the collection of more data.

Uncertainty about a quantity's true value can be shown by calculating error bounds on a point estimate. All of the above information is synthesized to reach a conclusion about the likelihood that exposure will occur. The exposure profile is one of the products of the analysis phase and is combined with the stressor-response profile during risk characterization.

Stressor-Response Analysis. Evaluating ecological risks requires an understanding of the relationships between stressor levels and resulting ecological responses. The stressor-response relationships used in a particular assessment depend on the scope and nature of the ecological risk assessment as defined in problem formulation and reflected in the analysis plan. For example, an assessor may need a point estimate of an effect (such as an LC_{50}) to compare with point estimates from other stressors. The shape of the stressor-response curve may be critical for determining the presence or absence of an effects threshold or for evaluating incremental risks, or stressor-response curves may be used as input for ecological effects models. If sufficient data are available, the risk assessor may construct cumulative distribution functions using multiple point estimates of effects. Or the assessor may use process models that already incorporate empirically derived stressor-response relationships.

In simple cases, the response will be one variable (e.g., mortality, incidence of abnormalities), and most quantitative techniques have been developed for univariate analysis. If the response of interest is composed of many individual variables (e.g., species abundances in an aquatic community), multivariate statistical techniques may be useful. These techniques have a long history of use in ecology but have not yet been extensively applied in risk assessment. Stressor-response relationships can be described using any of the dimensions of exposure (intensity, time, or space). Intensity is probably the most familiar dimension and is often used for chemicals as dose or concentration. The duration is also commonly used for chemical stressor-response relationships. The spatial dimension is more often of concern for physical stressors than for chemical stressors.

Data from individual experiments can be used to develop curves and point estimates both with and without associated uncertainty estimates. The advantages of curve-fitting approaches include using all of the available experimental data and the ability to interpolate to values other than the data points measured. If extrapolation outside the range of experimental data is required, risk assessors should justify that the observed experimental relationships remain valid. A disadvantage of curve fitting is that the number of data points required to complete an analysis may not always be available. For example, while standard toxicity tests with aquatic organisms frequently contain sufficient experimental treatments to permit regression analysis, frequently this is not the case for toxicity tests with wildlife species. Risk assessors sometimes use curve-fitting analyses to determine particular levels of effect for evaluation. These point estimates are interpolated from the fitted line. Point estimates may be adequate for simple assessments or comparative studies of risk and are also useful if a decision rule for the assessment was identified during the planning phase. Median effect levels are frequently selected because the level of uncertainty is minimized at the midpoint of the regression curve. While a fifty percent effect for an endpoint such as survival may not be appropriately protective for the assessment endpoint, median effect levels can be used for preliminary assessments or comparative purposes, especially when used in combination with uncertainty modifying factors. Selection of a different effect level (10%, 20%, etc.) can be arbitrary unless some clearly defined benchmark exists for the assessment endpoint. Thus, it is preferable to carry several levels of effect or the entire stressor-response curve forward to risk estimation.

When risk assessors are particularly interested in effects at lower stressor levels, they may seek to establish the no-effect level of a stressor based on comparisons between experimental treatments and controls. Statistical hypothesis testing is frequently used for this purpose. An advantage of statistical hypothesis testing is that the risk assessor is not required to pick a particular effect level of concern. The no-effect level is determined instead by experimental conditions such as the number of replicates as well as the variability inherent in the data. Thus it is important to consider the level of effect detectable in the experiment, called its power, in addition to reporting the no-effect level. Another drawback of this approach is that it is difficult to evaluate effects associated with stressor levels other than the actual treatments tested. Regression analysis has also been used as an alternative to statistical hypothesis testing.

Data available from multiple experiments can be used to generate multiple point estimates that can be displayed as cumulative distribution functions. These distributions facilitate identification of stressor levels that affect a minority or majority of species. A limiting factor in the use of cumulative frequency distributions is the amount of data needed as input. Cumulative effects distribution functions can also be derived from models that use Monte Carlo or other methods to generate distributions based on measured or estimated variation in input parameters for the models.

When multiple stressors are present, stressor-response analysis is particularly challenging. Stressor-response relationships can be constructed for each stressor separately and then combined. Alternatively, the relationship between response and the suite of stressors can be combined in one analysis. EPA prefers to directly evaluate complex chemical mixtures present in environmental media (wastewater effluents, contaminated soils, air emissions), but consider the relationship between the samples tested and the potential spatial and temporal variability in the mixture. The approach taken for multiple stressors depends on the feasibility of measuring the suite of stressors and whether an objective of the assessment is to project different stressor combinations.

In some cases, multiple regression analysis can be used to empirically relate multiple stressors and a response. Multiple regression analysis can be difficult to interpret if the explanatory variables (the stressors) are not independent. Principal components analysis can be used to extract independent explanatory variables formed from linear combinations of the original variables.

Establishing Cause and Effect Relationships (Causality). Causality is the relationship between cause (one or more stressors) and effect (assessment endpoint response to one or more stressors). Without a sound basis for linking cause and effect, uncertainty in the conclusions of an ecological risk assessment is likely to be high. Developing causal relationships is especially important for risk assessments driven by observed adverse ecological effects.

Evidence of causality may be derived from observational evidence or experimental data, and causal associations can be strengthened when both types of information are available. But since not all situations lend themselves to formal experimentation, scientists have looked for other criteria, based largely on observation rather than experiment, to support a plausible argument for cause and effect. While data to support some criteria may be incomplete or missing for any given assessment, these criteria offer a useful way of evaluating available information.

The strength of association between stressor and response is often the main reason that adverse effects are first noticed. A stronger response to a hypothesized cause is more likely to indicate true causation. Additional strong evidence of causation is when a response follows after a change in the hypothesized cause (predictive performance).

A cause-effect relationship that is demonstrated repeatedly (consistency of association) provides strong evidence of causality. Consistency may be shown by a greater number of instances of association between stressor and response, occurrences in diverse ecological systems, or associations demonstrated by diverse methods. In ecoepidemiology the occurrence of an association in more than one species and species population is very strong evidence for causation. Causality is supported if the same incident is observed by different persons under different circumstances and at different times.

Conversely, inconsistency in association between stressor and response is strong evidence against causality. Temporal incompatibility and incompatibility with experimental or observational evidence (factual implausibility) are also indications against a causal relationship.

Two other criteria may also be of some help in defining causal relationships: specificity of an association and probability. The more specific the effect, the more likely it is to have a consistent cause though some argue that effect specificity does little to strengthen a causal claim. In general, the more specific or localized the effects, the easier it is to identify the cause. Sometimes, a stressor may have a distinctive mode of action that suggests its role. For chemicals, ecotoxicologists have slightly modified Koch's postulates to provide evidence of causality. The injury, dysfunction, or other putative effect of the toxicant must be regularly associated with exposure to the toxicant and any contributory causal factors. Indicators of exposure to the toxicant must be found in the affected organisms. The toxic effects must be seen when normal organisms or communities are exposed to the toxicant under controlled conditions, and any contributory factors should be manifested in the same way during controlled exposures. The same indicators of exposure and effects must be identified in the controlled exposures as in the field.

These modifications are conceptually identical to Koch's postulates. While useful, this approach may not be practical if resources for experimentation are not available or if an adverse effect may be occurring over such a wide spatial extent that experimentation and correlation may prove difficult or yield equivocal results.

Experimental techniques are frequently used for evaluating causality in complex chemical mixtures. Options include evaluating separated components of the mixture, developing and testing a synthetic mixture, or determining how the toxicity of a mixture relates to the toxicity of individual components. The choice of method depends on the goal of the assessment and the resources and test data that are available.

Laboratory toxicity identification evaluations (TIEs) can be used to help determine which components of a chemical mixture are causing toxic effects. By using fractionation and other methods, the TIE approach can help identify chemicals responsible for toxicity and show the relative contributions to toxicity of different chemicals in aqueous effluents and sediments.

Risk assessors may utilize data from synthetic chemical mixtures if the individual chemical components are well characterized. This approach allows for manipulation of the mixture and investigation of how varying the components that are present or their ratios may affect mixture toxicity but also requires additional assumptions about the relationship between effects of the synthetic mixture and those of the environmental mixture.

When the modes of action of chemicals in a mixture are known to be similar, an additive model has been successful in 0predicting combined effects. In this situation, the contribution of each chemical to the overall toxicity of the mixture can be evaluated. However, the situation is more complicated when the modes of action of the chemical constituents are unknown or partially known.

Linking Measures of Effect to Assessment Endpoints. Assessment endpoints express the environmental values of concern for a risk assessment, but they cannot always be measured directly. When measures of effect differ from assessment endpoints, sound and explicit linkages between the two are needed. Risk assessors may make these linkages in the analysis phase or, especially when linkages rely on expert judgment, risk assessors may work with measures of effect through risk estimation (in risk characterization) and then make the connection with the assessment endpoints.

The scope and nature of the risk assessment and the environmental decision to be made help determine the degree of uncertainty and type of extrapolation that is acceptable. At an early stage of a tiered risk assessment, extrapolations from minimal data that involve large uncertainties are acceptable when the primary purpose is to determine whether a risk exists given worst-case exposure and effects scenarios. To define risk further at later stages of the assessment, additional data and more sophisticated extrapolation approaches are usually required.

The scope of the risk assessment also influences extrapolation through the nature of the assessment endpoint. Preliminary assessments that evaluate risks to general trophic levels, such as fish and birds, may extrapolate among different genera or families to obtain a range of sensitivity to the stressor. On the other hand, assessments concerned with management strategies for a particular species may employ population models.

Analysis phase activities may suggest additional extrapolation needs. Evaluation of exposure may indicate different spatial or temporal scales than originally anticipated. If spatial scales are broadened, additional receptors may need to be included in extrapolation models. If a stressor persists for an extended time in the environment, it may be necessary to extrapolate short-term responses over a longer period of exposure, and population level effects may become more important.

Whatever methods are employed to link assessment endpoints with measures of effect, apply the methods in a manner consistent with sound ecological principles and the availability of an appropriate database. For example, it is inappropriate to use structure-activity relationships to predict toxicity from chemical structure unless the chemical under consideration has a similar mode of toxic action to the reference chemicals. Similarly, extrapolations from upland avian species to waterfowl may be more credible if factors such as differences in food preferences, body mass, physiology, and seasonal behavior such as mating and migration habits are considered. Extrapolations made in a rote manner or that are biologically implausible erode the overall credibility of the assessment.

Finally, many extrapolation methods are limited by the availability of suitable databases. Although these databases are generally largest for chemical stressors and aquatic species, data do not exist for all taxa or effects. Chemical effects databases for mammals, amphibians, or reptiles are extremely limited. Risk assessors should be aware that extrapolations and models are only as useful as the data on which they are based and should recognize the great uncertainties associated with extrapolations that lack an adequate empirical or process-based rationale.

Approaches used by risk assessors to link measures of effect to assessment endpoints are based on expert judgment or on empirical or process models. Linkages based on expert judgment are not as desirable as empirical or process-based approaches, but sometimes provide the only option when data are lacking. Empirical extrapolations use experimental or observational data that may or may not be organized into a database. Process-based approaches are based on some level of understanding of the underlying operations of the system under consideration.

Expert judgment approaches rely on the professional expertise of risk assessors, expert panels, or others to relate changes in measures of effect to changes in the assessment endpoint. They are essential when databases are inadequate to support empirical models and process models are unavailable or inappropriate. Expert judgment linkages between measures of effect and assessment endpoints can be just as credible as empirical or process-based expressions, provided they have a sound scientific basis. Because of the uncertainties in predicting the effects of biological stressors such as introduced species, expert judgment approaches are commonly used.

Risks to organisms in field situations are best estimated from studies at the site of interest. However, such data are not always available. Frequently, risk assessors must extrapolate from laboratory toxicity test data to field effects. Variations in direct chemical effects between laboratory tests and field situations may not contribute as much to the overall uncertainty of the extrapolation.

In addition to single-species tests, laboratory multiple species tests are sometimes used to predict field effects. While these tests have the advantage of evaluating some aspects of a real ecological system, they also have inherent scale limitations (e.g., lack of top trophic levels) and may not adequately represent features of the field system important to the assessment endpoint.

Extrapolations based on expert judgment are frequently required when assessors wish to use field data obtained from one geographic area and apply them to a different area of concern, or to extrapolate from the results of laboratory tests to more than one geographic region. In either case, risk assessors should consider variations between regions in environmental conditions, spatial scales and heterogeneities, and ecological forcing functions.

Variations in environmental conditions in different geographic regions may alter stressor exposure and effects. If exposure to chemical stressors can be accurately estimated and are expected to be similar, the same species in different areas may respond similarly. Nevertheless, the influence of environmental conditions on stressor exposure and effects can be substantial.

Spatial scales and heterogeneities affect comparability between regions. Effects observed over a large scale may be difficult to extrapolate from one geographical location to another mainly because the spatial heterogeneity is likely to differ. Factors such as number and size of land-cover patches, distance between patches, connectivity and conductivity of patches, and patch shape may be important. Extrapolations can be facilitated by using appropriate reference sites, such as sites in comparable ecoregions.

Ecological forcing functions may differ between geographic regions. Forcing functions are critical abiotic variables that exert a major influence on the structure and function of ecological systems. Examples include temperature fluctuations, fire frequency, light intensity, and hydrologic regime. If these differ significantly between sites, it is inappropriate to extrapolate stressor effects from one system to another.

A variety of empirical and process-based approaches are available to risk assessors depending on the scope of the assessment and the data and resources available. Empirical and process-based approaches include numerical extrapolations between effects measures and assessment endpoints. These linkages range in sophistication from applying an uncertainty factor to using a complex model requiring extensive measures of effects and measures of ecosystem and receptor characteristics as input. But even the most sophisticated quantitative models involve qualitative elements and assumptions and thus require professional judgment for evaluation. Individuals who use models and interpret their results should be familiar with the underlying assumptions and components contained in the model.

Empirically based uncertainty factors or taxonomic extrapolations may be used when adequate effects databases are available but the understanding of underlying mechanisms of action or ecological principles is limited. When sufficient information on stressors and receptors is available, process-based approaches such as pharmacokinetic/pharmacodynamic models or population or ecosystem process models may be used. Regardless of the options used, risk assessors should justify and adequately document the approach selected.

Uncertainty factors are used to ensure that effects measures are sufficiently protective of assessment endpoints. Uncertainty factors are empirically derived numbers that are divided into measure of effects values to give an estimated stressor level that should not cause adverse effects to the assessment endpoint. Uncertainty factors have mostly been developed for chemicals because of the extensive ecotox-icologic databases available, especially for aquatic organisms. Uncertainty factors are useful when decisions must be made about stressors in a short time and with little information.

Uncertainty factors have been used to compensate for assessment endpoint/effect measures differences between endpoints (acute to chronic effects), between species, and between test situations (laboratory to field). Typically, uncertainty factors vary inversely with the quantity and type of effects measures data available. Uncertainty factors have been used in screening-level assessments of new chemicals, in assessing the risks of pesticides to aquatic and terrestrial organisms, and in developing bench-mark dose levels for human health effects.

In spite of their usefulness, uncertainty factors can also be misused, especially when used in an overly conservative fashion, as when chains of factors are multiplied together without sufficient justification. Like other approaches to bridging data gaps, uncertainty factors are often based on a combination of scientific analysis, scientific judgment, and policy judgment. Differentiate among these three elements when documenting the basis for the uncertainty factors used.

Empirical data can be used to facilitate extrapolations between species to species, genera, families, or orders or functional groups. Methods have been developed to extrapolate toxicity among freshwater and marine fish and arthropods. The uncer-tainties associated with extrapolating between orders, classes, and phyla tend to be very high. However, extrapolations can be made with fair certainty between aquatic species within genera and genera within families. Further applications of this approach for chemical stressors and terrestrial organisms are limited by a lack of suitable databases.

Dose-scaling or allometric regression has been used to extrapolate the effects of a chemical stressor to another species. The method is used for human health risk assessment but has not been applied extensively to ecological effects.

Allometric regression has been used with avian species and to a limited extent for estimating effects to marine organisms based on their length. For chemical stressors, allometric relationships can enable an assessor to estimate toxic effects to species not commonly tested, such as native mammalian species. Consider the taxonomic relationship between the known species and the species of interest. The closer the two are related, the more likely that the toxic response will be similar. Allometric approaches should not be applied to species that differ greatly in uptake, metabolism, or depuration of a chemical.

Process models for extrapolation are representations or abstractions of a system or process that incorporate causal relationships and provide a predictive capability that does not depend on the availability of existing stressor-response information as empirical models do. Process models enable assessors to translate data on individual effects, such as mortality, growth, and reproduction, to potential alterations in spe-cific populations, communities, or ecosystems. Such models can be used to evaluate risk hypotheses about the duration and severity of a stressor on an assessment endpoint that cannot be tested readily in the laboratory.

Two major types of models are single-species population models and multispe-cies community and ecosystem models. Population models describe the dynamics of a finite group of individuals through time and have been used extensively in ecology and fisheries management and to assess the impacts of power plants and toxicants on specific fish populations. Population models are useful in answering questions related to short- or long-term changes of population size and structure and can be used to estimate the probability that a population will decline below or grow above a specified abundance.

Proper use of the population models requires a thorough understanding of the natural history of the species under consideration, as well as knowledge of how the stressor influences its biology. Model input can include somatic growth rates, physiological rates, fecundity, survival rates of various classes within the population, and how these change when the population is exposed to the stressor and other environmental factors. In addition, the effects of population density on these parameters may be important and should be considered in the analysis of uncertainty.

Community and ecosystem models are particularly useful when the assessment endpoint involves structural (community composition) or functional (primary production) elements of the system potentially at risk. These models can also be useful when secondary effects are of concern. Changes in various community or ecosystem components such as populations, functional types, feeding guilds, or environmental processes can be estimated. By incorporating subordinate models that describe the dynamics of individual system components, these models permit evaluation of risk to multiple assessment endpoints within the context of the larger environmental system.

Risk assessors should evaluate the degree of aggregation in population or multispecies model parameters that is appropriate based both on the input data available and on the desired output of the model. For example, if a decision is required about a particular species, a model that lumps species into trophic levels or feeding guilds is not very useful. Assumptions concerning aggregation in model parameters should be included in the discussion of uncertainty.

The final product of ecological response analysis is a summary profile of what has been learned. Depending on the risk assessment, the profile may be a written document, or a module of a larger process model. Alternatively, documentation may be deferred until risk characterization. In any case, the objective is to ensure that the information needed for risk characterization has been collected and evaluated. A useful approach in preparing the stressor-response profile is to imagine that it will be used by someone else to perform the risk characterization. Using this approach, the assessor may be better able to extract the information most important to the risk characterization phase. In addition, compiling the stressor-response profile provides an opportunity to verify that the assessment and measures of effect identified in the conceptual model were evaluated.

Risk assessors should address several questions in the stressor-response profile. Depending on the type of risk assessment, affected ecological entities could include single species, populations, general trophic levels, communities, ecosystems, or landscapes. The nature of the effects should be germane to the assessment endpoints. Thus if a single species is affected, the effects should represent parameters appropriate for that level of organization. Examples include effects on mortality, growth, and reproduction. Short- and long-term effects should be reported as appropriate. At the community level, effects could be summarized in terms of structure or function depending on the assessment endpoint. At the landscape level, there may be a suite of assessment endpoints and each should be addressed separately.

Other information such as the spatial area or time to recovery may be appropriate, depending on the scope of the assessment. Causal analyses are important, especially for assessments that include field observational data.

Ideally the stressor-response profile should express effects in terms of the assessment endpoint, but this will not always be possible. Especially where it is necessary to use qualitative extrapolations between assessment endpoints and measures of effect, the stressor-response profile may only contain information on measures of effect. Under these circumstances, risk will be estimated using the measures of effects, and extrapolation to the assessment endpoints will occur during risk characterization.

Risk assessors need to be descriptive and candid about any uncertainties associated with the ecological response analysis. If it was necessary to extrapolate from measures of effect to the assessment endpoint, describe both the extrapolation and its basis. Similarly, if a benchmark or similar reference dose or concentration was calculated, discuss the extrapolations and uncertainties associated with its development. Finally, the assessor should clearly indicate major assumptions and default values used in models.

At the end of the analysis phase, the stressor-response and exposure profiles are used to estimate risks. These profiles provide the opportunity to review what has been learned and to summarize this information in the most useful format for risk characterization. Whatever form the profiles take, they ensure that the necessary information is available for risk characterization. Risk characterization is the topic for another book.

THE ITERATIVE NATURE OF ECOLOGICAL RISK ASSESSMENT

The ecological risk assessment process is by nature iterative. For example, it may take more than one pass through problem formulation to complete planning for the risk assessment, or information gathered in the analysis phase may suggest further problem formulation activities such as modification of the endpoints selected.

To maximize efficient use of limited resources, ecological risk assessments are frequently designed in sequential tiers that proceed from simple, relatively inexpensive evaluations to more costly and complex assessments. Initial tiers are based on conservative assumptions, such as maximum exposure and ecological sensitivity. When an early tier cannot define risk to support a management decision, a higher assessment tier is used that may require either additional data or applying more refined analysis techniques to available data. Iterations proceed until sufficient information is available to support a sound management decision, within the constraints of available resources.

WHO ARE RISK ASSESSORS?

Risk assessors are a diverse group of professionals who bring a needed expertise to a risk assessment. When a specific risk assessment process is well defined through regulations and guidance, one trained individual may be able to complete a risk assessment if needed information is available. However, as more complex risk assessments become common, an individual will rarely be able to provide the necessary breadth of expertise. Every risk assessment team should include at least one professional who is knowledgeable and experienced in using the risk assessment process. Other team members bring specific expertise relevant to the location, the stressors, the ecosystem, and the scientific issues and other expertise as determined by the type of assessment.

QUESTIONS ADDRESSED BY RISK ASSESSORS

- What is the scale of the risk assessment?
- What are the critical ecological endpoints and ecosystem and receptor characteristics?
- How likely is recovery and how long will it take?
- What is the nature of the problem: past, present, future?
- What is our state of knowledge on the problem?
- What data and data analyses are available and appropriate?
- What are the potential constraints (e.g., limits on expertise, time, availability of methods and data)?

QUESTIONS CONCERNING SOURCE, STRESSOR, AND EXPOSURE CHARACTERISTICS, ECOSYSTEM CHARACTERISTICS, AND EFFECTS

Source and Stressor Characteristics. What is the *source*? Is it anthropogenic, natural, point source, or diffuse nonpoint? What *type* of stressor is it: chemical, physical, or biological? What is the *intensity* of the stressor (e.g., the dose or concentration of a chemical, the magnitude or extent of physical disruption, the density or population size of a biological stressor)? What is the *mode of action*? How does the stressor act on organisms or ecosystem functions?

Exposure Characteristics. With what *frequency* does a stressor event occur (e.g., is it isolated, episodic, or continuous; is it subject to natural daily, seasonal, or annual periodicity)? What is its *duration*? How long does it persist in the environment (e.g., for chemical, what is its half-life, does it bioaccumulate; for physical, is habitat alteration sufficient to prevent recovery; for biological, will it reproduce and proliferate)? What is the *timing* of exposure? When does it occur in relation to critical organism life cycles or ecosystem events (e.g., reproduction, lake overturn)? What is the *spatial scale* of exposure? Is the extent or influence of the stressor local, regional, global, habitat-specific, or ecosystemwide? What is the *distribution*? How does the stressor move through the environment (e.g., for chemical, fate and transport; for physical, movement of physical structures; for biological, life history dispersal characteristics)?

Ecosystems Potentially at Risk. What are the *geographic boundaries*? How do they relate to functional characteristics of the ecosystem? What are the key *abiotic factors* influencing the ecosystem (e.g., climatic factors, geology, hydrology, soil type, water quality)? Where and how are *functional characteristics* driving the ecosystem (e.g., energy source and processing, nutrient cycling)? What are the *structural characteristics* of the ecosystem (e.g., species number and abundance, trophic relationships)? What *habitat* types are present? How do these characteristics influence the *susceptibility* (sensitivity and likelihood of exposure) of the ecosystem to the stressors? Are there *unique features* that are particularly valued (e.g., the last representative of an ecosystem type)? What is the *landscape context* within which the ecosystem occurs?

Ecological Effects. What are the *type and extent* of available ecological effects information (e.g., field surveys, laboratory tests, or structure-activity relationships)? Given the nature of the stressor (if known), which *effects* are expected to be elicited by the stressor? Under what *circumstances* will effects occur?

Cascading Adverse Effects: Primary (Direct) and Secondary (Indirect)

The interrelationships among entities and processes in ecosystems result in the potential for cascading effects: as one population, species, process, or other entity in the ecosystem is altered, other entities are affected as well. *Primary*, or direct, effects occur when a stressor acts directly on the assessment endpoint and causes an adverse response. *Secondary*, or indirect, effects occur when the *response* of an ecological entity to a stressor becomes a stressor to another entity. Secondary effects are not limited in number. They often are a series of effects among a diversity of organisms and processes that cascade through the ecosystem. For example, application of an herbicide on a wet meadow results in direct toxicity to plants. Death of the wetland plants leads to secondary effects such as loss of feeding habitat for ducks, breeding habitat for red-winged black birds, alteration of wetland hydrology that changes spawning habitat for fish, and so forth.

Questions for Source Description

- Where does the stressor originate?
- What environmental medium first receives stressors?
- Does the source generate other constituents that will influence a stressor's eventual distribution in the environment?
- Are there other sources of the same stressor?
- Are there background sources?
- Is the source still active?
- Does the source produce a distinctive signature that can be seen in the environment, organisms or communities?

Questions to Ask in Evaluating Stressor Distribution

- What are the important transport pathways?
- What characteristics of the stressor influence transport?
- What characteristics of the ecosystem will influence transport?
- What secondary stressors will be formed?
- Where will they be transported?

GENERAL MECHANISMS OF TRANSPORT AND DISPERSAL

Chemical stressors are transported by air current, in surface water (rivers, lakes, streams), over and/or through the soil surface, through ground water, and through the food web.

QUESTIONS TO ASK IN DESCRIBING CONTACT OR CO-OCCURRENCE

- Must the receptor actually contact the stressor for adverse effects to occur?
- Must the stressor be taken up into a receptor for adverse effects to occur?
- What characteristics of the receptors will influence the extent of contact or co-occurrence?
- Will abiotic characteristics of the environment influence the extent of contact or co-occurrence?
- Will ecosystem processes or community-level interactions influence the extent of contact or co-occurrence?

MEASURING INTERNAL DOSE USING BIOMARKERS AND TISSUE RESIDUES

Biomarkers, tissue residues, or other bioassessment methods may be useful in estimating or confirming exposure in cases where bioavailability is expected to be a significant issue, but the factors influencing it are not known. They can also be very useful when the metabolism and accumulation kinetics are important factors. These methods are most useful when they can be quantitatively linked to the amount of stressor originally contacted by the organism. In addition, they are most useful when the stressor-response relationship expresses the amount of stressor in terms of the tissue residues or biomarkers.

QUESTIONS ADDRESSED BY THE EXPOSURE PROFILE

- How does exposure occur?
- What is exposed?
- How much exposure occurs?
- When and where does it occur?
- How does exposure vary?
- How uncertain are the exposure estimates?
- What is the likelihood that exposure will occur?

QUESTIONS FOR STRESSOR-RESPONSE ANALYSIS

- Does the assessment require point estimates or stressor-response curves?
- Does the assessment require the establishment of a no-effect level?
- Would cumulative effects distributions be useful?

QUALITATIVE STRESSOR-RESPONSE RELATIONSHIPS

The relationship between stressor and response can be described qualitatively, for instance, using categories of high, medium, and low, to describe the intensity of response given exposure to a stressor. For example, one study assumed that seeds would not germinate if they were inundated with water at the critical time. This stressor-response relationship was described simply as a yes or no. In most cases, however, the objective is to describe quantitatively the intensity of response associated with exposure, and in the best case, to describe how intensity of response changes with incremental increases in exposure.

GENERAL CRITERIA FOR CAUSALITY

Criteria strongly affirming causality: strength of association, predictive performance, demonstration of a stressor-response relationship, and consistency of association.

Criteria providing a basis for rejecting causality: inconsistency in association, temporal incompatibility, and factual implausibility.

Other relevant criteria: specificity of association and theoretical and biological plausibility.

QUESTIONS ADDRESSED BY THE STRESSOR-RESPONSE PROFILE

- What ecological entities are affected?
- What is the nature of the effect(s)?
- What is the intensity of the effect(s)?
- Where appropriate, what is the time scale for recovery?
- What causal information links the stressor with any observed effects?
- How do changes in measures of effects relate to changes in assessment endpoints?

OSHA

Literal guidelines may be found in OSHA's Field Manual. The true OSHA guidelines are the safety standards (OSHA regulations) that apply to the control of chemical stressors in the workplace. For the sake and time and space, these will not be included here, even in paraphrased form. However, they should be consulted closely whenever conducting an exposure assessment in a workplace.

COOPERATION

The EPA and OSHA issued a memorandum of understanding which pledges the fullest possible cooperation and coordination between the two agencies including the development of joint training, data and information exchange, technical and professional assistance, referals of alleged violations and related matters concerning compliance and law enforcement.

REFERENCES

EPA. *Guidelines on Exposure Assessment.* 57 FR 22890. May 29, 1992.

EPA/630/R-95/002Ba. *Proposed Guidelines for Ecological Risk Assessment.* August 1996.

EPA & OSHA. *Memorandum of Understanding between The Occupational Safety and Health Administration and The Environmental Protection Agency on Minimizing Workplace and Environmental Hazards.* Nov. 23, 1990.

Glossary

a- (prefix) absence of.

Abduction a movement away from the body.

Aboral away from the mouth.

Absorbed Dose the amount of a substance penetrating across the absorption barriers (exchange boundaries) of an organism, by means of either physical or biological processes; synonym internal dose.

Absorption Barrier any of the exchange barriers of the body that allow differential diffusion of various substances across a boundary such as skin, lung tissue, and gastrointestinal tract wall.

Acceptable Risk risk that is low enough to be considered insignificant or *de minimis*.

Accuracy the measure of the correctness of data expressed as the difference between the measured value and the true or standard value.

ACGIH American Conference of Governmental Industrial Hygienists; a technical society devoted to the development of administrative and technical aspects of worker health protection.

Acetylcholine a substance produced by nerve fibers that causes muscles to contract.

Achilles' Tendon the thickest and strongest tendon of the human body.

Acne an inflammatory disease of the sebaceous glands, occurring mostly on the face, chest, and back.

Acoelous without a cavity; vertebra having the centrum flat on both ends.

Acrania lacking a brain case; the protochordates.

Actin one of two proteins involved in muscle contraction.

Action Current an electrical current that accompanies the activity of excitable tissue.

Acute Exposure a one-time exposure to a chemical; an exposure lasting less than 24 hours.

Acute Toxicity (CERCLA) measure of toxicological responses that result from a single exposure to a substance or from multiple exposures within a short period of time (typically several days or less). Specific measures of acute toxicity inlcude lethal dose$_{50}$ (LD$_{50}$) and lethal concentration$_{50}$ (LC$_{50}$), typically measured within a 24-hour to 96-hour period.

Adduction a movement toward the body.

Adenosine Diphosphate (ADP) an enzyme consisting of an adenine molecule, a ribose molecule, and two molecules of phosphoric acid.

Adenosine Triphosphate (ATP) an enzyme consisting of an adenine molecule, a ribose molecule, and three molecules of phosphoric acid.

Adipose Tissue tissue in which each cell contains a large vacuole filled with fat.

Administered Dose   the amount of a substance given to a test subject (human or animal) in order to determine dose-response relationships; synonym potential dose.

ADP   adenosine diphosphate.

Adrenal Gland   an endocrine gland located above each kidney that produces the corticosteroids and adrenaline.

Aerosol   dispersed microscopic particles in a gaseous medium; particles may be liquid, such as mist or fog, or solid such as dust, fume or smoke.

Afferent   carrying toward; for example, the direction of an afferent nerve impulse.

Agent   a chemical (or physical or biological) entity that is capable of inducing adverse response; synonym stressor.

Agglutination   the clumping together of antigens, caused by the chemical action of agglutinins on the surface of the antigens.

Agglutinin   an antibody that causes antigens to stick together.

Agglutinogen   an antigen toward which a particular antibody is antagonistic.

AIHA   American Industrial Hygiene Association

Albino   an individual with skin that has an absence of pigment.

Albumin   a protein that is the chief component of animal tissues and is insoluble in pure water.

Albuminuria   presence of albumin in urine.

Alecithal   without yolk.

Algae   chlorophyll-containing water plants that vary in size from microscopic to many feet in length.

-algia   (suffix) pain.

Allantois   an embryonic respiratory and excretory organ in the amniotes.

Alveoli   plural of alveolus.

Alveolus   an air sac in the lung where the blood exchanges carbon dioxide for fresh oxygen to carry to body tissues.

ambi-   (prefix) both.

Ambient   the conditions surrounding a person or sampling location.

Ambient Measurement   a measurement of the concentration of a chemical or pollutant taken in an ambient medium, normally with the intent of relating the measured value to the exposure of an organism that contacts that medium.

Ambient Medium   one of the basic categories of material surrounding or contacting an organism, such as outdoor air, indoor air, water, or soil, through which chemicals or pollutants can move and reach the organism.

Amino Acid   a class of organic chemicals, with the general formula $RCH(NH_2)COOH$, containing amine (NH_2) and organic acid (COOH) radicals.

Amnion   a membrane covering the embryo of reptiles, birds, and mammals.

Amniote   an animal possessing an amnion.

Amphiarthrosis   a joint that permits only slight movement.

Amphicoelus   having a cavity at each end; a veretebra with the centrum concave at both ends.

Amphiplatyan   acoelus.

Ampulla   a small, flask-shaped cavity.

an-   (prefix) absence of.

Anabolism cellular chemical processes that produce protoplasm.

Anamniote a vertebrate lacking an amnion; a fish or amphibian.

Anaphase the stage in mitosis in which daughter chromosomes migrate to opposite poles of the cell.

Anastomosis a joining of nerves, vessels, or other structures.

Androgen a male sex hormone.

Anemia a reduction of erythrocytes, hemoglobin, or hematocrit below normal.

angio- (prefix) tube or blood vessel.

Ankylosis a fusion of a joint through bone deposition.

Anlage the first embryonic indication of a structure.

Anoxia deficiency of oxygen in tissues.

ANS automatic nervous system.

ANSI American National Standards Institute

Antagonistic Muscles opposing members of a pair of skeletal muscles acting together to change directions of a limb.

ante- (prefix) before.

Anterior toward the front of the body.

anti- (prefix) against.

Antibodies substances in the blood that either naturally or through deliberate immunization counteract xenobiotic activity.

Antigen a foreign substance that stimulates the formation of a specific antibody.

anthro- (prefix) joint.

APF (symbol) (EPA) Assigned Protection Factor

Apocrine glands that undergo partial disintegration while secreting.

Aponeurosis an apronlike tendon that serves for attachment for muscles.

Apophysis a process or outgrowth, especially of a bone.

Appendicular Skeleton the arms, hands, legs, feet, shoulders, and pelvic bones.

Applied Dose the amount of a substance in contact with the primary absorption boundaries of an organism and available for absorption.

Arachnoid the middle membrane covering the brain and spinal cord.

Arboreal tree-dwelling.

Archenteron the original embryonic gut formed of the inner germ layer of the gastrula.

Archinephros the primitive excretory organ of the vertebrates.

Arithmetic Mean the sum of all the measurements in a data set divided by the number of measurements in the data set.

Arteriole the largest blood vessels, except the arteries, that carry blood away from the heart; a fine branch of an artery.

Artery the largest class of blood vessels, which carry blood away from the heart.

Ascorbic Acid vitamin C.

Asphyxia a deficiency of oxygen and accumulation of carbon dioxide in the body.

Assessment Endpoint the environmental value that is being protected by the exposure assessment/risk assessment process.

Assimilation the formation or replacement of protoplasm within cells from non-living materials.

Aster a structure of cytoplasmic fibers that appears during mitosis.

-asthenia   (suffix) weakness.

ASTM   American Society for Testing and Materials

ATP   adenosine triphosphate.

Atrioventricular Bundle   a bundle of nerve fibers that transmits impulses from the atria to the ventricles.

Atrium   the chamber of either side of the heart that receives blood from the veins.

Atrophy   a decrease in size of some organ or structure of the body.

Autonomic Nervous System (ANS)   the specialized nerves of the PNS that control the smooth muscles of the internal organs.

AV   atrioventricular.

Avogradro's Principle   equal volumes of gases at the same temperature and pressure contain the same number of molecules.

Axial Skeleton   the bones that make up the spinal column and the ribs.

Axon   the fiber extending from the neuron, which is longer than the other fibers, and carries impulses away from the cell body it serves.

B

Background Level   the concentration of a substance in a defined control area during a fixed period of time before, during, and after a data-gathering operation.

Basal Metabolic Rate (BMR)   the energy expended by the body per unit time, measured when a fasting subject is in a warm atmosphere, conscious, and at rest.

Basophil   a type of WBC in which the cytoplasmic granules stain blue with methylene blue.

Benign not malignant; a neoplasm that remains at its site of origin.

bi-   (prefix) two.

Bias   a systematic error inherent in a method or caused by some feature of the measurement system.

Bioavailability   the state of being capable of being absorbed and available to interact with the metabolic processes of an organism. Bioavailability is typically a function of chemical properties, physical state of the material to which an organism is exposed, and the ability of the individual organism to physiologically take up the chemical.

Biological Marker of Exposure   exogenous chemicals, their metabolites, or products of interactions between a xenobiotic chemical and some target molecule or cell that is measured in a compartment within an organism.

Biological Measurement   a measurement taken in a biological medium. For the purpose of exposure assessment via reconstruction of dose, the measurement is usually of the concentration of a chemical or metabolite or the status of a biomarker, normally with the intent of relating the measured value to the internal dose of a chemical at some time in the past. Biological measurements are also taken for purposes of monitoring health status and predicting effects of exposure.

Biological Medium   one of the major categories of material within an organism, such as blood, adipose tissue, or breath, through which chemicals can move, be stored, or be biologically, physically, or chemically transformed.

Biologically Effective Dose   the amount of a deposited or absorbed chemical that reaches the cells or target site where an adverse effect occurs, or where that chemical interacts with a membrane surface.

Blank   an unexposed sampling medium, or an aliquot of the reagents used in an analytical procedure, in the absence of an added analyte.

Blank Value   the measured value of a blank sample.

Blastocoele   the cavity of the blastula.

Blastomere   a cell formed in the primary divisions of an egg.

Blastula   the early stage of embryonic growth in which the cells are arranged in a sphere.

Blood Platelets   small grayish discs in blood that collect at the site a wound to assist the formation of clots.

Blood Pressure   the pressure exerted against the inner walls of the blood vessels as blood is pumped through the circulatory system.

BMR   basic metabolic rate.

Body Burden   the amount of a particular chemical stored in the body at a particular time, especially a potentially toxic chemical in the body as a result of exposure.

Bolus   a ball of food prepared by the mouth for swallowing.

Bounding Estimate   an estimate of exposure, dose, or risk that is higher than that incurred by the person in the population with the highest exposure, dose, or risk.

Boyle's Law   at a fixed temperature, the volume of any mass of air varies inversely with pressure.

Brachial   anterior appendage.

Branchial   of the gills.

Brain Stem   the lowest part of the brain, composed of the midbrain, pons, and medulla oblongata.

Brain Waves   tiny, but measurable, electrical impulses reflecting brain activity.

brady-   (prefix) slow.

Breathing Zone   a zone of air in the vicinity of an organism from which respired air is drawn.

Bronchi   two cartilaginous branches of the trachea that go to the right and left lung, respectively.

Bronchioles   tiniest branches of the bronchi, which have walls of muscle instead of cartilage and which terminate in the alveoli.

Buccal   of the mouth.

C

C   (symbol) (OSHA/ACGIH) Ceiling Limit TLV; the concentration that should not be exceeded during any part of the working day.

Callus   (1) new growth of bone tissue that surrounds a break during the process of repair. (2) a thickening of tissue caused by the normal growth of cells.

Calorie   (1) Calorie with a capital C is the amount of heat required to raise the temperature of one thousand grams of water one degree Celsius. (2) calories with a small case c is also called a gram-calorie and is the amount of heat required to raise the temperature of one gram of water one degree Celsius.

Cancellous   the spongy tissue in bones resembling lattice-work.

Cancer   a malignant neoplasm.

Capillary   the minutest branches of the arterial system, which diffuse blood to the tissues and eventually link up with the smallest branches of the venous system.

Carbohydrates   sugar, starch, and cellulose compounds.

Cardiac Sphincter   the muscle surrounding the entrance to the stomach, which permits ingested matter to enter and retains it after it has been admitted.

cardio-   (prefix) heart.

Carnivorous   flesh-eating.

Cartilage   a flexible supporting tissue composed of a nonliving matrix within which are living, nucleated cells.

CAS Number   Chemical Abstract Services Registry Number; a numeric designation assigned by the American Chemical Society's Chemical Abstract Service which uniquely identifies a specific chemical compound allowing for the conclusive identification of the chemical regardless of the nomenclature convention used.

Catabolism   chemical processes that result in the breaking down of protoplasm, releasing energy.

Catalyst   a substance that induces or speeds up a chemical or biochemical reaction by its mere presence, but which itself remains intact and nonreactive.

Caudal   toward the feet or tail.

Central Nervous System   the brain and spinal cord.

Centriole   one of two small bodies within the centrosome with a major role in cell division.

Centromere   the point on a chromosome to which the spindle is attached during mitosis.

Centrosome   the granular area near the nucleus of an animal cell that contains the centrioles.

Centrum   the body of a vertebrate.

Cephalad   toward the head.

Cephalic   of the head.

cephalo-   (prefix) head.

Cerebellum   the region of the brain where muscular activity is coordinated.

Cerebral Hemisphere   one of the two principle divisions of the cerebellum.

cerebro-   (prefix) brain.

Cerebrum   the largest of the three principle parts of the brain.

Cervical   of the neck.

Cervix   constricted part of an organ.

CFR   Code of Federal Regulations

Characterization of Ecological Effects   the part of the analysis phase of an ecological risk assessment that evaluates the ability of a stressor to cause adverse effects under a particular set of circumstances.

Characterization of Exposure   the part of the analysis phase of an ecological risk assessment that evaluates the interaction of the stressor with one or more ecological entities.

Charles' Law at a fixed pressure, the volume of a given mass of gas varies proportionally with the temperature.

Chemical Bond the link between atoms due to transfer, or sharing, of electrons.

Chiasma a crossing of fibers.

CHIP (EPA) Chemical Hazard Information Profile

chole- (prefix) biled.

Cholinesterase an enzyme that neutralizes the action of acetylcholine.

Chondro- (prefix) pertaining to cartilage.

Choroid Membrane a pigmented, highly vascular layer of the eye lying between the retina and the sclera.

CHRIS the Chemical Hazards Response Identification System; developed by the U.S. Coast Guard in cooperation with the National Academy of Sciences to provide information on the handling and disposal of toxic substances.

Chromatid one of the pair of spiral threads that make up a chromosome.

Chromatin the genetic material within the cell nucleus that is easily stained with basic dyes.

Chronic Exposure repeated exposures over the lifetime of the test species; 2 years for rodents; some scientists count any exposure lasting more than 3 months.

Chronic Value the geometric mean between the NOAEL and LOAEL.

Chyme the fluid form of food in the stomach.

Cilium a whiplike projection of a cell.

circum- (prefix) around.

Circumduction the movement of a limb so that its distal part rotates in a circle and its proximal end remains fixed.

CNS central nervous system.

Coenzyme the nonprotein component of an enzyme.

Collagen a glue-like substance found in various body tissues, especially bone and cartilage.

Colon the part of the large intestine that begins at the cecum and terminates at the sigmoid flexure.

Community an assemblage of populations of different species within a specified location of time and space.

Comparability the ability to describe likenesses and differences in the quality and relevance of two or more data sets.

Comparative risk assessment a risk assessment that uses an expert judgment approach.

Complex Exposure exposure to multiple chemicals or both chemical and nonchemical stressors.

Conceptual Model describes a series of working hypotheses of how a stressor might affect ecological entities and which describes the ecosystem potentially at risk, the relationship between measures of effect and assessment endpoints, and exposure scenarios.

Cones specialized cells in the retina of the eye that are able to distinguish colors.

Conjunctiva the delicate layer of epithelium that covers the outer surface of the cornea and the lining of the eyelids.

Connective Tissue   binding tissue that holds other tissues together or in proper relations to each other.

Contact Receptors   nerve receptors that transmit stimuli only on contact, such as touch.

contra-   (prefix) against.

Convolution   a twist or coil of any organ; especially, one of the prominent convex parts of the brain.

Cornea   the transparent portion of the eyeball, the outer surface of which is covered by the conjunctiva.

Cortex   the surface layer of an organ, such as the kidney or adrenal gland.

CPSC   Consumer Products Safety Commission

Craniad   toward the head.

Cranial Nerves   nerves arising directly from the brain and existing through an opening in the skull, passing to other body parts.

Craniata   animals possessing a brain case.

Cranium   the cavity that contains the brain; the skull.

Crossing-Over   the exchange of genes resulting from a breakage and recombination of chromosomes during meiosis.

Cumulative Distribution Function (CDF)   describe the likelihood that a variable will fall within different ranges of x. F(x), the value of y at x in a CDF plot, is the probability that a variable will have a value less than or equal to x.

Cumulative Ecological Risk Assessment   considers the aggregate ecological risk to the target entity caused by accumulation of risk from multiple stressors.

Cystic   pertaining to a bladder, particularly the gall bladder.

cyt-   (prefix) sac.

-cyte   (suffix) cell.

Cytoplasm   material within the cell membrane, outside the nucleus.

D

dactyl-   (prefix) digit.

Dalton's Law   the total pressure of a mixture of gases equals the sum of the partial pressures of the component gases individually.

Data Quality Objectives (DQO)   qualitative and quantitative statements of the overall level of uncertainty that a decision-maker is willing to accept in results or decisions derived from environmental data. DQO's are the statistical framework for planning and managing environmental data operations consistent with the data user's needs.

Deamination   removal of the amino ($-NH_2$) group from an organic compound, especially from an amino acid.

Deciduous   falling off or shedding periodically, such as primary human teeth, the skin of a reptile, or the leaves of a deciduous tree.

Deep   remote from the surface.

Dendrite   a fiber extending from the nerve cell, other than the axon, that carries impulses toward the cell.

Deoxyribonucleic Acid (DNA) the nucleic acid that bears coded genetic information.

derma- (prefix) skin.

Dermis the corium, or true skin, lying just beneath the epidermis or surface skin.

Diabetes Mellitus a disease in which the body cannot metabolize sugars properly because the pancreas is not producing adequate insulin.

Diaphragm the main muscle of breathing; separates the chest, or thoracic cavity, from the abdominal cavity.

Diaphysis the central shaft of a bone.

Diarthrosis a joint allowing free movement.

Diastole a period of relaxation of the heart muscles after contraction to bring blood back into the heart, filling the atrium and ventricles.

Differentiation the process during embryonic development when cells become specialized in structure and physiological function.

Diffusion the migration of molecules from a region of relatively high concentration to a region of lower concentration.

Diploid Number the number of chromosomes present in body cells of an organism, not counting the reproductive sex cells.

Disaccharide a carbohydrate formed by the union of two monosaccharides.

Dispersion the transport of a compound of concern through space and time.

Distal part furthest (from the heart).

Distance Receptors nerve receptors that receive stimuli transmitted over a distance, such as the eye receives light, or the ear receives sound.

Disturbance disruption of an ecosystem, community, or species, an ecosystem function, or characteristic, or a specific habitat.

DNA deoxyribonucleic acid.

Dorsal toward the back of the body.

Dose the amount of a substance available for interaction with metabolic processes or biologically significant receptors after crossing the outer boundary of an organism.

Dose Rate dose per unit time; also called dosage.

Dose-Response Assessment the determination of the relationship between the magnitude of the exposure to a chemical and the probability of occurrence of the health effect.

Dose-Response Curve a graphical representation of the quantitative relationship between administered, applied, or internal dose of a chemical or agent, and a specific biological response to that chemical or agent.

Dose-Response Relationship the resulting biological responses in an organ or organism expressed as a function of a series of different doses.

Dosimeter an instrument to measure dose, though most dosimeters actually measure exposure, not dose.

Dosimetry the process of measuring or estimating dose.

DRE (symbol) (EPA) Destruction and Removal Efficiency

Duodenum the first portion of the small intestine, beginning at the pyloric sphincter.

Dura Mater the outermost of the meninges.

Dust airborne solid particles ranging from 0.1 to 50 microns
and larger in diameter; dusts larger than 5 microns settle out in relatively still air
due to the force of gravity.

dys- (prefix) disordered, painful, difficult.

E

Ecological Entity a species, group of species, ecosystem function or characteristic,
or a specific habitat.

Ecological Exposure the exposure of a nonhuman receptor or organism to a
chemical, or a radiological or biological agent.

Ecological Risk Assessment an evaluation of the likelihood that adverse ecological effects may occur or are occurring as the result of exposure to one or more
stressors.

Ecosystem the biotic community and abiotic environment within a specified location in space and time.

-ectomy (suffix) surgical removal.

Edema the excessive accumulation of fluid in tissue spaces.

Effector a muscle, gland, or other tissue that responds to a definite stimulus.

Effluent waste material being discharged to the environment, either treated or
untreated.

Elastic Fibers fibers that are capable of returning to their original form after stretch-
ing or compressing.

Electrolyte a compound that ionizes upon being dissolved.

Embryo an organism in the earliest stages of its development; in human development, the time before four months.

-emia (suffix) blood.

en- (prefix) in.

Endocrine Gland a gland that secretes hormones directly into the blood stream.

Endoplasmic Reticulum a network of channels passing through a cell on whose
outer walls ribosomes are located.

Endoskeleton bony framework on the inside of the body, which grows with the
body, such as the human skeleton.

Endosteum a thin membrane lining the medullary canal.

Endothelium the squamous epithelial tissue lining the blood and lymph vessels
and the heart.

entero- (prefix) intestine.

Environmental Fate the destiny of a chemical or biological pollutant after release
into the environment. Environmental fate includes temporal and spatial considerations of transport, transfer, storage, and transformation.

Environmental Fate Model a mathematical abstraction of a physical system used
to predict the concentration of specific chemicals as a function of space and
time subject to transport, intermedia transfer, storage, and degradation in the
environment.

Environmental Medium one of the major categories of material found in the
physical environment that surrounds or contacts organisms. The media include

surface water, groundwater, soil, or air. The media are the means by which chemicals move and reach organisms.

Enzymes   substances that are created by living cells, whose presence aids another process within the body or speeds it to completion; a biochemical catalyst.

Eosinophil   a type of WBC in which the cytoplasmic granules are stained by eosin.

EPCRA   Emergency Planning and Community Right-to-Know Act

Epidermis   the outer, nonvascular, non-sensitive layer of horny cells that protects the dermis.

Epiglottis   a flap-like structure that lies over the opening to the larynx and prevents ingested material out of the air passage.

Epimysium   the sheath of fibrous connective tissue covering skeletal muscles.

Epiphyseal Plate   the cartilage mass between the epiphysis and the diaphysis.

Epiphysis   a portion of bone attached to another bone by a layer of cartilage.

Epithelium   tissue covering body surfaces, lining organs.

Erythema   abnormal redness of the skin due to local congestion due to sun-burn or infection.

erythro-   (prefix) red.

Erythrocyte   red blood corpuscle.

Esophagus   the musculo-membranous canal, about nine inches long, that extends from the pharynx to the stomach.

ESP   Electrostatic Precipitator

-esthesia   (suffix) feeling, sensation.

Estrogen   a female sex hormone.

Eustachian Tube   a canal running from the middle ear to the pharynx.

Excretion   the elimination of waste from cellular activity.

Exocrine Gland   a gland with a duct or ducts through which secretions pass.

Exoskeleton   a bony framework carried on the outside of the body, such as the skeleton of the crab.

Expiration   expelling of breath from the lungs.

Exposure   the contact or co-occurrence of a stressor with a receptor.

Exposure Assessment   the determination of the extent of human exposure to a chemical under various conditions.

Exposure Concentration   the concentration of a chemical in its transport or carrier medium at the point of contact.

Exposure Pathway   the physical course a chemical or pollutant takes from the source to the organism exposed.

Exposure Profile   the product of characterization of exposure in the analysis phase of ecological risk assessment and which summarizes the magnitude and spatial and temporal patterns of exposure for scenarios described in the conceptual model.

Exposure Route   the way a chemical or pollutants enters an organism after contact; ingestion, inhalation, or dermal absorption.

Exposure Scenario   assumptions concerning how an exposure may take place, including assumptions about the exposure setting, stressor characteristics, and activities that may lead to exposure.

Exposure to Mixtures exposure to multiple chemicals or to both chemical and nonchemical stressors.

Extension the movement that brings the parts of a limb toward a straight condition.

Extensor the muscle that straightens a joint once it is bent.

External outside.

External Respiration breathing; the visible act of inhaling and exhaling.

Exteroceptor a sensory organ that receives stimuli from external surroundings.

Extrinsic Eye Muscles muscles attached to the outer layer of the eyeball.

F

Fasciculus (1) a bundle of nerve, muscle, or tendon fibers separated by connective tissue. (2) a bundle or tract of nerve fibers with common connections and functions.

Fat an organic compound composed of glycerol and fatty acid.

Fat Soluble capable of dissolving in fats or lipids.

Fate the ultimate location of a released compound of concern; sometimes, an intermediate location when the ultimate location is not of particular interest or concern.

Fatigue the inability of nerves and muscles to respond to a stimulus following overactivity.

Fatty Acid an organic acid found in lipids.

Feces the waste products of animal digestion; excreted by the lower bowels.

Fertilization the union of gametes.

Fetus the unborn offspring of most mammals in the later stages of development.

Fibrillation an irregular, usually rapid, rate of excitation of the muscle tissue, particularly of the heart muscle.

Fibrinogen a plasma protein required for the formation of a blood clot.

Fixed-Location Monitoring sampling an environmental or ambient medium for pollutant concentration at one location continuously or repeatedly over some length of time.

Flagellum a small whiplike cell projection similar to a cilium, but longer.

Flexion the act of bending, or the condition of being bent.

Flexor the muscle that bends a joint.

Fog a visible aerosol of condensed liquid.

Foramen a normal opening in a tissue through which nerves, fluids, or blood vessels pass.

Fovea Centralis the most sensitive region of the retina, containing only cones.

Frontal Plane a plane that divides the body into anterior and exterior halves.

Fume an aerosol of solid particles from the condensation of vapors from molten metals; fumes are usually less than 1.0 micron in diameter and react with oxygen in the air to form a metal oxide.

G

Gamete an egg or sperm cell.

Gametogenesis the origin and formation of gametes.

Ganglion   a collection of neurons and fibers forming a subsidiary nerve center within a main nerve system.

Gas   a fluid that completely occupies a container at 77°F (25°C) and one atmosphere (760 torr).

gastro-   (prefix) stomach.

Gene   a factor in determining inheritance of a trait.

-genic   (suffix) causing, origin.

Genotype   the genetic makeup of an organism.

Geometric Means   the nth root of the product of n values.

Gland   a cell, tissue, or organ that manufactures and secretes a substance used elsewhere in the body.

Globin   a protein.

Globulins   a group of proteins in animal tissues that are soluble in pure water.

Glucose   sugar made from glycogen in the body; also found in nature in many fruits.

glyco-   (prefix) sugar.

Glycogen   starch stored in the body to be converted to glucose as needed.

Glycosuria   presence of sugar in urine.

Goblet Cell   a columnar cell that secretes mucus that lubricates epithelial surfaces.

Golgi Apparatus   a small body in the cytoplasm of most animal cells that produces secretions.

-graphy   (suffix) visualization.

Gray Matter   the substance forming the outer part of the brain and the inner part of the spinal cord, containing specialized cells, lacking myelin.

Greenstick Fracture   the fracture of a bone that is like the breaking of a twig, where cracking occurs, but not complete severance of the parts.

Guidelines   principles and procedures issued by EPA and OSHA that set basic requirements for general limits of acceptability for exposure assessments.

H

Haploid Number   the number of chromosomes in mature gametes.

Haversian Canals   openings running through the compact substance of the bone in a longitudinal direction and connecting with one another by transverse branches.

Hazard   (with respect to chemical stressors) the likelihood that a stressor will damage biological tissue in a given environment or situation.

Hazard Assessment   (1) evaluation of intrinsic effects of a stressor, or (2) defining a margin of safety or quotient by comparing toxicological effects concentration with an exposure estimate.

Hazard Identification   the determination of whether or not a particular chemical substance is causally linked to a particular health effect.

hem-   (prefix) blood.

hema-   (prefix) blood.

Hematin   an iron compound that combines with globin to form hemoglobin.

hemato-   (prefix) blood.

Hematocrit   the percentage of RBC in whole blood, normally around 45%.

Hematopoiesis   formation and maturation of blood cells.

hemi-   (prefix) half.

Hemisphere   either lateral half of the brain.

Hemoglobin   a red enzyme in red blood cells that has a strong affinity for oxygen and transports oxygen to the body tissues.

Henry's Law   the amount of gas dissolved in a liquid is proportional to the partial pressure of the gas.

HEPA   High Efficiency Particulate Absolute

hepa-   (prefix) liver.

hepato-   (prefix) liver.

Heterozygous   possessing two different alleles for the same trait.

HHE   (EPA) Health Hazard Evaluation.

High-End Exposure (Dose) Estimate   a plausible estimate of individual exposure or dose for those persons at the upper end of an exposure or dose distribution, conceptually above the 90th percentile, but not higher than the individual in the population who has the highest exposure or dose.

High-End Risk Descriptor   a plausible estimate of the individual risk for those persons at the upper end of the risk distribution, conceptually above the 90th percentile, but not higher than the individual in the population with the highest risk. Persons in the high end of the risk distribution have high risk due to high exposure, high susceptibility, or other reasons, and therefore persons in the high end of the exposure or dose distribution are not necessarily the same individuals as those in the high end of the risk distribution.

Homeostasis   the maintenance of a specific environment by the coordinated activities of various organ systems.

Hormone   a chemical messenger produced by an endocrine gland.

Horny Layer   the epidermis or non-sensitive outer skin layer that protects the dermis.

HSDB   Hazardous Substances Data Bank; part of the National Library of Medicine System; contains health and safety profiles for over 4,100 chemicals including use information, substance identification, animal and human toxicity, environmental fate, standards, personal protective equipment, fire and physical and chemical properties.

Humerus   bone of the upper arm.

hydro-   (prefix) water.

Hydrolysis   a reaction in which water combines with another compound, reducing it to a simpler form.

hyper-   (prefix) above, excess.

Hypertrophy   an increase in the size of an organ or body region due to enlargement of cells.

hypo-   (prefix) below, deficient.

Hypothalamus   a region of the brain that serves as the link between the CNS and endocrine system.

Hypoxia   inadequate oxygenation of air, blood, or cells.

hystero-   (prefix) uterus.

I

IARC   International Agency for Research on Cancer authorized by the World Health Organization (WHO) in 1966; studies causes of cancer in the human environment and publishes lists and monographs on seriously toxic chemicals and those which present carcinogenic risk.

IDLH   Immediately Dangerous to Life and Health

IH   Industrial Hygienist

Immunity   the condition of infection resistance.

in-   (prefix) in, inside.

Infarction   localized dead tissue, caused by inadequate blood supply.

Inferior   lower.

Insectivorous   insect-eating.

Inspiration   intake of breath.

Intake   the process by which a substance crosses the outer boundary of an organism without passing an absorption barrier.

Integument   the skin and its appendages.

Internal   inside.

Internal Dose   the amount of a substance that penetrates an absorption barrier, the exchange boundary of an organism, either by physical or biological processes; synonym absorbed dose.

Internal Respiration   the exchange of oxygen and carbon dioxide that takes place within the body.

Interoceptor   a sensory receptor stimulated in the viscera that receives stimuli connected with digestion, excretion, or secretion.

Interphase   the period in the life of a cell when it is not dividing.

intra-   (prefix) in, inside.

Involuntary Muscle   unstriated muscle tissue that acts independently of the will to govern automatic physical functions.

Iodopsin   the visual pigment in cones.

Ion   electrically charged atoms.

Irritability   the ability to respond to a stimulus.

Islets of Langerhans   small, isolated groups of cells embodied in the pancreas that form insulin.

Isotope   an element that has the same number of protons as another, but a different number of neutrons.

-itis   (suffix) inflammation.

J

Jaundice   the yellow color of skin, mucous membranes, and secretions due to an excess of bile pigment in the blood.

K

Karolymph   fluid within the membranes of a cell.

Karotyping   arranging and mapping chromosomes according to standard classification.

Kinetic Energy   the energy of motion.

L

Lacrimal Gland   the gland that secretes tears.

Lacteal   a lymph gland in a villus.

Lactose   a sugar found in milk.

Lacuna   a small depression or space.

Lamella   a thin scale or plate.

Larva   an immature animal that differs structurally from the adult.

Larynx   the organ of voice situated at the back of the neck, beginning below the base of the tongue, continuous with the windpipe, allowing breathing air to pass through.

Lateral   away from the center of the body; to the side.

LC$_{50}$ (symbol) the atmospheric concentration expected to be lethal for 50% of an exposed population (all having the same airborne exposure).

LD$_{50}$ (symbol) the dose expected to be lethal in 50% of a dosed population (all receiving the same dose).

LEL   (symbol) Lower Explosive Limit.

Leucocytes   white blood cells.

Leukemia   a disease of the blood that kills RBC, favoring WBC.

leuko-   (prefix) white.

Leukopenia   a decrease in the normal number of WBC, typically due to marrow damage or infection.

LEV   Local Exhaust Ventilation.

Ligament   a tough band of connective tissue that connects bones to each other.

Limit of Detection (LOD)   the minimum concentration of an analyte that, in a given matrix and with a specific analytical method, has a 99% probability of being identified, qualitatively or quantitatively measured, and reported to be greater than zero.

Lines of Evidence   information derived from different sources or by different techniques that are used to interpret and compare risk estimates.

Lipid   a group of organic compounds consisting of carbon, hydrogen, and oxygen, the simplest being fats and oils.

Lowest Observed Adverse Effect Level (LOAEL)   the lowest level of a stressor evaluated in a test that causes statistically significant differences from the controls.

Lymph   a clear, yellowish fluid from blood that bathes all cells and that returns to the circulatory system via the lymphatic vessels.

Lymph Cells   lymphocytes.

Lymphocyte   a leucocyte formed in the lymph.

Lysosome   a membrane-enclosed vesicle within the cytoplasm that disposes of unwanted foreign material from a cell.

M

macro- (prefix) large.

mal- (prefix) disordered, bad.

mamma- (prefix) breast.

Material (OSHA) used interchangeably with substance.

Matrix a mold; the cavity in which anything is formed; a specific type of medium (surface water, drinking water, soil, air) in which the analyte of interest may be contained.

Maximum Accepted Toxic Concentration (MATC) a term that means, for a particular ecological effects test, either 1) the range between the NOAEL and LOAEL, or 2) the geometric mean of the NOAEL and LOAEL.

Maximum Tolerated Dose (MTD) the highest dose in an animal bioassay which the animal tolerates without showing acute toxicity.

Maximally Exposed Individual (MEI) the single individual with the highest exposure in a given population; also, most exposed individual. EPA warns assessors to look for contextual definitions when encountering this term in literature as it has historically been defined in several ways.

Maximum Exposure Range a semiquantitative term that refers to the extreme uppermost portion of the distribution of exposures. This term should refer to the part of the individual exposure distribution that falls above the 98th percentile, but below the exposure received by the MEI.

Measure of Effect a measurable ecological characteristic that is related to the value characteristic chosen as the assessment endpoint.

Measure of Exposure a measurable stressor characteristic that is used to quantify exposure.

Measurement Endpoint measure of effect.

Medial toward the center of the body.

Median Value the value in a measurement data set such that half the measured values are greater than it is and half are less.

Median Lethal Concentration (LC$_{50}$) see LC$_{50}$.

Medulla Oblongata the part of the brain that extends from the pons to the spinal cord.

Meiolecithal pertaining to an ovum having very little yolk.

Meiosis cell division that forms gametes resulting in a haploid number of chromosomes.

Melanin a dark pigment in skin, the eyes, or hair.

Membrane a thin layer of tissue surrounding a body part or separating adjacent cavities.

Meninges coverings of the brain and spinal cord.

meno- (prefix) monthly, menstrual.

Meroblastic a partial cleavage of the ovum, in the region of the animal pole.

Merocrine those glands that do not injure themselves or destroy cells during secretion.

Mesencephalon the middle division of the embryonic brain; the midbrain.

Mesenchyme loosely organized, embryonic tissue.

Mesentery   a thin sheet of tissue that suspends an organ in a cavity.

Mesodaeum   part of the embryonic gut that is lined with endoderm.

Mesoderm   the middle embryonic germ layer.

Mesodermal Somite   a metameric division of the epimere.

Mesolecithal   pertaining to an ovum with a moderate amount of yolk.

Mesomere   a primary division of the embryonic mesoderm initiating the development of the urogenital organs and ducts.

Mesonephros   a kidney structure developed from the middle part of the archinephros, serving as the embryonic kidney of amniotes.

Mesothelium   the mesodermal lining of the body cavity.

Metabolism   the process of transforming ingested materials, food, into tissue elements and energy for use in body growth, maintenance, repair, and general function.

Metamerism   the repetition of similar body segments along the axis of an animal.

Metamorphosis   structural change of an animal, particularly when an animal passes from the larval to the adult stage.

Metanephros   a kidney structure developed from the posterior part of the archinephros, the functional adult kidney of amniotes.

Metaphase   the stage in mitosis when the chromosomes separate.

Metencephalon   the anterior division of the hindbrain.

mg/m³   (symbol) milligrams of a substance per cubic centimeter of air.

Micelle   a large chain molecules in parallel arrangement.

micro-   (prefix) small.

Microenvironments   well-defined surroundings such as home, office, automobile, kitchen, store, that can be treated as homogeneous, or at least well characterized, in the concentrations of a chemical or other agent.

Microenvironments Method   a method used in predictive exposure assessments to estimate exposures by sequentially assessing exposure for a series of areas, or microenvironments, that can be approximated by constant or well-characterized concentrations of a chemical or other agent.

Microvillus   a structure of epithelial cells lining the intestines.

Midbrain   the middle portion of the embryonic brain.

Midsagittal Plane   an imaginary line passing through the skull and spinal cord, dividing the body in half.

Mist   an aerosol of suspended liquid droplets generated by condensation from the gaseous to the liquid state or by the mechanical breaking up of a liquid into a dispersed state.

Mitochondrion   a small living structure within cytoplasm, associated with energy transformation.

Mitosis   cell division that results in two daughter cells containing the same number and type of chromosomes.

MNS   motor nervous system.

Mode   the value in the data set that occurs most frequently.

Monosaccharide   a carbohydrate that cannot be broken down chemically to a simple carbohydrate; a simple sugar.

Monte Carlo Technique   a repeated random sampling from the distribution of values for each of the parameters in a generic exposure or dose equation to derive an estimate of the distribution of exposures or doses in the population.

Motor Nerve   a nerve that causes or pertains to motion and moving the body.

mppcf (symbol) millions of particles per cubic feet of air.

Most Exposed Individual (MEI)   the single individual with the highest exposure in a given population; also, maximally exposed individual. EPA warns assessors to look for contextual definitions when encountering this term in literature as it has historically been defined in several ways.

MSDS   (OSHA) Material Safety Data Sheet.

MSHA   Mine Safety and Health Administration.

Mucosa   the membrane lining cavities and canals of the body that communicate with the outside; it secretes mucus.

Mucus   a complex protein-carbohydrate compound that is secreted by mucous glands to lubricate the internal body surfaces.

Multiple Exposure   exposure to multiple chemicals or to both chemical and nonchemical stressors.

Mutation   a change in the characteristics of an organism that can be inherited.

Myasthenia Gravis   a disorder characterized by weakness of certain voluntary muscles, thought to be caused by an excess of a chemical that blocks acetylcholine at the myoneural junction.

Myelencephalon   the posterior division of the hindbrain.

Myelin Sheath   a covering of fatty material over a nerve fiber or process.

Myelon   the spinal cord.

myo-   (prefix) muscle.

Myocomma   the connective tissue septum between two myotomes.

Myosin   one of the two proteins involved in muscle contraction.

Myotome   the median part of the mesodermal somite; gives rise to the skeletal muscles.

N

Naris   the nostril.

Neoteny   retention of larval characteristics in the adult resulting from environmental factors.

nephri-   (prefix) kidney.

nephro-   (prefix) kidney.

Nephron   the functional unit of the kidneys.

Neural   of the nervous system.

Neurectoderm   the division of the epiblast that gives rise to the nervous system.

neuro-   (prefix) nerve.

Neurocranium   the part of the skull that surrounds the brain.

Neurofibril   a strand in a nerve cell that crosses the nerve cell body to connect the dendrites and axons.

Neuroglial Cell   a nonconducting cell that is essential for the protection, structural support, and metabolism of the nerve cell.

Neuron   a nerve cell.

Neuroplasm   the protoplasm in the neuron that fills the spaces between the fibrils.

NIOSH   National Institute of Occupational Safety & Health.

No Observed Adverse Effect Level (NOAEL)   the highest level of a stressor evaluated in a test that does not cause statistically significant differences from the controls.

Nonparametric Statistical Methods   these methods do not assume a functional form with identifiable parameters for the statistical distribution of interest.

Notochord   the primitive or embryonic supporting column of the Chordata.

NSPS   (EPA) New Source Performance Standards

Nuclear Membrane   a semipermeable membrane surrounding the nucleus of a cell, separating it from the cytoplasm.

Nucleic Acid   a group of organic acids found in chromatin.

Nucleolus   a body found within the nucleus of cell that appears to be associated with the formation of nucleoproteins.

Nucleoprotein   a protein with nucleic acid chains, found in the nucleus of a cell.

Nucleotide   a compound consisting of a purine or pyrimidine base with a sugar and phosphoric acid.

Nucleus   (1) the control center of a cell, containing the chromosomes. (2) the center of an atom, containing protons and neutrons. (3) a group of specialized nerve cells lying within the white matter of the CNS.

O

Occipital   pertaining to the back of the skull.

Occipital Lobe   one of the sections, or lobes, of the cerebrum; located toward the back of the brain.

Odoriferous   offering stimulus to the olfactory nerve.

Oil   a relatively simple lipid that is liquid at ordinary temperatures.

Oligodendroglia   small spheroidal cells of the nervous system that are found around nerve cells as supporting structures.

Oligolecithal   an ovum having a large amount of yolk.

-oma   (suffix) tumor.

Omnivorous   eating both plants and animal flesh.

Ontogeny   development of the individual.

Opisthocoelus   concave behind; a vertebra with the centrum convex anteriorly and concave posteriorly.

Opisthonephros   kidney structure developed from the middle and posterior parts of the archinephros; the functional adult kidney of most anamniotes.

Organelle   a structure within a cell's protoplasm that has a specific function.

OSHA   Occupational Safety and Health Administration.

Osmosis   the passage of a solvent through a semipermeable membrane.

Osmotic Pressure   the pressure exerted on one side of a semipermeable membrane resulting from differences in concentration of a substance between the two sides.

Osteoblast   a cell that forms new bone material or repairs old bone material.

Osteoclast   a cell that eliminates bone tissue not needed for skeletal strength and efficiency. After an injury or fracture has healed and left superfluous bony ridges.

Osteocyte   a bone cell carrying on continuous maintenance activity within the bone.

Osteomyelitis   an infection of bone tissue.

-ostomy   (suffix) opening.

oto-   (prefix) ear.

Ovary   the egg-producing organ in the female.

Oviparous   reproduction in which the young hatch from eggs outside the mother's body.

Ovoviviparous   reproduction in which the eggs are retained in the oviduct during the developmental period.

Ovulation   growth and discharge of an ovum.

Ovum   a female reproductive cell; an egg.

Oxidation   a chemical reaction in which electrons are removed from one molecule or atom, releasing energy.

Oxyhemoglobin   a bright red iron compound containing oxygen.

P

Pacinian Corpuscles   elliptical, semi-transparent bodies that occur along the nerves supplying the skin, especially of the hands and feet, and record deep pressure stimuli.

Paedogenesis   genetically fixed retention of larval characteristics in the adult.

Pain Receptors   bare nerve endings in the tissue cells of skin and other organs, though not present everywhere in the body.

Pancreas   a glandular organ secreting digestive juices and delivering them through a duct into the duodenum; special areas in the pancreas, called the Islands of Langerhans, manufacture insulin.

Papilla   any projection of tissue above a normal surface.

para-   (prefix) side.

Parasympathetic Nerves   nerves that are part of the automatic nervous system and that reverse the action of the sympathetic nerves.

-paresis   (suffix) weakness.

Parietal   situated in the outer layer or wall.

Pathway   the physical course a chemical or pollutant takes from the source to the organism exposed.

-pathy   (suffix) disease.

Peduncle   a stem or supporting structure.

PEL   (symbol) (OSHA) Permissible Exposure Limits developed by OSHA to indicate maximum airborne concentration of a contaminant to which an employee may be exposed over the duration specified.

Pentadactyl   having five digits, the basic pattern of the tetrapod limb, including the human limb.

Pentose   a five carbon sugar, such as ribose.

peri-   (prefix) around.

Perichondrium   the fibrous membrane covering cartilage.

Periosteum   a membrane covering the surface of bones, except where ligaments are attached or where bone becomes cartilaginous.

Peripheral Nervous System   the nervous system that continues on or from the central nervous system, and includes nereves running to the sense organs, heart, and internal organs, and the skeletal muscles.

Peristalsis   wave-like motion of esophagus, stomach, and intestines that keeps food and its residue moving along.

Peritoneum   a layer of tissue that lines the abdominal cavity and many of its organs.

Personal Measurement   a measurement collected from an individual's immediate environment, using active or passive devices to collect the samples.

Phagocyte   a type of WBC that devours bacteria and other foreign bodies.

Pharmacokinetic   the study of the time course of absorption, distribution, metabolism, and excretion of a foreign substance, such as a drug or a pollutant, in an organism's body.

Pharynx   the space at the back of the mouth, extending upward to meet the nasal cavities, continuous with the esophagus and the larynx going downward.

-phobia   (suffix) fear.

Photosynthesis   a process by which green plants convert solar energy to carbohydrates from carbon dioxide and water in the air.

Phylogeny   the evolutionary history of a group of organisms.

Pia Mater   the innermost of the meninges.

Placoid Scale   a type of dermal scale present in elasmobranchs.

Plantar   pertaining to the sole of the foot.

Plasma   the fluid part of blood.

Plexus   a network of interlacing nerves, blood vessels, or lymphatics.

PMN (EPA) Premanufacture Notification.

-pnea   (suffix) breathing.

pneumo-   (prefix) air, lung.

PNS   peripheral nervous system.

Poikilothermic   having variable body temperature.

Point-of-contact Measurement of Exposure   an approach to quantifying exposure by taking measurements of concentration over time at or near the point of contact between the chemical and an organism while the exposure is taking place.

poly-   (prefix) many.

Polylecithal   having a large amount of yolk.

Polysaccharides   a carbohydrate composed of linked monosaccharides.

Population   an aggregate of individuals of a species within a specified location in space and time.

Portal Vein   a large vein that passes from the digestive organs to the liver.

post-   (prefix) after.

Posterior   situated in back; at or toward the tail end of an animal.

Potential Dose   the amount of chemical contained in a material ingested, air breathed, or bulk material applied to the skin.

Potential Energy   stored energy in a body based on its location.

POTW   (EPA) Publicly Owned Treatment Works.

PPE   Personal Protective Equipment.

ppm (symbol) parts of vapor or gas per million parts of contaminated air by volume.

pre- (prefix) before.

Precision a measure of the reproducibility of a measured value under a given set of conditions.

Prehensile adapted for grasping and squeezing.

Primary Effect an effect where the stressor acts on the ecological component directly, not through the effects on other components within the ecosystem; synonym direct effect.

Probability Density Function (PDF) describe the relative likelihood that a variable will have different particular values of x. The probability that a variable will have a value within a small interval around x can be approximated by multiplying f(x), the value of y at x in a PDF plot, by the width of the interval.

Probability Samples samples collected from a statistical population such that each sample has a known probability of being selected.

Procoelus concave anteriorly; vertebra with the centrum concave anteriorly and convex posteriorly.

Proctodaeum the ectoderm lined, posterior portion of the embryonic gut.

Pronation movement of a hand or foot to a palm or sole downward position.

Prone lying horizontal, face down.

Prophase the first stage of mitotic division, during which the chromosomes appear.

Propioceptor a receptor whose function is connected with locomotion or posture.

Prosencephalon the most anterior embryonic division of the brain; the forebrain.

Protein a major component of living matter composed of carbon, hydrogen, nitrogen, and other elements.

Protein-bound Iodine (PBI) iodine present in protein molecules of the blood, reflecting thyroid hormone concentration.

Protoplasm the living matter of cells or tissues.

Proximal part nearest (the heart); nearest to the point of origin.

pseudo- (prefix) false.

Pterygium a little wing, describing fins and bones.

Pulmonary of the lungs.

Pulmonary Circulation circulation of blood through the lungs.

pulmuno- (prefix) lung.

pyelo- (prefix) kidney.

Pyloric Sphincter the muscle surrounding the opening at the lower end of the stomach.

pyo- (prefix) pus.

Q

Quadrupedal walking on four legs.

Quality Assurance (QA) an integrated system of activities involing planning, quality control, quality assessment, reporting, and quality improvement to ensure that a product or service, in this case the exposure assessment, meets defined standards of quality with a stated level of confidence.

Quality Control (QC) the overall system of technical activities whose purpose is to measure and control the quality of a product or service, in this case the exposure assessment, so that it meets the needs of the users. The aim of QC is to provide quality that is satisfactory, adequate, dependable, and economical.

Quantification Limit (QL) the concentration of an analyte in a specific matrix for which the probability of producing analytical values above the detection limit is 99%. QL describes the capability of the analyte of being detected, whereas the similar term LOD the ability of the analytical method or instrument, or both, to detect.

R

Ramus a branch of a vessel, nerve, or bone.

Random Samples samples selected from a statistical population such that each sample has an equal probability of being selected.

Range the difference between the largest and smallest values in a measurement data set.

RBC (abbreviation) red blood cells.

RCRA Resource Conservation and Recovery Act

Reasonable Worst Case a semiquantitative term referring to the lower portion of the high end of the exposure, dose, or risk distribution. The reasonable worst case has historically been loosely defined, including synonymously with maximum exposure, or worst case, and assessor's are cautioned by EPA to look for contextual definitions when encountering this term in the literature. Reasonable worst case should refer to a range that can be described as the 90th percentile in the distribution, but below about the 98th percentile.

Receptor (1) peripheral nerve endings in the skin, and special sense organs. (2) an entity that is exposed to a stressor.

Receptor Cells specialized cells that collect information for the organism.

Reconstruction of Dose an approach to quantifying exposure from internal dose, which is in turn reconstructed after exposure has occurred, from evidence within an organism such as chemical levels in tissues or fluids or from evidence of other biomarkers of exposure.

Recovery the rate and extent of return of a population or community to a condition that existed before the introduction of a stressor.

Red Blood Cells minute, circular discs in the blood that carry oxygen to the tissues and carbon dioxide away from them.

Red Corpuscles red blood cells.

Red Marrow bone marrow found in the interstices of cancellous bones.

Reflex an involuntary response in which a stimulus is received by a nerve, transmitted, and finally translated into muscular activity.

Reflex Arc the mechanism by which a reflex action occurs.

REL (symbol) (NIOSH) Recommended Exposure Limits issued by NIOSH to aid in controlling hazards in the workplace; generally expressed as 8- or 10-hour TWAs for a 40-hour workweek and/or ceiling levels with time limits ranging from instantaneous to 120 minutes.

Relative Risk Assessment estimating risks associated with different stressors or management actions.

Renal of the kidney.

Representativeness the degree to which a sample is, or samples are, characteristic of the whole medium, exposure, or dose for which the samples are being used to make an inference.

Respiration the total process by which oxygen is absorbed into the system and the oxidation products, carbon dioxide and water vapor, are released; the outward signs of this process are inhalation and exhalation.

retro- (prefix) behind.

rhino- (prefix) nose.

Rhombencephalon the posterior-most section of the embryonic brain; the hindbrain.

rhombo- (prefix) having a parallelogram or kite-shaped figure.

Ribonucleic Acid (RNA) nucleic acids found principally in the cytoplasm of cells.

Ribose a five-carbon sugar found in ribonucleic acid.

Ribosome a granule found in cytoplasm containing RNA.

Risk the probability that an untoward health effect may occur or an emergency situation develop as a result of a specified exposure or release of a compound of concern.

Risk Assessment a complex series of steps which describe the process by which scientific data are used to define individual health effects caused by exposure to chemicals (in this case).

Risk Cascade the series of interactions of exposures and effects resulting in secondary exposures, secondary effects, and ultimate effects, or causal chain, pathway, or network.

Risk Characterization the description of the nature and in many cases the magnitude of human risk with its uncertainties; a phase of ecological risk assessment that integrates the exposure and stressor response profiles to evaluate the likelihood of adverse ecological effects associated with exposure to a stressor.

Risk Management the process of setting and implementing policies to integrate the results of a risk assessment with engineering data and social, regulatory and legal concerns.

Rods specialized cells in the retina of the eye that are sensitive to fine degrees of light and dark.

Rostrum a beaklike structure or process.

Route the way a chemical or pollutant enters an organism after contact; ingestion, inhalation, dermal exposure, injection.

S

Sacrum the region of the vertebral column consisting of fused vertebrae that form part of the hip girdle.

Sagittal Plane a vertical line that divides the body into right and left halves.

Salivary Glands six glands, three on either side of the mouth, that secrete saliva, which moistens food during chewing and begins the digestive process.

Sample   a small part of something designed to show the nature or quality of the whole.

Sampling Frequency   the time interval between the collection of successive samples.

Sampling Plan   a set of rules or procedures specifying how a sample is to be selected and handled.

Scenario Evaluation   an approach to quantifying exposure by measurement or estimation of both the amount of a substance contacted, and the frequency/duration of contact, and subsequently linking these together to estimate exposure or dose.

-scopy   (suffix) see.

Sebaceous Gland   a gland that secretes an oil substance.

Sebum   the oily substance secreted by sebaceous glands.

Secondary Effect   an effect where the stressor acts on supporting components of the ecosystem, which in turn have an effect on the ecological component of interest; synonym indirect effect.

semi-   (prefix) half.

Sensory Nerve   a nerve that conveys sensations from the periphery of the organism to the central nervous system.

Sentinel Compound   a specific compound of a chemical class that is used to estimate the properties of other chemicals in the same class.

Septum   the wall of muscle separating the left and right portions of the heart; the partition between the two nostrils.

Serum   the clear, yellowish fluid that separates from blood after it clots; plasma that lacks fibrinogen.

Serum Albumin   a plasma protein that maintains osmotic pressure.

Serum Globulin   a plasma protein that contains antibodies.

SIC   Standard Industrial Classification

Signal   instruction or direction sent from the brain.

Skeletal Muscle   muscle that is attached to the skeleton.

Skeleton   the bony framework of the body.

Sinuses   hollow spaces or cavities in the skull.

Smoke   an aerosol of carbon or soot less than 0.1 microns in diameter from the incomplete combustion of carbonaceous materials; typically smoke contains both liquid droplets and solid (dry) particles.

Smooth Muscle   unstriated muscle tissue that acts independently of the will to govern automatic physical functions.

SNS   sensory nervous system.

SNUR   (EPA) Significant New Use Rule

Soft Palate   the rear portion of the roof of the mouth.

Solar Energy   the light or heat from the sun, which is capable, when harnessed, of doing work, providing fuel.

Source   an entity or action that releases to the environment or imposes on the environment a chemical (or physical or biological) stressor or stressors.

Source Characterization Measurements   measurements made to characterize the rate of release of agents into the environment from a source of emission such as an incinerator, landfill, industrial or municipal facility or sampling operation.

Source Rate the rate (amount per unit time) of a compound of concern that enters the atmosphere from a particular origin.

Source Regime (1) characterizes exposure to multiple chemicals or to both chemical and nonchemical stressors; (2) a synonym for exposure that is intended to avoid overemphasis on chemical exposures; or (3) a description of the series of interactions of exposures and effects resulting in secondary exposures, secondary effects, and, finally, ultimate effects or causal chain, pathway, or network.

Source Term the type, magnitude, and patterns of chemical stressors released.

Sperm a male reproductive cell.

Sphincters muscles surrounding certain parts of the intestinal tract that widen or narrow the lumen of the gut as required.

Spinal Column the bony vertebral column.

Spinal Cord the neerve structure running within the canal of the spinal collumn.

Standard Operating Procedure (SOP) a procedure adopted for repetitive use when performing a specific measurement or sampling operation.

Statistical Control the process by which the variability of measurements or data outputs of a system is controlled to the extent necessary to produce stable and reproducible results. Measurements are under statistical control when statistical evidence demonstrates that the critical variables in the measurement process are being controlled to such an extent that the system yields data that are reproducible within well-defined limits.

Statistical Significance an inference that the probability is low that the observed difference in quantities being measured could be due to artificial variability in the data rather than actual difference in the quantities themselves.

STEL (symbol) (OSHA/ACGIH) Short-Term Exposure Limit TLV; the highest concentration to which workers can be exposed to for a short period of time without suffering from either irritation, chronic or irreversible tissue change or narcosis of sufficient degree to increase accident proneness, impair self-rescue or materially reduce work efficiency provided that no more than four excursions above the TWA per day are allowed within this STEL limit.

Stimuli a goad to incite action, exertion, or response from an organism.

Stratum Lucidum the translucent layer of the epidermis.

Stressor a physical, chemical, or biological entity that induces an adverse response; synonym agent.

Stressor-Response Profile the product of characterization of ecological effects in the analysis phase of ecological risk assessment and which summarizes the data on effects of a stressor and the relationship of the data to the assessment endpoint.

Striated Muscle muscle that is attached to the skeleton.

Striped Muscle muscle that is attached to the skeleton.

sub- (prefix) under.

Subacute Exposure repeated exposures over a period of typically no more than 14 days; some scientists say up to one month.

Subchronic Exposure repeated exposures or continuous exposure lasting from one to 3 months.

Substance (OSHA) used interchangeably with material.

super- (prefix) above, greater.

Superficial near the surface.

Superior upper.

Supination turning the palm of the hand or sole of the foot upward.

Supine lying horizontally on the back, face upward.

supra- (prefix) above, greater.

Surrogate Data substitute data or measurements on one substance used to estimate analogous or corresponding values of another substance.

Suture line of joining or closure.

Sympathetic Nerves the portion of the automatic nervous system that stimulates the smooth muscles of the body to activity.

Synapse the intertwining of terminal branches of neurons so that nerve impulses may pass from one to the other.

Synovial Fluid the clear fluid normally present in joint cavities.

System a group of organs serving a common function.

Systemic of the body, taken as a whole.

Systole contraction of the heart muscles to squeeze blood out.

T

tachy- (prefix) fast.

Tachycardia a very rapid heart beat.

TCRI (EPA) Toxic Chemical Release Inventory (see TRI)

Telencephalon the anterior division of the brain.

Telolecithal yolk concentrated in one hemisphere of an egg.

Telophase the final stage in mitosis, in which a new cell membrane appears and the parent cell divides into half.

Tendon a bank of dense, fibrous tissue forming the end of a muscle that attaches the muscle to a bone.

Tetrapod having four appendages based on the pentadactyl plan; amphibians, reptiles, birds, mammals.

Thalamus either of two large masses of gray matter above the midbrain serving to relay incoming sensory impulses.

TLV (symbol) Threshold Limit Value a registered trademark for an exposure limit developed by the ACGIH; stated as a TWA, STEL or ceiling.

Total Fluid Intake the consumption of all types of fluids including tapwater, milk, soft drinks, alcoholic beverages, and the added moisture content of home-prepared or purchased foods.

Total Tapwater food and beverages that are prepared or reconstituted with tapwater, such as coffee, tea, frozen juices, and soups.

Toxicity the ability of a substance to damage biological
tissue.

Toxicology the study of the noxious effects of chemical and physical agents on the human body.

Trachea the windpipe; air from the nose or mouth passes into the trachea on the way to the larynx.

trans- (prefix) across.

trema- (prefix) opening.

TRI (EPA) Toxic Chemical Release Inventory (a variation).

Trophic Levels a functional classification of taxa within a community that is based on feeding relationships.

TSCA Toxic Substances Control Act, a law passed by Congress to protect human health and the environment by requiring testing and necessary use restrictions to regulate the commerce of certain chemical substances.

Tumor an abnormal mass resulting from a growth disturbance of cells.

TWA (symbol) (OSHA/ACGIH) time-weighted average TLV; the time-weighted average concentration for a normal 8-hour workday or a 40-hour workweek to which nearly all workers may be repeatedly exposed, day after day, with no effect.

U

Unguis non-living, corneous layer of a nail, hoof, or claw.

Upper Bound Estimate of Risk the risk assumed in the absence of detail toxicity data.

Uptake the process by which a substance crosses an absorption barrier and is absorbed into the body.

ur- (prefix) exretory system.

Uremia a disorder due to kidney failure.

-uria (suffix) urine.

urino- (prefix) excretory system.

uro- (prefix) excretory system.

V

Vacuole a space within a cell.

Valve a device that permits control of an opening so as to allow free passage in one direction, but deny passage in the other direction.

Vapor the gaseous phase of a material which is otherwise liquid at 77°F (25°C) and one atmosphere (760 torr).

Vas vessel or duct.

Vasomotor Nerves nerves that act upon blood vessels.

Veins the largest category of blood vessels returning blood to the heart.

Ventral toward the front of the body.

Venules the largest blood vessels returning blood to the heart except the veins.

Vertebra skeletal segments of the spinal column.

Vesicle small sac or bladder; blister.

Villus a small, fingerlike projection from the mucosa of the small intestine, through which digested food is absorbed by the blood.

Viviparous reproduction in which the young develop in the uterus of the female, receiving nourishment from the maternal blood stream.

VOC (EPA) Volatile Organic Chemical

Vocal Chords two transverse parallel folds of mucous membrane at the upper end of the larynx, which can either be relaxed toward the sides of the larynx during breathing or tightened and pulled toward each other to vibrate and produce sound.

Voluntary Muscle muscle that is attached to the skeleton; subject to the will.

W

WBC white blood cells.

White Blood Cells minute bodies in the blood, one-third larger than RBCs, which attack infection.

White Corpuscles white blood cells.

White Matter the part of the CNS having neurons with myelin.

WHO World Health Organization, part of the United Nations, with programs in occupational health information systems, early detection of health impairment, and international health-based permissible exposure limits for occupational exposure.

Whorl circular arrangement of like parts around a point on an axis, such as leaves, flowers, or skin ridges on fingerprints.

Worst Case a semiquantitative term referring to the maximum possible exposure, dose, or risk, that can conceivably occur, whether or not this exposure, dose, or risk actually occurs or is observed in a specific population. EPA cautions assessors to look for contextual definitions when encountering this term in literature as it has been loosely defined, historically.

Y

Yellow Marrow bone marrow that is found in the hollow, center shaft of cancellous bones.

Yolk nutritive material within the ovum.

Z

zygo- (prefix) yoke or pair.

Zygote a fetilized egg cell.

REFERENCES

American Medical Association. *The Wonderful Human Machine.* Chicago AMA, 1961.

Cornett, Frederick D. and Pauline Gratz. *Modern Human Physiology.* New York Holt, Rinehart and Winston, Publishers, 1982.

EPA. 40 CFR 300 — National Oil and Hazardous Substances Pollution Contingency Plan. *47 FR 31203.* July 16, 1982.

EPA. *Exposure Factors Handbook.* Rockville, MD Government Institutes, Inc., 1996.

EPA. "Proposed Guidelines for Ecological Risk Assessments." *61 FR 47552.* Sept. 9, 1996.

EPA. "EPA Guidelines on Exposure Assessment." *57 FR 22890.* May 29, 1992.

Goin, Coleman J. and Olive B. Goin. *Comparative Vertebrate Anatomy.* New York Barnes & Noble, Inc., 1965.

Hathaway, Gloria, Nick H. Proctor, James P. Hughes, and Michael L. Fischman, eds *Proctor and Hughes' Chemical Hazards of the Workplace.* 3d ed. New York Van Nostrand Reinhold, 1991.

Jayjock, Michael A. and Neil C. Hawkins. "A Proposal for Improving the Role of Exposure Modeling in Risk Assessment." *Am. Industrial Hygiene Assn. J.* December 1993.

OSHA. *OSHA Preamble and Proposed Rule to Revise Air Contaminants Standards for Construction, Maritime and Agriculture.* 57 FR 26002. June 12, 1992.

Talty, John T. ed. *Industrial Hygiene Engineering Recognition, Measurement, Evaluation, and Control.* 2nd ed. Park Ridge, NJ Noyes Data Corp., 1988.

Appendix I
Solubility of Common Contaminants

Chemical	Water, ppm	Chemical	Water, ppm
A		2-Butanone	249,000
Acenaphthene	3.42	Butyl Alcohol	74,600
Acetaldehyde	Infinite	secButyl Alcohol	184,000
Acetic Acid	Infinite	tertButyl Alcohol	Infinite
Acetic Anhydride	150,000	Butylamine	Infinite
Acetone	Infinite	secButylamine	Infinite
Acetonitrile	Infinite	Butylbenzene	13.8
Acetylene	1,000	Butylcyclohexane	0.178
Acrylic Acid	Infinite	in salt water	0.124
Acrylonitrile	73,500	Butylcyclopentane	0.431
Allyl Alcohol	Infinite	in salt water	0.301
Aniline	34,200	1-Butyne (Ethylacetylene)	2,870
Arochlor 1254	0.012	Butyraldehyde	83,700
		Butyronitrile	31,900
B			
Benzene	1,780	**C**	
Benzonitrile	10,000		
Benzo(a)pyrene	0.012	Carbon Dioxide	1,950
Benzo(g,h,i)perylene	0.0007	Carbon Disulfide	1,880
Benzoic Acid	3,390	Carbon Monoxide	23.8
Biphenyl	7	Carbon Tetrachloride	800
Bromobenzene	410	Carbonyl Sulfide	1,150
Bromobutane	616	Chlorobenzene	500
Bromodichloromethane	4,400	1-Chlorobutane	740
Bromoethane	9,000	2-Chlorobutane	1,000
Bromoform	3,010	Chlorodifluoromethane	2,770
Bromomethane	13,400	Chloroethane	9,050
2-Bromo-2-methylpropane	600	Chloroethene	2,700
1-Bromopentane	127	Chlorofluoromethane	10,500
1-Bromopropane	2,450	Chloroform	8,200
2-Bromopropane	2,870	Chloromethane	5,900
3-Bromo-1-propene	3,820	2-Chloro-2-methylbutane	3,320
1,3-Butadiene	735	1-Chloro-2-methylpropane	924
Butane	61.4	Chloropentafluoroethane	58
1-Butanethiol	600	1-Chloropentane	89.3
1-Butene	222	2-Chlorophenol	29,000
Butadyne	9,450	1-Chloropropane	2,700
		2-Chloropropane	3,100

3-Chloro-1-propene	4,000	1,2-Dichloroethane	8,680
Chlorotrifluoromethane	90	1,1-Dichloroethylene	3,350
m-Cresol	21,800	cis-1,2-Dichloroethylene	3,500
o-Cresol	24,500	trans-1,2-Dichloroethylene	6,300
p-Cresol	19,400	Dichlorofluoromethane	18,800
Cumene	50	Dichloromethane	19,400
Cyanogen	9,490	2,4-Dichlorophenoxyacetic Acid	620
Cycloheptane	30	1,2-Dichloropropane	2,750
Cycloheptatriene	620	1,3-Dichloropropane	2,730
Cyclohexane	56.1	Dichlorotetrafluoroethane	137
in salt water	39.19	Diethylamine	Infinite
Cyclohexanol	38,200	3,3-Diethylhexane	0.0659
Cyclohexanone	93,200	3,4-Diethylhexane	0.0744
Cyclohexene	213	3,3-Diethyl-2-methylpentane	0.0555
1-Cyclohexylheptane	0.014	m-Difluorobenzene	114
in salt water	0.01	o-Difluorobenzene	1,140
1-Cyclohexylhexane	0.025	p-Difluorobenzene	1,220
in salt water	0.018	1,1-Difluoroethane	2,500
1-Cyclohexylnonane	0.0095	Difluoroethene	165
in salt water	0.0066	Difluoromethane	4,390
1-Cyclohexyloctane	0.0101	Diiodomethane	1,240
in salt water	0.0071	Dimethyl Phthalate	4,320
Cyclooctane	7.9	Dimethylamine	620,000
Cyclopentane	160	2,2-Dimethylbutane	23.8
in salt water	111.8	2,3-Dimethylbutane	20.5
Cyclopentene	535	1,1-Dimethylcyclohexane	5.99
1-Cyclopentyldecane	0.0059	in salt water	4.18
in salt water	0.0041	cis-1,2-Dimethylcyclohexane	6
1-Cyclopentylheptane	0.016	in salt water	4.19
in salt water	0.011	trans-1,2-Dimethylcyclohexane	6
1-Cyclopentylhexane	0.038	cis-1,3-Dimethylcyclohexane	5.8
in salt water	0.027	in salt water	4.05
1-Cyclopentylnonane	0.0065	trans-1,3-Dimethylcyclohexane	4.49
in salt water	0.0045	in salt water	3.14
1-Cyclopentyloctane	0.009	cis-1,4-Dimethylcyclohexane	4.5
in salt water	0.0063	in salt water	3.15
1-Cyclopentylpentane	0.115	trans-1,4-Dimethylcyclohexane	3.84
in salt water	0.08	in salt water	2.68
Cyclopropane	538	1,1-Dimethylcyclopentane	22.53
		in salt water	15.74
D		cis-1,2-Dimethylcyclopentane	11.91
		in salt water	8.32
Decane	0.052	trans-1,2-Dimethylcyclopentane	18.16
Decyl Alcohol	37	in salt water	12.69
1,2-Dibromomethane	4,170	cis-1,3-Dimethylcyclopentane	19.27
1,2-Dibromopropane	1,430	in salt water	13.46
m-Dichlorobenzene	123	trans-1,3-Dimethylcyclopentane	18.3
o-Dichlorobenzene	92.3	in salt water	12.79
p-Dichlorobenzene (1,4-)	80	2,2-Dimethylheptane	0.381
Dichlorodifluoromethane	300	2,3-Dimethylheptane	0.252
1,1-Dichloroethane	8,700		

2,4-Dimethylheptane	0.377	Ethylcyclohexane	2.92
2,5-Dimethylheptane	0.319	in salt water	2.04
2,6-Dimethylheptane	0.333	Ethylcyclopentane	9.56
3,3-Dimethylheptane	0.303	in salt water	6.68
3,4-Dimethylheptane	0.25	3-Ethyl-2,2-dimethylhexane	0.111
3,5-Dimethylheptane	0.319	4-Ethyl-2,3-dimethylhexane	0.179
4,4-Dimethylheptane	0.333	3-Ethyl-2,3-dimethylhexane	0.0752
2,2-Dimethylhexane	1.5	4-Ethyl-2,3-dimethylhexane	0.0867
2,3-Dimethylhexane	0.945	3-Ethyl-2,4-dimethylhexane	0.0903
2,4-Dimethylhexane	1.29	4-Ethyl-2,4-dimethylhexane	0.0858
2,5-Dimethylhexane	1.37	3-Ethyl-2,5-dimethylhexane	0.123
3,3-Dimethylhexane	1.15	4-Ethyl-3,3-dimethylhexane	0.0783
3,4-Dimethylhexane	0.802	3-Ethyl-3,4-dimethylhexane	0.0816
2,4-Dimethyl-3-isopropylpentane	0.106	3-Ethyl-2,2-dimethylpentane	0.358
1,3-Dimethylnaphthalene	8	3-Ethyl-2,3-dimethylpentane	0.202
1,4-Dimethylnaphthalene	11.4	3-Ethyl-2,4-dimethylpentane	0.308
1,5-Dimethylnaphthalene	3.37	Ethylene	131
2,3-Dimethylnaphthalene	3	Ethylene Chloride	9,200
2,6-Dimethylnaphthalene	2	Ethylene Glycol	Infinite
2,2-Dimethyloctane	0.107	Ethylene Oxide	Infinite
2,3-Dimethyloctane	0.0729	Ethyleneimine	Infinite
2,4-Dimethyloctane	0.112	3-Ethylheptane	0.22
2,5-Dimethlyoctane	0.0981	4-Ethylheptane	0.242
2,6-Dimethyloctane	0.0891	3-Ethylhexane	0.809
2,7-Dimethyloctane	0.0914	bis(2-Ethylhexyl)phthalate	0.285
3,3-Dimethyloctane	0.0854	3-Ethyl-2-methylheptane	0.0854
3,4-Dimethyloctane	0.0763	4-Ethyl-2-methylheptane	0.11
3,5-Dimethyloctane	0.0936	5-Ethyl-2-methylheptane	0.0922
3,6-Dimethyloctane	0.0872	3-Ethyl-3-methylheptane	0.0748
4,4-Dimethyloctane	0.103	4-Ethyl-3-methylheptane	0.0811
4,5-Dimethyloctane	0.0814	3-Ethyl-5-methylheptane	0.0996
2,2-Dimethylpentane	4.4	3-Ethyl-4-methylheptane	0.0779
2,3-Dimethylpentane	5.25	4-Ethyl-4-methylheptane	0.0872
2,4-Dimethylpentane	4.41	3-Ethyl-2-methylhexane	0.287
3,3-Dimethylpentane	5.94	4-Ethyl-2-methylhexane	0.359
2,2-Dimethylpropane	33.2	3-Ethyl-3-methylhexane	0.25
2,6-Dinitrotoluene	1,320	3-Ethyl-4-methylhexane	0.253
1,4-Dioxane	431,000	3-Ethyl-2-methylpentane	0.943
para-Dioxane	Infinite	3-Ethyl-3-methylpentane	0.821
Dodecane	0.0037	1-Ethylnaphthalene	10.7
Dodecyl Alcohol	4	2-Ethylnaphthalene	8
		3-Ethyloctane	0.0652
E		4-Ethyloctane	0.0754
Eicosane	0.0019	3-Ethylpentane	3.02
Ethane	60.4	Ethylsulfide	3,120
Ethanethiol	14,800	o-Ethyltoluene	93.1
Ethyl Acetate	73,700	p-Ethyltoluene	94.9
Ethyl Alcohol	Infinite	3-Ethyl-2,2,3-trimethylpentane	0.0561
Ethyl Benzene	165	3-Ethyl-2,2,4-trimethylpentane	0.116
Ethyl Ether	60,900	3-Ethyl-2,3,4-trimethylpentane	0.0563

F

Fluorobenzene	1,550
Fluoroethane	2,160
Fluoromethane	2,390
1-Fluoropropane	3,860
2-Fluoropropane	3,660
Formic Acid	Infinite
Furan	9,900

H

Heptachlor	0.18
Heptadecane	0.0012
1-Heptadecanol	0.008
Heptanal	1,520
Heptane	2.24
Heptene	18.2
Heptyl Alcohol	1,740
Heptyne	94
Hexachlorobenzene	0.11
Hexachloroethane	50
Hexadecane	0.0009
1-Hexadecanol	0.035
Hexafluoroethane	7.9
Hexanal	5,640
Hexane	13.3
2-Hexanone	14,000
1-Hexene	69.7
Hexyl Alcohol	5,870
Hexylbenzene	1.02
1-Hexyne	688

I

Iodobenzene	201
Iodoethane	3,900
Iodomethane	26,100
1-Iodopropane	1,070
2-Iodopropane	1,400
Isophorone	12,000
4-Isopropylheptane	0.0961
3-Isopropyl-2-methylhexane	0.0646

M

Mesitylene	48.2
Methane	24.4
Methanethiol	24,000
Methanol	Infinite
2-Methyl 1,3-Butadiene	642
3-Methyl-1-Butene	130
Methyl tert-Butyl Ether	52,100

Methyl Ether	65,200
Methyl Ethyl Ketone	268,000
Methyl Formate	23,800
Methyl Naphthalene	25.4
2-Methyl-1-pentene	78
4-Methyl-1-pentene	48
Methyl Propyl Ether	30,500
Methyl isoPropyl Ether	65,100
Methyl Sulfide	19,600
Methylamine	548,000
2-Methylbutane (isoPentane)	47.8
Methylcyclohexane	16
in salt water	11.18
Methylcyclopentane	42.6
in salt water	29.2
Methylene Chloride	20,000
2-Methylheptane	0.849
3-Methylheptane	0.792
4-Methylheptane	0.843
2-Methylhexane	2.54
3-Methylhexane	2.64
1-Methylnaphthalene	28
2-Methylnaphthalene	25.4
2-Methylnonane	0.0636
3-Methylnonane	0.0611
4-Methylnonane	0.0679
5-Methylnonane	0.07
2-Methyloctane	0.22
3-Methyloctane	0.207
4-Methyloctane	0.115
2-Methylpentane	13.8
3-Methylpentane	17.9
2-Methylpropane (isoButane)	48.9
2-Methylpropene	263
m-Methylstyrene	89
p-Methylstyrene	89

N

Naphthalene	32.1
Nitrobenzene	1,900
Nitroethane	50,100
Nitromethane	10,000
1-Nitropropane	13,800
2-Nitropropane	16,700
Nonadecane	0.0016
Nonanal	105
Nonane	0.122
1-Nonene	1.12
Nonyl Alcohol	130
1-Nonyne	7.2

O

Octadecane	0.0021
1-Octadecanol	0.0011
Octafluorocyclobutane	50
Octanal	370
Octane	4.31
1-Octene	4.1
Octyl Alcohol	540
1-Octyne	24

P

Pentachloroethane	500
Pentachlorophenol	14
Pentadecane	0.0015
1-Pentadecanol	0.089
1,4-Pentadiene	558
2,2,3,3,4-Pentamethylpentane	0.0667
2,2,3,4,4-Pentamethylpentane	0.0924
Pentane	38.5
2-Pentanone	55,400
1-Pentene	148
cis-2-Pentene	203
trans-2-Pentene	203
Pentyl Alcohol	22,000
tertPentyl Alcohol	110,000
Pentylbenzene	3.84
Pentylcyclohexane	0.059
in salt water	0.041
Pentyne	786
Phenol	93,000
2-Picoline	Infinite
3-Picoline	Infinite
Propane	62.4
Propene	200
Propionaldehyde	405,000
Propionitrile	93,400
Propyl Alcohol	Infinite
isoPropyl Alcohol	Infinite
Propyl Ether	3,820
isoPropyl Ether	11,200
Propylamine	Infinite
Propylbenzene	52.2
Propylcyclohexane	0.677
in salt water	0.473
Propylcyclopentane	2.04
in salt water	1.425
Propylene Oxide	259,000
4-Propylheptane	0.103
Propyne (Methylacetylene)	3,640

Pyridine	Infinite
Pyrrolidine	Infinite

S

Styrene	322

T

1,1,2,2-Tetrachloroethane	3,000
Tetrachloroethylene (Perchloroethylene; Tetrachloroethene)	150
Tetradecane	0.0022
1-Tetradecanol	0.3
Tetrafluoroethene	158
Tetrahydrofuran	0.3
1,2,4,5-Tetramethylbenzene	3.48
2,2,3,3-Tetramethylbutane	1.53
2,2,3,3-Tetramethylhexane	0.0893
2,2,3,4-Tetramethylhexane	0.0966
2,2,3,5-Tetramethylhexane	0.166
2,2,4,4-Tetramethylhexane	0.125
2,2,4,5-Tetramethylhexane	0.17
2,2,5,5-Tetramethylhexane	0.295
2,3,3,4-Tetramethylhexane	0.0718
2,3,3,5-Tetramethylhexane	0.13
2,3,4,4-Tetramethylhexane	0.0837
2,3,4,5-Tetramethylhexane	0.11
3,3,4,4-Tetramethylhexane	0.0547
2,2,3,3-Tetramethylpentane	0.255
2,2,3,4-Tetramethylpentane	0.374
2,2,4,4-Tetramethylpentane	0.663
2,3,3,4-Tetramethylpentane	0.238
Thiophene	3,020
Toluene	542
1,2,4-Trichlorobenzene	30
1,1,1-Trichloroethane	4,400
1,1,2-Trichloroethane	4,500
Trichloroethylene	1,100
Trichlorofluoromethane	1,080
2,4,6-Trichlorophenol	800
1,2,3-Trichloropropane	1,900
1,1,2-Trichlorotrifluoroethane	170
Tridecane	0.0035
Triethylamine	7,290
Trifluoromethane	900
Trifluorotoluene	450
Triiodomethane	100
Trimethylamine	291,000
1,2,3-Trimethylbenzene	65.5
1,2,4-Trimethylbenzene	57

2,2,3-Trimethylbutane	5.67	2,2,5-Trimethylhexane	0.54
cis-cis-1,3,5-Trimethylcyclohexane	1.93	2,3,3-Trimethylhexane	0.292
in salt water	1.37	2,3,4-Trimethylhexane	0.272
cis-trans-1,3,5-Trimethylcyclohexane	1.74	2,3,5-Trimethylhexane	0.409
in salt water	1.22	2,4,4-Trimethylhexane	0.425
2,2,3-Trimethylheptane	0.103	3,3,4-Trimethylhexane	0.252
2,2,4-Trimethylheptane	0.167	2,2,3-Trimethylpentane	1.27
2,2,5-Trimethylheptane	0.146	2,2,4-Trimethylpentane	2.22
2,2,6-Trimethylheptane	0.161	2,3,3-Trimethylpentane	0.99
2,3,3-Trimethylheptane	0.0899	2,3,4-Trimethylpentane	2.30
2,3,4-Trimethylheptane	0.0913		
2,3,5-Trimethylheptane	0.0876	**U**	
2,3,6-Trimethylheptane	0.112	Undecane	0.044
2,4,4-Trimethylheptane	0.145		
2,4,5-Trimethylheptane	0.109	**V**	
2,4,6-Trimethylheptane	0.173		
2,5,5-Trimethylheptane	0.132	Valeraldehyde	22,000
3,3,4-Trimethylheptane	0.0824	Vinyl Chloride	2,670
3,3,5-Trimethylheptane	0.113		
3,4,4-Trimethylheptane	0.0858	**X**	
3,4,5-Trimethylheptane	0.0799	m-Xylene	174
2,2,3-Trimethylhexane	0.363	o-Xylene	221
2,2,4-Trimethylhexane	0.529	p-Xylene	202

REFERENCES

Ayers, Kenneth W. *et al. Environmental Science and Technology Handbook.* Rockville, MD: Government Institutes, Inc. 1994.

Kaczmar, Swistoslav W., Edwin C. Tifft, Jr. and Cornelius B. Murphy, Jr. "Site Assessment under CERCLA: The Importance of Distinguishing Hazard from Risk." *Hazardous Materials Control Monograph Series: Health Assessment.* Silver Spring, MD: Hazardous Materials Control Research Institute.

Lide, David R. ed. *Basic Laboratory and Industrial Chemicals: A CRC Quick Reference Handbook.* Boca Raton, FL: CRC Press. 1993.

"Tech Resources." *Environmental Technology.* Vol. 6, Issue 7, 1997 Resource Guide.

Yaws, Carl L. and Haur-Chung Yang. "Water Solubility Data for Organic Compounds." *Pollution Engineering.* October 1990, pp. 70-5.

Yaws, Carl L., Haur-Chung Yang, Jack R. Hopper, and Keith C. Hanson. "232 Hydrocarbons: Water Solubility Data." *Chemical Engineering.* April 1990, pp. 177–180.

Yaws, Carl L. and Xiaoyan Lin. "Solubility of Hydrocarbons in Salt Water." *Pollution Engineering.* January 1994, pp. 70–2

Appendix II
Vapor Pressure of Common Contaminants

Chemical	Vapor Pressure, mmHg				
	0°C	20°C	25°C	100°C	BP °C
A					
Acetaldehyde		740			21.0
Acetic Acid		11			117.7
Acetic Anhydride		4			138.9
Acetone		180			56.1
Acetonitrile		73			81.7
Acetylene Tetrabromide		0.02			245.6
Acrolein		210			52.8
Acrylamide		0.007			237.5
Acrylonitrile	29.8	83	105.8	1,455	77.2
Allyl Alcohol		17			96.1
Allyl Chloride		295			45.0
Allyl Glycidyl Ether		2			153.9
n-Amyl Acetate		5			149.4
sec-Amyl Acetate		7			120.6
Aniline		0.6			
B					
Benzene	24.7	75	95.3	1,343	80
Benzyl Chloride		1[A]			178.9
Bromine		172			59.4
Bromoform		5			149.4
2-Butxoyethanol		0.8			170.6
n-Butyl Acetate		15			125.6
sec-Butyl Acetate			24		112.2
tert-Butyl Acetate					97.8
n-Butyl Alcohol		6			117.2
sec-Butyl Alcohol			24[B]		100
tert-Butyl Alcohol			42		82.2
Butylamine		82			77.8
n-Butyl Glycidyl Ether		3			163.9
Butyl Mercaptan		35			98.3
p-tert-Butyl Toluene		0.7			192.8
C					
Carbon Disulfide		297			46.7
Carbon Tetrachloride		91			76.7
Chloroacetaldehyde		100			85.6
α-Chloroacetophenone		0.01			244.4

Chemical	Vapor Pressure, mmHg				
	0°C	20°C	25°C	100°C	BP °C
Chlorobenzene			12		132.2
o-Chlorobenzylidene malononitrile		1			312.5
Chlorobromomethane			160		71.1
Chlorodiphenyl (42% chlorine)		0.001			345.6
Chlorodiphenyl (54% chlorine)		0.00006			377.5
Chloroform		160			61.7
bis-Chloromethyl Ether		30^A			106.1
1-Chloro-1-nitropropane			6		142.8
Chloropicrin		20			112.2
β-Chloroprene		188			59.4
Cresol (all isomers)			1^C		196.9
Crotonaldehyde		19			103.9
Cumene			5		152.2
Cyclohexane			98		80.6
Cyclohexanol		1			161.1
Cyclohexanone			5		155.6
Cyclohexene			160^C		82.8
D					
Decaborane			0.05		212.8
Diacetone Alcohol		1			167.8
1,2-Dibromo-3-chloropropane		0.8			195.6
Dibutyl Phosphate		1			100
o-Dichlorobenzene		1			180.6
p-Dichlorobenzene			0.4		173.9
1,1-Dichloroethane		71.6	228.8		57.2
cis-1,2-Dichloro-ethylene			201		82.2
trans-1,2-Dichloro-ethylene	102.3		331.5		
Dichloroethyl Ether		0.7			177.8
1-Dichloro-1-nitro-ethane		15			123.9
Diethylamine		192			55.6
2-Diethylamino-ethanol		21			162.8
Difluorodibromo-methane		620			24.4
Diglycidyl Ether			0.09		260
Diisobutyl Ketone		2			167.8
Diisopropylamine		70			83.9
Dimethylaniline			1^C		192.2
Dimethylformamide			4		152.8
1,1-Dimethylhydrazine			157		63.9

Chemical	Vapor Pressure, mmHg				
	0°C	20°C	25°C	100°C	BP °C
Dimethylphthalate		0.01			238.9
Dimethyl Sulfate		0.1			187.8
Dinitro-o-cresol		0.00005			312.2
Dinitrotoluene		1			300
Dioxane		29			101.1
Diphenyl			1[C]		253.9
Dipropylene Glycol Methyl Ether			0.4[D]		190
E					
Epichlorohydrin		13			116.7
Ethanolamine		0.4			170.6
2-Ethoxyethanol		4			135
2-Ethoxyethylacetate		2			156.1
Ethyl Acetate		74			77.2
Ethylacrylate		29			99.4
Ethyl Benzene			9.6[D]	256.5	136.1
Ethyl Bromide		400[E]			38.3
Ethyl Butyl Ketone		4			147.8
Ethylene Chlorohydrin		5			127.8
Ethylenediamine		11			116.1
Ethylene Dibromide		12			131.1
Ethylene Dichloride		64			83.3
Ethylene Glycol Dinitrate		0.05			197.2
Ethyleneimine		160			56.1
Ethyl Ether		440			34.4
Ethyl Formate		200			54.4
Ethyl Mercaptan		442			35
N-Ethylmorpholine		6			138.3
Ethyl Silicate		1			168.9
F					
Fluorotrichloromethane		690			23.9
Formic Acid		35			106.7
Furfural		1[F]			161.7
Furfural Alcohol			1[G]		170
G					
Glycidol			0.9		160
H					
n-Heptane		40[A]			98.3
n-Hexane			150		68.9
2-Hexanone			4		127.8
sec-Hexyl Acetate		3			147.2
Hydrazine		10			113.3
Hydroquinone		0.00001			285
I					
Iodine			0.3		185
Isoamyl Acetate		4			142.2

Chemical	Vapor Pressure, mmHg				
	0°C	20°C	25°C	100°C	BP °C
Isobutyl Acetate		13			117.2
Isobutyl Alcohol		9			108.3
Isophorone			0.4		215
Isopropyl Acetate		42			90
Isopropyl Alcohol		33			82.8
Isopropylamine		460			32.8
Isopropyl Glycidyl Ether			9		137.2
M					
Maleic Anhydride		0.2			202.2
Mercury		0.0012			356.7
Mesityl Oxide		9			130
2-Methoxyethanol		6			124.4
2-Methoxyethanol Acetate		2			145
Methyl Acetate		173			57.2
Methyl Acrylate		65			80
Methylal		330			43.9
Methyl Alcohol		92			63.9
Methyl n-Amyl Ketone		3			151.7
Methylcyclohexane		43[A]			101.1
Methylcyclohexanol			2[B]		167.5
o-Methylcyclohexanone		1			162.8
Methylene bis-Phenyl Isocyanate			0.001[H]		172.2
Methylene Chloride		350			40
Methyl Ethyl Ketone		71			79.4
Methyl Formate		476			31.7
3-Methyl-3-heptanone		2			157.2
Methylhydrazine			50		87.8
Methyl Iodine			400		42.8
Methyl Isobutyl Carinol		3			132.8
Methyl Isobutyl Ketone		16			116.7
Methyl Isocyanate		348			59.4
Methyl Methacrylate			40		101.1
α-Methyl Styrene		2			165.6
Monomethyl Aniline			1[I]		195.6
Morpholine		6			128.9
N					
Naphtha (Petroleum Distillates)		40			133.9
Naphthalene		0.08			217.8
α-Naphthylamine				1[J]	300.6
β-Naphthylamine				1[K]	306.1
Nicotine		0.08			250
Nitric Acid		48			82.8
p-Nitroaniline			1[L]		332.2
Nitrobenzene			1[M]		210.6
Nitroethane		16			113.9

Chemical	Vapor Pressure, mmHg				
	0°C	20°C	25°C	100°C	BP °C
Nitrogen Dioxide		720			21.1
Nitromethane		28			101.1
1-Nitropropane		8			131.7
2-Nitropropane		13			120.6
N-Nitrosodimethylamine		3			152.2
m-Nitrotoluene			1^N		232.2
o-Nitrotoluene			1^N		222.2
p-Nitrotoluene			1^N		237.8
O					
Octane		10			125.6
Osmium Tetroxide			11^P		130
P					
Pentaborane			200		60
Pentachlorophenol		0.001			308.9
n-Pentane		400^R			36.1
2-Pentanone		16			101.7
Perchloromethyl					
Mercaptan		65			147.2
Phenol		0.4		39.9	181.7
Phenyl Ether			0.02		258.9
Phenyl Glycidyl Ether		0.01			245
Phenylhydrazine			1^S		243.3
Phosphoric Acid		0.03			212.8
Phosphorus (Yellow)		0.03			280
Phosphorus Trichloride		100^E			76.1
β-Propiolactone			3		161.7
n-Propyl Acetate			40^T		101.7
n-Propyl Alcohol			21		97.2
Propylene Dichloride		40			96.7
Propyleneimine		112			66.7
Propylene Oxide		445			34.4
n-Propyl Nitrate		18			110.6
Pyridine			20		115.6
S					
Styrene		5	6.1	191.3	145
Sulfur Monochloride		7			137.8
Sulfur Pentafluoride		561			28.9
T					
1,1,2,2-Tetrachloro-1,2-					
difluoroethane		40			92.8
1,1,2,2-Tetrachloroethane			4.7	186.8	146.7
Tetrachloroethylene		14	18.2	406.5	121.1
(Perchloroethylene)					
Tetraethyl Lead		0.2			108.9
Tetrahydrofuran		132			66.1
Tetramethyl Lead		23			100

	Vapor Pressure, mmHg				
Chemical	0°C	20°C	25°C	100°C	BP °C
Tetranitromethane		8			126.1
Toluene		20R	28.4	559.5	111.1
Toluene-2,4-diisocyanate			0.01		251.1
o-Toluidine		0.3			200
1,1,1-Trichloroethane		100			73.9
1,1,2-Trichloroethane		19	23.3	499.5	113.9
Trichloroethylene	30.3	58	71.3		87.2
1,2,3-Trichloropropane			10U		328.9
1,1,2-Trichloro-1,2,2-trifluoroethane		285			47.8
Triethylamine		54			89.4
Tri-o-cresol Phosphate			0.00002		410
Turpentine			5		161.9
V					
Vinyl Toluene		1			170.6
X					
m-Xylene		9			131.7
o-Xylene		7			144.4
p-Xylene		9			138.3

Legend:

A: @ 22.2°C	J: @ 104.4°C
B: @ 30°C	K: @ 107.8°C
C: @ 37.8°C	L: @ 142.2°C
D: @ 26.1°C	M: @ 44.4°C
E: @ 21.1°C	N: @ 50.0°C
F: @ 18.9°C	P: @ 27.2°C
G: @ 31.7°C	R: @ 18.3°C
H: @ 40°C	S: @ 71.7°C
I: @ 36.1°C	T: @ 28.8°C
	U: @ 46.1°C

REFERENCES

Ayers, Kenneth W. *et al. Environmental Science and Technology Handbook.* Rockville, MD: Government Institutes, Inc. 1994.

Kaczmar, Swistoslav W., Edwin C. Tifft, Jr. and Cornelius B. Murphy, Jr. "Site Assessment under CERCLA: The Importance of Distinguishing Hazard from Risk." *Hazardous Materials Control Monograph Series: Health Assessment.* Silver Spring, MD: Hazardous Materials Control Research Institute.

Lide, David R. ed. *Basic Laboratory and Industrial Chemicals: A CRC Quick Reference Handbook.* Boca Raton, FL: CRC Press. 1993.

NIOSH Pocket Guide to Chemical Hazards. U.S. Department of Health and Human Services. Public Health Service. Centers for Disease Control. National Institute for Occupational Health and Safety. June 1990.

Appendix III
Henry's Law Constant of Common Contaminants

Chemical	Mol. Wt.	Henry's Law Constant
A		
Acenaphthene	154.210	0.00510
Acetaldehyde	44.053	0.0001
Acetone	58.080	0.0017
Acetonitrile	41.052	0.00002
Acetylene (Ethyne)	26.038	0.0252
Arochlor 1254		0.15
B		
Benzene	78.114	0.23
Benzo(a)pyrene	252.320	0.0001
Benzonitrile	103.123	0.0005
Biphenyl	154.211	0.0009
Bromobenzene	157.010	0.0021
1-Bromobutane	137.019	0.012
Bromodichloromethane	163.830	0.127
Bromoethane	108.966	0.0074
Bromoform	252.75	0.035
Bromomethane	94.939	0.0068
2-Bromo-2-methylpropane	137.019	0.031
1-Bromopentane	151.046	0.02
1-Bromopropane	122.992	0.0072
2-Bromopropane	122.992	0.0097
3-Bromo-1-propene	120.977	0.0059
1,3-Butadiene	54.091	0.071
Butadiyne	50.060	0.51
Butane	58.123	0.917
1-Butanethiol	90.183	0.009
1-Butene	56.107	0.245
Butylbenzene	134.221	0.013
1-Butyne (Ethylacetylene)	54.091	0.018
C		
Carbon Dioxide	44.010	0.022
Carbon Disulfide	76.131	0.019
Carbon Monoxide	28.010	1.14

Chemical	Mol. Wt.	Henry's Law Constant
Carbon Tetrachloride	153.823	1.282
Carbon Tetrafluoride	88.005	5.33
Carbonyl Sulfide	60.070	0.051
Chlorobenzene	112.558	0.17
1-Chlorobutane	92.568	0.017
2-Chlorobutane	92.568	0.018
Chlorodifluoromethane	86.469	0.030
Chloroethane	64.514	0.034
Chloroethene	62.499	0.022
Chlorofluoromethane	68.478	0.0065
Chloroform	119.378	0.171
Chloromethane	50.488	0.0083
2-Chloro-2-methylbutane	106.595	0.0032
1-Chloro-2-methylpropane	92.568	0.016
Chloropentafluoroethane	154.467	2.58
1-Chloropentane	106.595	0.049
2-Chlorophenol	128.560	0.00093
1-Chloropropane	78.541	0.011
2-Chloropropane	78.541	0.014
3-Chloro-1-propene	76.525	0.0091
Chlorotrifluoromethane	104.459	1.12
Cumene	120.194	0.0145
Cyanogen	52.035	0.0053
Cycloheptane	98.188	0.0936
1,3,5-Cycloheptatriene	92.140	0.0046
Cyclohexane	84.161	0.193
Cyclohexene	82.145	0.045
Cyclooctane	112.214	0.104
Cyclopentane	70.134	0.188
Cyclopentene	68.118	0.064
Cyclopentylpentane	140.268	1.83
Cyclopropane	42.080	0.076

D

Chemical	Mol. Wt.	Henry's Law Constant
Decane	142.284	4.72
1,2-Dibromomethane	187.862	0.0007
1,2-Dibromopropane	201.888	0.0015
m-Dichlorobenzene	147.004	0.0032
o-Dichlorobenzene	147.004	0.003
p-Dichlorobenzene (1,4-)	147.004	0.104
Dichlorodifluoromethane	120.914	0.390
1,1-Dichloroethane	98.959	0.24
1,2-Dichloroethane	98.959	0.051
1,1-Dichloroethylene	96.943	1.841
cis-1,2-Dichloroethylene	96.943	0.31
trans-1,2-Dichloroethylene	96.943	0.429
Dichlorofluoromethane	102.923	0.0052
Dichloromethane	84.933	0.0025
1,2-Dichloropropane	112.986	0.0027

Chemical	Mol. Wt.	Henry's Law Constant
1,3-Dichloropropane	112.986	0.001
1,2-Dichlorotetrafluoroethane	170.922	1.21
3,3-Diethylhexane	142.284	7.73
3,4-Diethylhexane	142.284	7.19
3,3-Diethyl-2-methylpentane	142.284	8.59
3,3-Diethylpentane	128.257	6.55
m-Difluorobenzene	114.094	0.078
o-Difluorobenzene	114.094	0.007
p-Dichlorobenzene	114.094	0.0076
1,1-Difluoroethane	66.050	0.026
Difluoroethene	64.035	0.388
Difluoromethane	52.024	0.0115
Diiodomethane	267.836	0.0004
2,2-Dimethylbutane	86.177	1.52
2,3-Dimethylbutane	86.177	1.30
cis-1,2-Dimethylcyclohexane	112.214	0.356
trans-1,2-Dimethylcyclohexane	112.214	0.477
trans-1,4-Dimethylcyclohexane	112.214	0.872
2,2-Dimethylheptane	128.257	4.79
2,3-Dimethylheptane	128.257	5.27
2,4-Dimethylheptane	128.257	4.63
2,5-Dimethylheptane	128.257	4.82
2,6-Dimethylheptane	128.257	4.66
3,3-Dimethylheptane	128.257	5.27
3,4-Dimethylheptane	128.257	5.34
3,5-Dimethylheptane	128.257	4.87
4,4-Dimethylheptane	128.257	5.22
2,2-Dimethylhexane	114.230	3.41
2,3-Dimethylhexane	114.230	3.73
2,4-Dimethylhexane	114.230	3.52
2,5-Dimethylhexane	114.230	3.34
3,3-Dimethylhexane	114.230	3.75
3,4-Dimethylhexane	114.230	4.05
2,4-Dimethyl-3-isopropylpentane	142.284	7.59
1,3-Dimethylnaphthalene	156.227	0.0007
1,4-Dimethylnaphthalene	156.227	0.0005
1,5-Dimethylnaphthalene	156.227	0.0006
2,3-Dimethylnaphthalene	156.227	0.0006
2,6-Dimethylnaphthalene	156.227	0.0012
2,2-Dimethyloctane	142.284	5.93
2,3-Dimethyloctane	142.284	6.39
2,4-Dimethyloctane	142.284	5.70
2,5-Dimethyloctane	142.284	6.00
2,6-Dimethyloctane	142.284	6.19
2,7-Dimethyloctane	142.284	5.71
3,3-Dimethyloctane	142.284	6.65
3,4-Diemthyloctane	142.284	6.52
3,5-Dimethyloctane	142.284	6.26
3,6-Dimethyloctane	142.284	6.08

Chemical	Mol. Wt.	Henry's Law Constant
4,4-Dimethyloctane	142.284	6.53
4,5-Dimethyloctane	142.284	6.57
2,2-Dimethylpentane	100.203	3.15
2,3-Dimethylpentane	100.203	1.73
2,4-Dimethylpentane	100.203	2.94
3,3-Dimethylpentane	100.203	1.84
2,2-Dimethylpropane	72.150	2.105
2,6-Dinitrotoluene	182.140	0.0002
1,4-Dioxane	88.106	0.0006
Dodecane	170.337	7.18
E		
Eicosane	282.552	0.0003
Ethane	30.069	0.482
Ethanethiol	62.129	0.0029
Ethyl Benzene	106.167	0.359
3-Ethyl-2,2-dimethylhexane	142.284	7.10
4-Ethyl-2,2-dimethylhexane	142.284	6.11
3-Ethyl-2,3-dimethylhexane	142.284	7.78
4-Ethyl-2,3-dimethylhexane	142.284	7.16
3-Ethyl-2,4-dimethylhexane	142.284	7.18
4-Ethyl-2,4-dimethylhexane	142.284	7.48
3-Ethyl-2,5-dimethylhexane	142.284	6.52
4-Ethyl-3,3-dimethylhexane	142.284	7.79
3-Ethyl-3,4-dimethylhexane	142.284	7.81
3-Ethyl-2,2-dimethylpentane	128.257	5.33
3-Ethyl-2,3-dimethylpentane	128.257	6.45
3-Ethyl-2,4-dimethylpentane	128.257	5.48
Ethylene	28.054	0.208
Ethylene Chloride	98.960	0.038
3-Ethylheptane	128.257	5.17
4-Ethylheptane	128.257	5.17
3-Ethylhexane	114.230	3.73
3-Ethyl-2-methylheptane	142.284	6.56
4-Ethyl-2-methylheptane	142.284	6.22
5-Ethyl-2-methylheptane	142.284	6.31
3-Ethyl-3-methylheptane	142.284	7.27
4-Ethyl-3-methylheptane	142.284	6.87
3-Ethyl-5-methylheptane	142.284	6.18
3-Ethyl-4-methylheptane	142.284	6.85
4-Ethyl-4-methylheptane	142.284	7.10
3-Ethyl-2-methylhexane	128.257	5.19
4-Ethyl-2-methylhexane	128.257	4.81
3-Ethyl-3-methylhexane	128.257	5.70
3-Ethyl-4-methylhexane	128.257	5.37
3-Ethyl-2-methylpentane	114.230	3.81
3-Ethyl-3-methylpentane	114.230	4.21
1-Ethylnaphthalene	156.227	0.0004
2-Ethylnaphthalene	156.227	0.0006

Chemical	Mol. Wt.	Henry's Law Constant
3-Ethyloctane	142.284	6.20
4-Ethyloctane	142.284	6.29
3-Ethylpentane	100.203	2.54
Ethylsulfide	90.183	0.0017
o-Ethyltoluene	120.194	0.0042
p-Ethyltoluene	120.194	0.0049
3-Ethyl-2,2,3-trimethylpentane	142.284	9.76
3-Ethyl-2,2,4-trimethylpentane	142.284	7.62
3-Ethyl-2,3,4-trimethylpentane	142.284	9.26

F

Fluorobenzene	96.104	0.0062
Fluoroethane	48.060	0.0222
Fluoromethane	34.033	0.014
1-Fluoropropane	62.087	0.016
2-Fluoropropane	62.087	0.017
Furan	68.075	0.0054

H

Heptachlor	0.046	
Heptadecane	240.471	0.053
Heptanal	114.187	0.0004
Heptane	100.203	2.700
1-Heptene	98.188	0.401
1-Heptyne	92.172	0.0707
Hexachlorobenzene	284.784	0.412
Hexachloroethane	236.740	0.138
Hexadecane	226.445	0.228
Hexafluoroethane	138.012	16.93
Hexanal	100.160	0.0005
Hexane	86.177	1.291
2-Hexanone	0.0016	
1-Hexene	84.161	0.296
Hexylbenzene	162.274	0.01
1-Hexyne	82.145	0.0214

I

Iodobenzene	204.010	0.0013
Iodoethane	155.966	0.0054
Iodomethane	141.939	0.0028
1-Iodopropane	169.993	0.0083
2-Iodopropane	169.993	0.0089
Isophorone	138.210	0.0003
Isopropyl Ether	102.176	0.0017
4-Isopropylheptane	6.54	
3-Isopropyl-2-methylhexane	142.284	8.91

M

Mesitylene	120.194	0.0079

Chemical	Mol. Wt.	Henry's Law Constant
Methane	16.043	0.637
Methyl Acetylene (Propyne)	40.065	0.011
Methyl tert-Butyl Ether	88.150	0.196
Methyl Ethyl Ketone	72.120	0.002
Methyl Naphthalene	142.200	0.0032
2-Methyl-1,3-butadiene	68.118	0.077
2-Methylbutane	72.150	0.137
3-Methyl-1-butene	70.134	0.522
Methylcyclohexane	98.188	0.427
Methylcyclopentane	84.161	0.357
Methylene Chloride	84.930	0.10
2-Methylheptane	114.230	3.65
3-Methylheptane	114.230	3.71
4-Methylheptane	114.230	3.66
2-Methylhexane	100.203	3.42
3-Methylhexane	100.203	0.308
1-Methylnaphthalene	142.200	0.0004
2-Methylnaphthalene	142.200	0.0005
2-Methylnonane	142.284	5.62
3-Methylnonane	142.284	5.87
4-Methylnonane	142.284	5.96
5-Methylnonane	142.284	5.97
2-Methyloctane	128.257	4.79
3-Methyloctane	128.257	5.04
4-Methyloctane	128.257	9.94
2-Methylpentane	86.177	1.74
3-Methylpentane	86.177	1.12
2-Methyl-1-pentene	84.161	0.277
4-Methyl-1-pentene	84.161	0.625
2-Methylpropane (Isobutane)	58.123	1.152
2-Methylpropene	56.107	0.207
m-Methylstyrene	118.178	0.0038
p-Methylstyrene	118.178	0.0028

N

Naphthalene	128.173	0.02
Nitrobenzene	123.110	0.0012
Nitromethane	61.040	0.0003
1-Nitropropane	89.094	0.0001
2-Nitropropane	89.094	0.0001
Nonanal	142.241	0.0014
Nonane	128.257	5.93
Nondecane	268.525	0.0029
1-Nonene	126.241	0.794
1-Nonyne	124.225	0.142

O

Octadecane	254.498	0.0088
Octafluorocyclobutane	200.031	3.87

Chemical	Mol. Wt.	Henry's Law Constant
Octane	114.230	4.93
1-Octene	112.214	0.627
1-Octyne	110.199	0.082
P		
Pentachloroethane	202.295	0.0018
Pentachlorophenol	266.340	0.00015
Pentadecane	212.418	0.479
1,4-Pentadiene	68.118	0.118
2,2,3,3,4-Pentamethylpentane	142.284	9.65
2,2,3,4,4-Pentamethylpentane	142.284	9.57
Pentane	72.150	0.126
1-Pentene	70.134	0.398
cis-2-Pentene	70.134	0.225
trans-2-Pentene	70.134	0.23
Pentylbenzene	148.247	0.017
1-Pentyne	68.118	0.049
Phenol	94.113	0.000017
Propane	44.096	0.685
Propene	42.080	0.204
Propylbenzene	120.194	0.0102
Propylcyclopentane	112.214	0.893
Propylether	102.176	0.0022
4-Propylheptane	5.71	
S		
Styrene	104.152	0.0026
T		
1,1,2,2-Tetrachloroethane	167.849	0.021
Tetrachloroethylene (Perchloroethylene)	165.833	1.035
Tetradecane	198.391	1.13
Tetrafluoroethene	100.016	0.612
Tetrahydrofuran	72.120	0.002
1,2,4,5-Tetramethylbenzene	134.221	0.025
2,2,3,3-Tetramethylbutane	114.230	3.78
2,2,3,3-Tetramethylhexane	142.284	8.14
2,2,3,4-Tetramethylhexane	142.284	7.89
2,2,3,5-Tetramethylhexane	142.284	6.32
2,2,4,4-Tetramethylhexane	142.284	8.57
2,2,4,5-Tetramethylhexane	142.284	6.61
2,2,5,5-Tetramethylhexane	142.284	5.54
2,3,3,4-Tetramethylhexane	142.284	8.37
2,3,3,5-Tetramethylhexane	142.284	7.03
2,3,4,4-Tetramethylhexane	142.284	8.15
2,3,4,5-Tetramethylhexane	142.284	6.85
3,3,4,4-Tetramethylhexane	142.284	9.68
2,2,3,3-Tetramethylpentane	128.257	6.25
2,2,3,4-Tetramethylpentane	128.257	5.69

Chemical	Mol. Wt.	Henry's Law Constant
2,2,4,4-Tetramethylpentane	128.257	5.09
2,3,3,4-Tetramethylpentane	128.257	6.27
Thiophene	84.136	0.0029
Toluene	92.141	0.29
1,2,4-Trichlorobenzene	314.820	0.128
1,1,1-Trichloroethane	133.404	0.77
1,1,2-Trichloroethane	133.404	0.31
Trichloroethylene	131.388	0.544
Trichlorofluoromethane	137.368	0.122
2,4,6-Trichlorophenol	197.450	0.0002
1,2,3-Trichloropropane	147.431	0.0003
1,1,2-Trichlorotrifluoroethane	187.376	0.485
Tridecane	184.364	2.31
Trifluoromethane	70.014	0.075
A,A,A-Trifluorotoluene	146.112	0.016
Triiodomethane	393.732	0.0029
1,2,3-Trimethylbenzene	120.194	0.0036
1,2,4-Trimethylbenzene	120.194	0.0056
2,2,3-Trimethylbutane	100.203	2.38
2,2,3-Trimethylheptane	142.284	6.79
2,2,4-Trimethylheptane	142.284	6.09
2,2,5-Trimethylheptane	142.284	6.18
2,2,6-Trimethylheptane	142.284	5.76
2,3,3-Trimethylheptane	142.284	7.12
2,3,4-Trimethylheptane	142.284	6.85
2,3,5-Trimethylheptane	142.284	7.15
2,3,6-Trimethylheptane	142.284	6.20
2,4,4-Trimethylheptane	142.284	6.55
2,4,5-Trimethylheptane	142.284	6.55
2,4,6-Trimethylheptane	142.284	5.49
2,5,5-Trimethylheptane	142.284	6.37
3,3,4-Trimethylheptane	142.284	7.46
3,3,5-Trimethylheptane	142.284	7.00
3,4,4-Trimethylheptane	142.284	7.48
3,4,5-Trimethylheptane	142.284	7.18
2,2,3-Trimethylhexane	128.257	5.24
2,2,4-Trimethylhexane	128.257	4.79
2,2,5-Trimethylhexane	128.257	5.17
2,3,3-Trimethylhexane	128.257	5.68
2,3,4-Trimethylhexane	128.257	5.60
2,3,5-Trimethylhexane	128.257	4.84
2,4,4-Trimethylhexane	128.257	5.31
3,3,4-Trimethylhexane	128.257	5.94
2,2,3-Trimethylpentane	114.230	3.81
2,2,4 — Trimethylpentane	114.230	3.34
2,3,3-Trimethylpentane	114.230	4.10
2,3,4-Trimethylpentane	114.230	1.76

U

Chemical	Mol. Wt.	Henry's Law Constant
Undecane	156.311	1.83

V

Valeraldehyde	86.133	0.0002
Vinyl Chloride	62.500	355

X

m-Xylene	106.167	0.0067
o-Xylene	106.167	0.266
p-Xylene	106.167	0.0061

The constant from various sources is listed as atm/mole-m^3 at 25°C.

REFERENCES

Ayers, Kenneth W. *et al. Environmental Science and Technology Handbook.* Rockville, MD: Government Institutes, Inc. 1994.

Fang, C.S. and Sok-Leng Khor. "Reduction of Volatile Organic Compounds in Aqueous Solutions through Air Stripping and Gas-Phase Carbon Absorption." *Environmental Progress.* November 1989, pp. 270–8.

Lide, David R. ed. *Basic Laboratory and Industrial Chemicals: A CRC Quick Reference Handbook.* Boca Raton, FL: CRC Press. 1993.

"Tech Resources." *Environmental Technology.* Vol. 6, Issue 7, 1997 Resource Guide.

Yaws, Carl, Haur-Chung Yang, and Xiang Pan. "Henry's Law Constants for 362 Organic Compounds in Water." *Chemical Engineering.* November 1991, pp. 179–85.

Appendix IV
Octanol–Water Partition Coefficient of Common Contaminants

Chemical	Kow	Chemical	Kow
A		4-Chlorobiphenyl	79,400
		p-Chloro-m-cresol	891
Acenaphthene	21,380	4-Chlorodiphenyloxide	12,000
Acenaphthylene	11,749	Chloroethane	34.67
Acetophenone	39.8	Bis(2-Chloroethoxy)methane	18
Acridine	5,140	Bis(2-Chloroethyl)ether	38
Acrylonitrile	8.32	Chloromethane	8
Alachlor	830	Bis(2-Chloromethyl)ether	2.4
Aldicarb	11.02	2-Chlorophenol	148
Aniline	7	2-Chlorophenyl phenyl ether	100,000
Anthracene	22,000	Chrysene	407,380
Atrazine	476	Crufomate	2,780
		Cyanazine	150
B			
		D	
Bentazon	220		
Benzene	135	2,4-D	37
Benzidine	64	Dalapon	6
Benzo[a]anthracene	407,380	DDD	1,047,000
Benzo[b]fluoranthene	3,715,352	DDE	583,000
Benzo[a]pyrene	1,096,478	DDT	960,000
Biphenyl	7,540	Dialifor	49,300
Bromobenzene	900	Dibromochloromethane	123
Bromodichloromethane	76	Dichlofenthion	137,000
Bromoform	200	o-Dichlorobenzene	2,399
Bromomethane	12.59	p-Dichlorobenzene	2,455
4-Bromophenylphenylether	141,254	4,4'-Dichlorobiphenyl	380,000
Butylbenzylphthalate	26,301	Dichlorodifluoromethane	144.54
trans-sec-Butyl-4-chlorodiphenyloxide	16,000	2,4-Dichlorophenol	562
		Dichlorovos	10,000
C		Diethylaniline	9
		Di-2-ethylhexyl phthalate	9,500
Captan	224	Diethyl phthalate	26,303
Carbaryl	230	Dimethoate	0.51
Carbofuran	40		
Chlorobenzene	692		

Dimethylnitrosamine	1.15
Dimethyl phthalate	2,630
4,6-Dinitro-o-cresol	708
2,4-Dinitrophenol	34
2,4-Dinitrotoluene	102
Dinoseb	4,900
Diphenylnitrosamine	372
Diphenyl oxide	15,800
Di-n-propylnitrosamine	20
Diuron	94
Dursban	97,700

E

Endrin	218,000
Ethyl Benzene	1,412.5
Ethylene dichloride	30

F

Fenuron	10
Fluoranthene	213,796
Fluorene	15,136

H

Hexachlorobenzene	168,000
Hexachlorobutadiene	5,495
Hexachloroethane	2,187

I

Imidan	677
Ipazine	2,900

L

Leptophos	2,020,000
Linuron	124

M

Malathion	780
Methomyl	2
Methoxychlor	47,500
2-Methoxy-3,5,6-trichloropyridine	18,500
9-Methylanthracene	117,000
3-Methylcholanthrene	3,333,000
Methylene Chloride	17.78
2-Methylnaphthalene	13,000
Methylparathion	82
Mexacarbate	1,370
Monolinuron	40
Monuron	29

N

Naphthalene	2,040
1-Naphthol	762
Nitrapyrin	2,590
Nitrobenzene	62
2-Nitrophenol	58
4-Nitrophenol	81

P

Parathion	6,400
Pentachlorobenzene	154,000
Pentachlorophenol	102,000
Phenanthrene	28,840
Phenol	28.8
Phosalone	20,100
Phthalic anhydride	0.24
Picloram	2
PCBs (generically)	21,700
Pyrene	135,000

R

Ronnel	46,400

S

Sevin	230
Silvex	2,600
Simazine	155
Styrene	891.3

T

2,4,5-T	4
Tetracene	800,000
1,2,4,5-Tetrachlorobenzene	47,000
1,1,2,2-Tetrachloroethane	363
Tetrachloroethylene (Perchloroethylene)	758
Toluene	489.8
1,2,4-Trichlorobenzene	18,197
1,1,1-Trichloroethane	148
Trichloroethylene	195
Trichlorofluoromethane	338.84
Trichlorofon	3
2,4,6-Trichlorophenol	2,399
Trichlopyr	3
Trichlopyr(butoxyethyl ester)	12,300
Trichlopyr(triethylamine salt)	3
Trietazine	2,300

U

Urea	0.001

REFERENCES

Ayers, Kenneth W. *et al. Environmental Science and Technology Handbook.* Rockville, MD: Government Institutes, Inc. 1994.

Kaczmar, Swistoslav W., Edwin C. Tifft, Jr. and Cornelius B. Murphy, Jr. "Site Assessment under CERCLA: The Importance of Distinguishing Hazard from Risk." *Hazardous Materials Control Monograph Series: Health Assessment.* Silver Spring, MD: Hazardous Materials Control Research Institute.

Lide, David R. ed. *Basic Laboratory and Industrial Chemicals: A CRC Quick Reference Handbook.* Boca Raton, FL: CRC Press. 1993.

Ney, Ronald E., Jr. *Fate and Transport of Organic Chemicals in the Environment: A Practical Guide.* 2d. Ed. Rockville, MD: Government Institutes, Inc., 1995.

Answers To End
of Chapter Questions

1. The American approach has been control by regulation.
2. The purpose of the MOU is to establish and improve the working relationship between EPA and OSHA and to improve efforts to protect the workers, the public, and the environment at facilities under the jurisdiction of either agency. The implication is that either agency can call in the other to a facility being inspected.
3. Opinions about the Rene Dubose prophecy may vary. Discussions could range from how clean is clean to are we over- or under-regulating ourselves. Whatever stand the reader/student takes is fine, so long as he or she separates opinion from scientific issues.
4. Many Americans feel the Superfund is unjust and un-American. These opinions, or more liberal ones, are fine, but everyone should be clear about separation of opinion from scientific issues. Neither conservative nor liberal opinions are right nor wrong, mostly they differ on how to implement policies. If the discussion is allowed to come to a consensus, it will. Time permitting.

CHAPTER 2

1. Exposure is potential for contact with a chemical. Dose is the amount received, or the amount that actually contacts the receptor. Since the assessor rarely has the luxury of understanding the pathway with confidence, exposure is usually used as the dose, in effect, assuming no loss of material between the source and receptor.
2. Acute and chronic exposure are terms that have to do with how long a person or the environment is potentially in contact with a chemical. Acute and chronic effects, on the other hand, are the results of a dose received, and are health effects requiring immediate medical attention in the former case, and presenting long-term medical problems in the latter case.
3. 1 ppm.
   $mg/m^3 \times 24.5 = ppm \times M$, or $ppm = mg/m^3 \times 24.5/M$
   $ppm = 4.16 \times 24.5/100.1 = 1$ (about)
4. 30.63 mg/m^3.
   $mg/m^3 \times 24.5 = ppm \times M$, or $mg/m^3 = ppm \times M/24.5$
   $mg/m^3 = (5)(150.1)/24.5 = 30.63$
5. Because the biological effective dose (BED) has been difficult to quantify.

6. 10 mg/Kg.
   D = CI/WF, therefore C = DWF/I.
   C = (0.1)(100)(1.0)/1 = 10.
7. The possibility of a false negative is 18%. In other words, a one in five chance exists that no effect will be observed though one should have been. In order to offset this potential for a false negative, the toxicologist typically increases the range of doses administered. To give the animals a reasonable dose, unless several hundreds of animals are tested, means that the probability is unacceptably high that no effect will be observed, when, in fact, an effect is feasible.
8. Each person accepts risk for himself or herself. Inconsistencies exist because some individuals prefer safety and security, while others enjoy a degree of danger. Bungee jumpers and sky divers are among the latter group. However, a bungee jumper may not accept the risk of living next door to a waste incinerator, even though it is less risky than jumping off tall structures with a thin elastic line tied around your waist.

CHAPTER 3

1. Delta cholinesterase.
2. Testing for mercury in urine.
3. DDT is a chlorinated hydrocarbon and does not inhibit cholinesterase. Therefore, the test would be ineffective in determining if a dose were received.
4. The recent ingestion of food containing arsenic.
5. Phenol.
6. Yes.
7. The 24-hour number represents an average for the day, whereas the spot number may represent a minimum or maximum exposure, and the exposure has not been ongoing long enough to level out the metabolite in the urine. One or both samples may have been defective. One or both analyses may have been inaccurate.
8. No damage has occurred to heart muscle so far. That is all you can say at this time, because the sensitizer may still be present, but have caused no damage yet.
9. Dichlorofluoroethane   96.1%
   Methanol   3.4%
   Dinitrobenzene   0.5%
   DCFE $= Cl_2FC_2H_3$ = 117 lb./lb.-mole
   Methanol $= CH_3OH$ = 32 lb./lb.-mole
   Dinitrobenzene $= C_6H_4(NO_2)_2$ = 168 lb./lb.-mole
   On a 100 lb. Basis:
      DCFE   96.1 lb.   = 0.821 lb.-mole = 88.3% v/v
      Methanol   3.4 lb.   = 0.106 lb.-mole = 11.4% v/v
      Dinitrobenzene   0.5 lb.   = 0.003 lb.-mole = 00.3% v/v

10. $88°F = 31.1°C$
    0.12 mmHg @ 25°C
    x mmHg @ 31.1°C
    760 mmHg @ 190°C
    46.05 mmHg/°C
    ∴ at 88°F, the vapor pressure is 281 mmHg.
11. Density = mass per unit volume.
    5.148 g/2.6 l = 1.98 g/l
12. $D = 0.1782 \ g/l \ l \times 273K/304K = 0.160 \ g/l$

CHAPTER 4

1. Cancer is a neoplasm, or series of neoplasms, that grow out of control and attack other cells. The student should further elaborate on mechanisms and stages.
2. About one in seven.
3. A cancer-causing agent is called a *carcinogen, tumorigen, oncogen,* or *blastomogen.*
4. A carcinoma.
5. A sarcoma.
6. Oncogenes are formed from protooncogenes by retroviral involvement, however, point mutations or DNA rearrangements, such as translocations or gene amplifications, also lead to oncogenes.
7. The answer should describe briefly initiation, promotion, progression, and metastasis.
8. The one-hit model.
9. The multi-hit model.
10. The first three do: acrylonitrile, arsenic, and chloroform.
11. Trichloroethylene has the lowest potency factor.
12. The toxic goo might have caused death or malformation if they had been developing as eggs in their mothers in which case the chemical would have been a developmental toxin. If they had been eggs when exposed and had been born with malformation, the goo would have been a teratogen. If they had been exposed at any time and their off-spring had been malformed, the goo would have been a mutagen. It may also have been a mutagen, if in later life, they developed abnormal growths. The concept of the mutations in the movie *Teenage Mutant Ninja Turtles*© does not fly in the face of science.
13. Any toxic substance that interferes with the control mechanism for the male pituitary hormones, *Luteinizing Hormone* (LH) and *Follicle Stimulating Hormone* (FSH), can lead to male infertility.
14. The cholinesterase inhibitor causes excited synapses to keep firing, leading to incoordination and other neuropathological symptoms.
15. Intervention almost always leads to recovery.
16. Benzene.

17. This response causes *angioedema* (fluid in the vessel), *urticaria* (hives), and *anaphylaxis* (shock or hypersensitivity).

CHAPTER 5

1. The atmosphere, lithosphere, and hydrosphere.
2. Terrestrial biomes are influenced by latitude, elevation, moisture, and temperature.
3. Forest, grassland, shrub land, desert, streams, lakes, ponds, wetlands, ocean, littoral or shallow water, benthic or ocean bottom zone, rocky shores, sandy shores, estuaries, tidal marshes, tundra, taiga, chaparral.
4. Primary consumers are the grazing animals and small prey animals that humans often eat for meat.
5. Biodegradation is any process that breaks down hazardous compounds into simpler, often less hazardous forms. Used as a sales pitch biodegradable is meant to give the impression that the purchaser will not have to worry about the fate of the product after use. Sometimes a product may be advertised as "completely safe" or "harmless" or "safe to put into the sewer." In fact, few chemical products can be put into the sewer without being treated in some manner first. And many cannot be put into the sewer even after being treated to render them less hazardous. Never assume that the manufacturer's chemist has done all the research necessary to determine the fate of the product.
6. By discharging phosphate detergents, ammonia compounds, fertilizer, and other nutrients into it.
7. The pH of the aqueous solution, its hardness, the concentration of free oxygen, and light stress on the biota.
8. Uptake by marine biota; currents.
9. Neoplasms in flatfish.
10. Answers will vary but ought to be consist with the discussion in the text. This question could be used as a major report project.

CHAPTER 6

1. Adverse consequences of productivity are ignored.
2. A system of commerce and production where each and every act is inherently sustainable and restorative.
3. Although an engineering term that is used for the expression of the energy laws, entropy in economics is reflected in the fact that more and more wealth is merely circulated among services and consumers rather than being truly created by the manufacture of products of value. This is also reflected in diminished resources in terms of either amount available or quality of the remainder.
4. EPA has not clearly demonstrated proof of environmental justice. In fact, data released by EPA so far indicates that the siting of incinerators, for

instance, is not particularly unjust. However, for the impoverished or minority communities that perceive environmental injustice, the issue is emotionally packed.

CHAPTER 7

1. $I = 0.25$ kg fish/meal $\times$ 1 meal per week $\times$ 1 week/7 days
   $= 0.0357$ kg/day.
2. The mechanism of chemical uptake is diffusion, which means that uptake depends on the concentration gradient across the boundary, and permeability of the barrier.
3. Because some other pathology may lead to death of the study subject before the tumor has a chance to progress to malignancy, therefore we assume the worst case. That is, we assume that if the tumor had time it would progress to malignancy.
4. Exposure response curves.
5. Type, intensity, duration, frequency, timing, and scale.
6. Acute, subacute, subchronic, and chronic.

CHAPTER 8

1. $C_1 = C_0 \dfrac{P_1 X_1}{P_0 X_0}$

   30% by weight toluene and 40% by weight benzene
   $X_0 = 0.30$; $X_1 = 0.40$
   $P_0 = 29.9$ mmHg; $P_1 = 95.9$ mmHg
   $C_1 = 150$ ppm$[(95.9)(0.40)/(29.9)(0.30)] = 641$ ppm

Index

T - #0075 - 101024 - C0 - 234/156/24 [26] - CB - 9781566702164 - Gloss Lamination